Communication Technology Update

6th Edition

August E. Grant & Jennifer Harman Meadows, Editors

In association with
Technology Futures, Inc.

**Focal
Press**

Boston • Oxford • Johannesburg • Melbourne • New Delhi • Singapore

Editors

August E. Grant
Jennifer Harman Meadows

Technology Futures, Inc.

Managing Editor   Julia A. Marsh
Production Editor   Debra R. Robison
Art Director   Helen Mary V. Marek

Focal Press is an imprint of Butterworth–Heinemann.

Copyright © 1998 by Butterworth–Heinemann

A member of the Reed Elsevier group

 Butterworth–Heinemann supports the efforts of American Forests and the Global ReLeaf program in its campaign for the betterment of trees, forests, and our environment.

Library of Congress Cataloging-in-Publication Data

ISBN 0-240-80326-4

British Library Cataloguing-in-Publication Data
A catalogue record for this book is available from the British Library.

The publisher offers special discounts on bulk orders of this book.
For information, please contact:

Manager of Special Sales
Butterworth–Heinemann
225 Wildwood Avenue
Woburn, MA 01801-2041
Tel: 781-904-2500
Fax: 781-904-2620

For information on all Butterworth–Heinemann publications available, contact our World Wide Web home page at: http://www.bh.com

10 9 8 7 6 5 4 3 2

Printed in the United States of America

Table of Contents

Section IV–Telephony & Satellite Technologies   **197**

Section V–Conclusions   **277**

Updates can be found on the
Communication Technology Update Home Page
on the Internet at

http://www.tfi.com/ctu/

Preface

In 1992, the first edition of the *Communication Technology Update* was published. As we put together this edition of the *Update*, we took a look back at the first edition. We expected massive changes in content, but we were a little surprised at how much the form of the *Update* had changed.

The 1992 edition of the *Communication Technology Update* was a slim 123 pages, containing brief synopses of recent developments in 37 different communication technologies. Feedback from readers in the intervening years encouraged us to provide more detail and include historical information on each technology, combine similar technologies into a single chapter, and expand the number of graphics to help explain each technology. The book you are reading today is thus a product of the combined vision of the editors, the chapter authors, and the readers.

This edition of the *Update* is changed in three respects from the fifth edition. As always, every chapter was completely rewritten to present the most recent developments in each technology. In addition, each chapter author has agreed to contribute periodic updates that will appear in the *Communication Technology Update* home page on the Internet—http://www.tfi.com/ctu. The home page will also continue to supplement the text with links to a wide variety of information available over the Internet.

Many of the chapter authors included Internet links in their bibliographies that would provide additional information. Because the URLs for these links change so often, we will publish them on the home page for the Update rather than in each chapter. We encourage you to suggest additional links (and to let us know if a link is out of date). We hope that these uses of the home page make this book more useful for you over a long period of time.

The third change is the promotion of Jennifer Meadows to co-editor of this volume. Jennifer was the assistant editor for the past two editions. Her knowledge of the technologies and attention to detail have improved the volume every year she has been involved with it. Her promotion to co-editor has given her the opportunity to further improve this edition, allowing more one-on-one attention to individual chapter authors during the writing, editing, and production phases.

This book uses a novel application of desktop publishing that allows us to have the book printed and available within two months after the last chapters are written. Because we know that changes in the technologies discussed will happen before the book is printed (no matter how fast we are), the *Communication Technology Update* home page has been created to provide up-to-date information no matter how long it has been since the chapter was written. We encourage you to contribute to the home page by sending Internet links and ideas for content.

This compilation is the product of dozens of people who have worked right up to the deadline date to provide the latest developments in all areas of communication technology. We are especially grateful to the staff at Technology Futures, Inc., including Managing Editor Julia Marsh, Production Editor Deb Robison, and Art Director Helen Mary Marek. Most of all, we are grateful to our authors for their continued involvement and enthusiasm for this project.

As always, we encourage you to suggest new topics, glossary additions, and possible authors for the next edition of this book by communicating directly with us via e-mail, fax, snail mail, or voice.

Augie Grant
College of Journalism & Mass Communications
Carolina Coliseum
University of South Carolina
Columbia, SC 29208
Phone: 803.777.4464
Fax: 803.777.0638
Email: augie@sc.edu

Jennifer H. Meadows
Department of Communication Design
California State University, Chico
Chico, CA 95929-0504
Phone: 530.898.4775
Email: jmeadows@oavax.csuchico.edu

The Umbrella Perspective On Communication Technology

August E. Grant, Ph.D.*

Communication technologies are the nervous system of contemporary society, transmitting and distributing sensory and control information, and interconnecting a myriad of interdependent units. Because these technologies are vital to commerce, control, and even interpersonal relationships, any change in communication technologies has the potential for profound impacts on virtually every area of society.

One of the hallmarks of the industrial revolution was the introduction of new communication technologies as mechanisms of control that played an important role in almost every area of production and distribution of manufactured goods (Beniger, 1986). These communication technologies have evolved throughout the past two centuries at an increasingly rapid rate. The evolution of these technologies shows no signs of slowing, so an understanding of this evolution is vital for any individual wishing to attain or retain a position in business, government, or education.

This text provides you with a snapshot of this evolutionary process. The individual chapter authors have compiled facts and figures from thousands of sources to provide the latest information on 24 sets of communication technologies. Each discussion explains the roots and evolution, the recent developments, and the current status of the technology as of mid-1998. In discussing each technology, we will deal not only with the

---

* Associate Professor and Director, Center for Mass Communications Research, College of Journalism and Mass Communications, University of South Carolina (Columbia, South Carolina).

hardware, but also with the software, the organizational structure, the political and economic influences, and the individual users.

Although the focus throughout the book is on individual technologies, these snapshots comprise a larger mosaic representing the communication networks that bind individuals together and enable us to function as a society. No single technology can be understood without comprehending the competing and complimentary technologies and the larger social environment within which these technologies exist. As discussed in the following section, all of these factors (and others) have been considered in preparing each chapter through application of the "umbrella perspective." Following this discussion, an overview of the remainder of the book is presented.

Defining Communication Technology

The most obvious aspect of communication technology is the hardware—the physical equipment related to the technology. The hardware is the most tangible part of a technology system, and new technologies typically spring from developments in hardware. However, understanding communication technology requires more than just studying the hardware. It is just as important to understand the messages communicated through the technology system. These messages will be referred to in this text as the "software." It must be noted that this definition of "software" is much broader than the definition used in computer programming. For example, our definition of computer software would include information manipulated by the computer (such as this text, a spreadsheet, or any other stream of data manipulated or stored by the computer), as well as the instructions used by the computer to manipulate the data.

The hardware and software must also be studied within a larger context. Rogers' (1986) definition of "communication technology" includes some of these contextual factors, defining it as "the hardware equipment, organizational structures, and social values by which individuals collect, process, and exchange information with other individuals" (p. 2). An even broader range of factors is suggested by Ball-Rokeach (1985) in her "Media System Dependency Theory," which suggests that communication media can be understood by analyzing dependency relations within and across levels of analysis, including the individual, organizational, and system levels. Within the system level, Ball-Rokeach (1985) identifies at least three systems for analysis: the media system, the political system, and the economic system.

These two approaches have been synthesized into the "Umbrella Perspective on Communication Technology" illustrated in Figure 1.1. The bottom level of the umbrella consists of the hardware and software of the technology (as previously defined). The next level is the organizational infrastructure—the group of organizations involved in the production and distribution of the technology. The top is the system level, including the political, economic, and media systems, as well as other groups of individuals or organizations serving a common set of functions in society. Finally, the "handle" for the umbrella is the individual user, implying that the relationship between the user and a technology must be examined in order to get a "handle" on the technology. The basic premise of the umbrella perspective is that all five areas of the umbrella must be examined in order to understand a technology.

(The use of an "umbrella" to illustrate these five factors is the result of the manner in which they were drawn on a chalkboard during a lecture in 1988. The arrangement of the five attributes resembled an umbrella, and the name stuck. Although other diagrams have since been used to illustrate these five factors, the umbrella remains the most memorable of the lot.)

Figure 1.1
The Umbrella Perspective on Communication
Technology

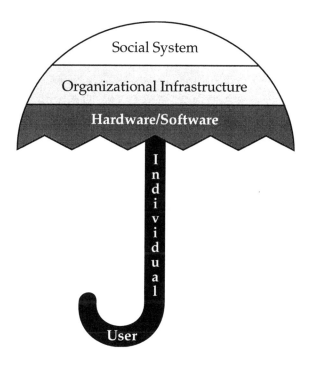

Source: A. E. Grant

Factors within each level of the umbrella may be identified as "enabling," "limiting," "motivating," and "inhibiting." *Enabling* factors are those which make an application possible. For example, the fact that coaxial cable can carry dozens of channels is an enabling factor at the hardware level, and the decision of policy makers to allocate a portion of the spectrum for cellular telephone is an enabling factor at the system level (political system).

Limiting factors are the opposite of enabling factors. Although coaxial cable increased the number of television programs which could be delivered to a home, most coaxial networks cannot transmit more than 54 channels of programming. To the viewer, 54 channels might seem to be more than is needed, but to the programmer of a new cable television channel who is unable to get space on a filled-up cable system, this hardware factor represents a definite limitation. Similarly, the fact that policy makers permitted only two companies to offer cellular telephone service in each market is a system-level limitation on that technology.

Motivating factors are those which provide a reason for the adoption of a technology. Technologies are not adopted just because they exist. Rather, individuals, organizations, and social systems must have a reason to take advantage of a technology. The desire of local telephone companies to increase profits, combined with the fact that growth in providing local telephone service is limited, is an organizational factor motivating the telcos to enter the markets for new communication technologies. Individual users who desire information more quickly can be motivated to adopt electronic information technologies.

Inhibiting factors are the opposite of motivating ones, providing a disincentive for adoption or use of a communication technology. An example of an inhibiting factor at the software level might be a new electronic information technology that has the capability to update information more quickly than existing technologies, but does not use that capability to provide continuously-updated messages. One of the most important inhibiting factors for most new technologies is the cost to individual users. Each potential user must decide whether the cost is worth the service, considering his or her budget and the number of competing technologies.

All four types of factors—enabling, limiting, motivating, and inhibiting—can be identified at the system, organizational, software, and individual user levels. However, hardware can only be enabling or limiting; by itself, hardware does not provide any motivating factors. The motivating factors must always come from the messages transmitted (software) or one of the other levels of the umbrella.

The final dimension of the umbrella perspective relates to the environment within which communication technologies are introduced and operate. These factors can be termed "external" factors, while ones relating to the technology itself are "internal" factors. In order to understand a communication technology or to be able to predict the manner in which a technology will diffuse, both internal and external factors must be studied and compared.

Each communication technology discussed in this book has been analyzed using the umbrella perspective to ensure that all relevant factors have been included in the discussions. As you will see, in most cases, organizational and system-level factors (especially political factors) are more important in the development and adoption of communication technologies than the hardware itself. For example, political forces have, to date, prevented the establishment of a world standard for high-definition television production and transmission. As individual standards are selected in countries and regions, the standard selected is as likely to be the product of political and economic factors as of technical attributes of the system.

Organizational factors can have similar powerful effects. For example, the entry of a single company, IBM, into the personal computer business resulted in fundamental changes in the entire industry. Finally, the individuals who adopt (or choose not to adopt) a technology, along with their motivations and the manner in which they use the technology, have profound impacts upon the development and success of a technology following its initial introduction.

Each chapter in this book has been written from the umbrella perspective. The individual writers have endeavored to update developments in each area to the extent possible in the brief summaries provided. Obviously, not every technology experienced developments in each of the five areas, so each report is limited to areas in which relatively recent developments have taken place.

Overview of Book

The technologies discussed in this book have been organized into three sections: electronic mass media, computers and consumer electronics, and satellites and telephony. These three are not necessarily exclusive; for example, direct broadcast satellites (DBS) could be classified as either an electronic mass medium or a satellite technology. The final decision regarding where to put each technology was made by determining which set of current technologies most closely resembled the technology from the user's perspective. Thus, DBS was classified with electronic mass media. This process also locates a cable television technology—cable telephony and data services—in the telephony section.

Each chapter is followed by a brief bibliography. These reference lists represent a broad overview of literally thousands of books and articles that provide details about these technologies. It is hoped that the reader will not only use these references, but will examine the list of source material to determine the best places to find newer information since the publication of this *Update*.

Most of the technologies discussed in this book are continually evolving. As this book was completed, many technological developments were announced but not released, corporate mergers were under discussion, and regulations had been proposed but not passed. Our goal is for the chapters in this book to establish a basic understanding of the structure, functions, and background for each technology, and for the supplementary Internet home page to provide brief synopses of the latest developments for each technology discussed. (The address for the home page is: http://www.tfi.com/ctu.) Each chapter author has agreed to submit periodic updates to the home page, allowing you to become familiar with many developments that have taken place since this book was published.

The final two chapters attempt to draw larger conclusions from the preceding discussions. The first of these two chapters presents a detailed statistical abstract of many of the technologies discussed, allowing you to more easily compare technologies. The final chapter then attempts to place these discussions in a larger context, noting commonalties among the technologies and trends over time. It is impossible for any text such as this to ever be fully comprehensive, but it is hoped that this text will provide the reader with a broad overview of the current developments in communication technology.

Bibliography

Ball-Rokeach, S. J. (1985). The origins of media system dependency: A sociological perspective. *Communication Research, 12* (4), 485-510.

Beniger, J. (1986). *The control revolution.* Cambridge, MA: Harvard University Press.

Rogers, E. M. (1986). *Communication technology: The new media in society.* New York: Free Press.

ELECTRONIC MASS MEDIA

Digital technologies are revolutionizing virtually all aspects of mass media. Digital video compression, interactivity, and new business opportunities are fueling an explosion in the number of mass media and the programming they provide.

The changes are most evident in multichannel video distribution services. As the following chapter indicates, cable television continues to reinvent itself, incorporating digital technology to increase channel capacity and provide new services. Chapter 4 then explains how direct broadcast satellite services (DBS) have emerged as the most aggressive competitors to cable television.

The factor shared by all of multichannel distribution services is programming. Most of these services will depend upon revenues from the pay television services explored in Chapter 3, including premium cable channels and various types of pay-per-view television.

Chapter 5 explores how digital technology is forcing the biggest change ever in broadcast television as broadcasters must choose whether to use their new digital frequencies to provide one channel of high-definition television, a "multicast" of up to five channels of standard definition programming, or some combination of the two.

Not all technologies have fared as well. "Wireless cable" services (MMDS, discussed in Chapter 7) have had a more difficult time competing with cable television. Similarly, Chapter 4 explores how interactive television efforts have continued to disappoint inventors and investors.

Finally, Chapter 8 explains how radio is preparing for its own digital revolution. That revolution may take longer than the television revolution, but digital technology promises the same degree of change in radio as it has offered to all areas of television broadcasting.

In reading these chapters, you should consider two basic communication technology theories. Diffusion theory helps us to understand that the introduction of innovations is a process that occurs over time among members of a social system (Rogers, 1983). Different types of people adopt a technology at different times, and for different reasons. The smallest group of adopters are the innovators who are first to adopt, but they usually adopt for reasons that are quite different from later adopters. Hence, it is dangerous to predict the ultimate success, failure, diffusion pattern, gratifications, etc. of a new technology by studying the first adopters.

Diffusion theory also suggests five attributes of an innovation that are important to its success: compatibility, complexity, trialability, observability, and relative advantage (Rogers, 1983). In studying or predicting diffusion of a technology, use of these factors suggests that analysis of competing technologies is as important as the attributes of the new technology.

A second theory to consider is the "Principle of Relative Constancy" (McCombs, 1972; McCombs & Nolan, 1992). This theoretical perspective suggests that, over time, the aggregate disposable income devoted to the mass media, as a proportion of gross national product, is constant. In simple terms, people spend a limited amount of their income on the media discussed in this section, and that amount rarely increases when new media are introduced. In applying this theory to the electronic mass media discussed in the following chapters, consider which media will win a share of audience income, and what will happen to the losers.

Bibliography

McCombs, M. (1972). Mass media in the marketplace. *Journalism Monographs*, 24.

McCombs, M., & Nolan, J. (1992). The relative constancy approach to consumer spending for media. *Journal of Media Economics, 5* (2), 43-52.

Rogers, E. M. (1983). *Diffusion of innovations*, 3rd Ed. New York: Free Press.

2

Cable Television

Robyn R. Booth, M.M.C.*

Cable television has changed drastically from its humble beginnings in 1949 as a system designed solely to retransmit existing broadcast signals to remote areas. Today, cable is a complex industry of more than 170 networks that provides original programming to 12,000 systems owned by 75 operators in more than 34,000 communities around the United States.

The cable industry has been able to keep up this growth because it consistently provides a variety of programming and improvements to cable hardware and system capacity. However, the relationship between cable companies and their subscribers has been troubled in the past due, in part, to poor customer service and perceived unfair pricing. The federal government even stepped in and passed legislation intended to increase competition and lower monthly rates for cable subscribers. Increasing competition for cable television comes in the form of satellite and wireless television systems, and cable is losing some of its best subscribers to these systems. Threatened with the loss of much-needed revenue, the cable industry is offering more and improved services at basic rates in order to maintain revenues and profitability.

Background

When broadcast television first emerged, people in many areas of the country had trouble receiving television signals from transmitter towers because mountains or trees blocked the signal or because their antennas were too weak to receive the signals from afar (Baldwin & McVoy, 1988; Bartlett, 1995). Cable television was developed in 1948 to solve this problem by bringing existing broadcast signals to rural areas with community

---

* Columbia Metropolitan Magazine (Columbia, South Carolina).

antennas placed at high elevations, usually on mountains or on top of tall poles (Baldwin & McVoy, 1988; Bartlett, 1995). Cable operators amplified weak signals and fed them through a coaxial cable-connected system of amplifiers (known as the trunk and cable system) into households (Baldwin & McVoy, 1988; Bartlett, 1995).

Several men have laid claim to building the first community antenna television (CATV) system. In Astoria, Oregon, Ed Parsons experimented with wires and an antenna to receive television signals from Seattle. John Walson, from Mahanoy City, Pennsylvania, also claimed to have developed a master antenna cable system (Baldwin & McVoy, 1988). These operations are believed to be the very first cable television systems (Crandall & Furchtgott-Roth, 1996).

In 1949, Robert J. Tarlton of Lansford, Pennsylvania, invested in a company that built a master antenna at the summit of the Allegheny Mountains which amplified television signals from Philadelphia. The signals were then distributed to homes in the community by coaxial cables hung on telephone poles (Baldwin & McVoy, 1988). The Lansford system became the first subscription cable system (Crandall & Furchtgott-Roth, 1996).

The number of cable systems has grown steadily since 1949. By 1960, there were nearly 700 CATV systems. In 1971, 2,750 systems were serving almost six million homes (Baldwin & McVoy, 1988). By the end of the 1970s, nearly 15 million households were cable subscribers, and by the end of the 1980s, that number had jumped as high as 50 million households (Crandall & Furchtgott-Roth, 1996). In 1994, the cable industry boasted more than 55 million subscribers (Crandall & Furchtgott-Roth, 1996), with that number rising to more than 65 million in 1998 (NCTA, 1998a).

Multiple system operators (MSOs) own approximately two-thirds of all cable systems. The top five MSOs at the beginning of 1998 were:

- Tele-Communications, Inc. (TCI), 15.7 million subscribers.
- Time Warner Cable, 12.6 million.
- MediaOne, 4.9 million.
- Comcast Corporation, 4.4 million.
- Cox Communications, Inc., 3.2 million.

The largest single system was Time Warner Cable's New York City system, with just over one million subscribers. (TCI's largest system was the Puget Sound, Washington system, ranked seventh with 424,500 subscribers [NCTA, 1998a].)

Cable Hardware

Traditional cable systems provide services through a trunk-and-feeder system, which uses three main elements: the headend, the distribution network (which includes the trunk-and-feeder), and the subscriber drop (see Figure 2.1) (Baldwin & McVoy, 1988).

The headend is the point at which all satellite- or microwave-transmitted program signals are received, assembled, and processed for transmission by the distribution network (Baldwin & McVoy, 1988).

Figure 2.1
Traditional Cable TV Network Tree and Branch
Architecture

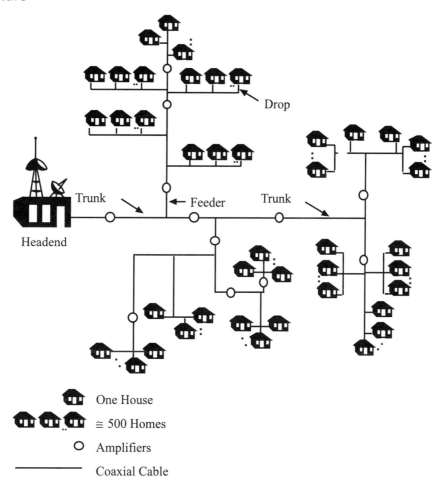

One House

≅ 500 Homes

○ Amplifiers

Coaxial Cable

Source: Technology Futures, Inc.

The distribution network carries the program signal through the community using a system of coaxial or fiber optic cables strung along telephone or power lines or buried underground. The distribution network, often called the distribution plant, consists of two elements: the trunk system and the feeder system. The trunk consists of a large-diameter cable that leaves the headend and travels through the community, splitting at various points along the route and ending at the end of the service area. The trunk's purpose is to deliver signals to subscriber neighborhoods; no subscribers are directly served from the trunk. Instead, bridger amplifiers are located along the trunk at intervals between one-quarter and one-half mile to feed the signal to the feeder system. The feeder system consists of smaller cables running along the streets in a neighborhood to which subscribers connect. The subscriber drop then takes the signal from the feeder system into the subscriber's home where it is picked up by the television receiver (Baldwin & McVoy, 1988).

This complex system of cables, amplifiers, and receivers is necessary to make sure the television signal stays strong. As the signal travels down the cable, it loses strength. Amplifiers are set up at equal intervals along the distribution network to make sure that a person living 10 miles away from the headend gets the same reception as the person living next door to it.

The earliest cable distribution systems consisted of coaxial cable one-half-inch in diameter with a copper conductor in the center to carry the TV signals. The conductor is surrounded by a plastic or foam insulator which is surrounded by a braided copper shield. A plastic jacket is placed around the cable to protect it from moisture, salt, and other damage (Baldwin & McVoy, 1988).

The development of fiber optic transmission technology has led cable providers to shift from purely coaxial cable systems to hybrid fiber/coax (HFC) cable systems. Fiber optic technology has several advantages over coaxial cable. The wider range of frequencies increases the bandwidth available for transmission, making it possible for signals to be transmitted for greater distances without the need for amplification (Bartlett, 1995). Fiber optic cables use light to transmit signals rather than copper wire. In cable systems, fiber optic cables are used as trunk lines to carry signals to coaxial distribution lines that, in turn, link to drop cables connected to the subscribers (Johnson, 1994).

The desire to provide additional services, including a variety of digital and two-way services, is prompting most cable operators to rebuild their systems and replace the traditional plant with an HFC system (see Figure 2.2). Instead of the traditional trunk-and-branch architecture, the service area is divided into a number of neighborhoods containing up to 500 houses, called "nodes." Each node contains an "optical network interface" that selects signals from a fiber optic cable and delivers that set of signals via coaxial cable to all homes in the neighborhood. The coaxial cable uses about half of its bandwidth to deliver the traditional array of cable channels, with most of the other half used to deliver other signals, including Internet service, telephone conversations, digital television, and pay-per-view movies. A portion of the cable's bandwidth is used for "upstream" communication—the signals from the house back to the node. Upstream communication is necessary for telephone calls, Internet access, and other interactive services.

Cable Software

When it was first introduced, cable was intended to supplement existing broadcast networks. Because cable companies focused mainly on retransmitting broadcast television and original cable networks showed primarily previously-aired broadcast programming, these first cable networks were not seen by program providers as a viable market for new shows.

Figure 2.2
Hybrid Fiber/Coax Cable TV System

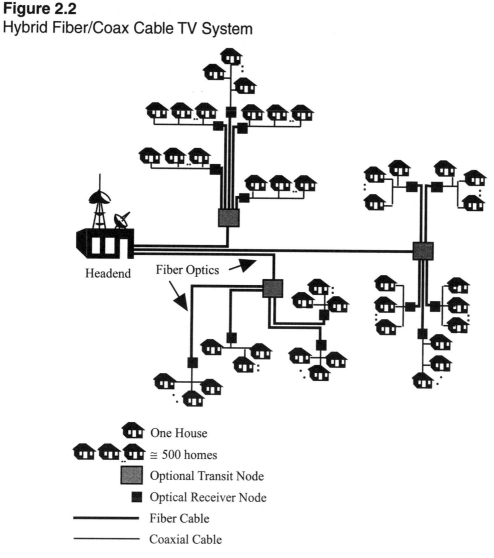

One House

≅ 500 homes

Optional Transit Node

Optical Receiver Node

Fiber Cable

Coaxial Cable

Headend   Fiber Optics

Source: Technology Futures, Inc.

As broadcast station signals grew stronger and more prolific, the need for cable waned. In order to survive, and to attract customers in urban areas that already had three or more broadcast television stations, the cable industry needed to offer something new. Original programming proved to be the something that the industry needed to stand out.

Home Box Office (HBO) began the premium-channel revolution in 1972, initiating satellite delivery in 1975. By 1979, 10 premium channels and 11 basic cable networks offered original programming (Crandall & Furchtgott-Roth, 1996). Over the past 23 years, the cable industry has grown to offer original programming on more than 170 networks, with each local cable system carrying an average of 50 channels.

Programming on cable channels now competes on an equal standing with broadcast channels. In 1997, 15 cable networks were honored with 138 prime-time Emmy nominations for programming, amounting to 35% of 391 nominations. Cable networks won nine of the 28 prime-time awards, a record number for the industry (NCTA, 1998a).

While programming on the broadcast networks appeals to a large audience with differing interests, cable can appeal to population segments, offering programming to a smaller, more homogenous audience. Viewers can turn on their televisions and see news, old movies, new movies, B-movies, adult movies, cartoons, westerns, dramas, comedies, rock music videos, country music videos, rap music videos, shows about science fiction, gardening, antiques, construction—the list goes on. Cable also provides more than 75% of all children's programming viewed in cable households (NCTA, 1998b).

Cable now has more programming than ever. An estimated 5,000 cable systems were originating programming in their own studios in 1996 (Kizar, 1997). With the increased number of available channels resulting from the development of digital television, we'll be looking at hundreds of new programming possibilities.

Regulations and Cable

During the 1960s and 1970s, public policy debate focused on the assumption that cable television companies were natural monopolies. "Into the foreseeable future, it was commonly argued, the home would be served by only one cable company for the same reason that it would be served by a single telephone company system: construction of a competitive parallel network would entail wasteful duplications" (Johnson, 1994, p. 2). Other companies did not find it profitable to break into markets with legacy cable systems. Because this lack of competition could create a monopoly cable industry, much of the policy enacted was intended to protect subscribers against monopoly pricing.

Cable basically faced little regulation until 1972, when the Federal Communications Commission (FCC) enacted rules to aid expansion. These rules included:

- Requirements that cable systems obtain approval from the local franchising authority to raise rates.

- Requirements for obtaining franchises.

- Limits on length of franchise periods.

- Channel capacity requirements.

- Must-carry provisions for broadcast signals.

- Creation of public access channels.

- Rules banning cross-ownership of cable and other utilities, including telephone.

Cable companies complained about the hassle of needing government approval for rate increases, which they thought discouraged expansion and development. They also thought that increased competition in the 1970s and early 1980s made government regulation unnecessary and inappropriate (Johnson, 1994). The 1984 Cable Communication Act was developed to free the cable industry from excessive regulation, mainly by eliminating the local government's power to regulate rates. The act also prohibited broadcast networks from owning cable systems and limited telephone company ownership of cable systems to systems outside of their own service regions (Albarran, 1996). The act permitted franchising authorities to regulate basic cable rates only in cases where the cable system did not have effective competition, leading to the unintentional formation of unregulated monopolies. Deregulation of cable rates became effective in November 1986, and most companies qualified (Johnson, 1994).

Following deregulation, cable rates rose rapidly and sharply. According to one study, between November 1986 and April 1991, the monthly rate for the lowest-priced cable service rose by 56%. Rates for the most popular service rose by 61%—nearly three times the rate of general inflation for that time period. These rate increases and consumer complaints about service led to the Cable Television Consumer Protection and Competition Act of 1992, which placed regulation authority into the hands of the federal government. The FCC ordered a reduction in rates for basic service by a total of 17% by 1994, but the complexity of the rate regulations actually resulted in rate increases for many subscribers (Johnson, 1994). The 1992 Act also required program providers to sell their services to direct broadcast satellite (DBS) systems, multipoint multichannel distribution services (MMDS), and satellite master antenna television (SMATV) systems (Albarran, 1996).

One of the most important provisions of the 1992 Cable Act gave local television broadcasters a choice of "must-carry" or "retransmission consent" regarding carriage on cable systems. A broadcaster choosing "must-carry" could force cable operators in their local service areas to carry their signal. The "retransmission consent" choice prohibited cable operators from carrying a television station without permission. Initially, most broadcasters chose retransmission consent, expecting to receive cash payments for their signals similar to the fees paid by cable operators for other cable networks. Cable operators refused to pay, and most deals were settled with broadcasters receiving other considerations, such as an extra channel or an opportunity to promote their programming during local advertising slots on cable channels.

The Telecommunications Act of 1996 deregulated rates once again. Most small cable systems were granted almost immediate rate deregulation. For large systems, extended basic tiers of service were scheduled for deregulation over a three-year period; rate regulation is scheduled to end on March 31, 1999 (Cole, et al., 1996).

Rate deregulation is permitted in markets where cable companies have effective competition. Effective competition includes any franchise area where another company provides comparable video programming services to subscribers by any means other than through digital broadcast satellite. There is no penetration minimum, and comparable service means at least 12 channels of programming, including broadcast signals (Cole, et al., 1996).

The new law also allowed cable companies to offer telephone service, and telephone companies to offer video services. It required cable operators to scramble any channels that they considered to be inappropriate for children, and they were given the right to refuse to accept any programming considered obscene or indecent (Cole, et al., 1996).

Recent Developments

Recent developments in cable television include high-speed Internet access via cable modems, high-definition digital television signals, interactive television, and wireless cable.

Most cable system signals are transmitted through a combination of fiber optics and coaxial cable, also called hybrid/fiber coax. This combination creates a high-capacity broadband network capable of carrying at least 1,000 times more information than ordinary twisted pair copper telephone wires, at a speed 100 times faster than Integrated Services Digital Network (ISDN) phone lines (NCTA, 1998b). Almost four million homes are in areas wired with HFC cable systems (McConville, 1997).

Cable modems access the Internet through these HFC cable systems, allowing for faster access without tying up phone lines. According to the National Cable Television Association (NCTA), a 28.8K modem using regular phone lines takes nearly 23 minutes to download a file of about five megabytes. A 128 Kb/s ISDN line takes about five minutes. A cable modem, on the other hand, takes only 26 seconds. More than 45,000 homes are already receiving access via cable modems (McConville, 1997).

Bill Gates and Microsoft invested $1 billion in Comcast Corporation to help build cable plants that will offer high-speed Internet access. Microsoft, Sun Microsystems, and Intel are building the key components of TCI's OpenCable box that will include a cable modem (Barthold, 1998a). The "Baby Bells" and GTE are teaming up with Intel, Microsoft, and Compaq to build their own cable modems in an attempt to give the cable companies a run for the money.

The cable industry is also developing and building digital set-top boxes that are needed to pick up digital programming signals. Digital service is designed to deliver more channels with better audio and video quality than regular analog service. The promise of digital video is that one analog channel carrying one program can be replaced with a digital signal that will initially carry four to six, then 10, and later up to 20 programs (Evans, 1996). Broadcasters and cable systems in limited areas should begin transmitting digital signals in 1998. Many television manufacturers are also gearing up to provide high-definition television sets to play those signals.

In December 1997, General Instrument Corporation said it would build 15 million digital set-top boxes for cable operators in the next three to five years (Barthold, 1998c). Scientific-Atlanta is claiming that their boxes will be out almost 18 months sooner, and that they will supply digital networks to nine cable providers with as many as 500,000 digital boxes. These boxes will provide video on demand, Internet access, e-mail, interactive channels, and other services that require real-time two-way interaction (Barthold, 1998c).

According to *Cable World*, cable modems and digital boxes were the top two technological accomplishments of 1997 (Barthold, 1998b). It is predicted that 1998 will be the year in which consumers will be able to buy their own cable modems and digital set-top boxes (Barthold, 1998b). A key feature in making digital television and high-speed Internet access through cable modems a reality is interoperability (Barthold, 1998a). It is important for subscribers and cable systems alike that people be able to buy cable modems and digital set-top boxes from dealers such as Radio Shack in much the same way that they buy their TVs and VCRs—without having to worry about compatibility.

Cable companies are also using more fiber optic cables, which are considered to be the ideal way to transport media for such applications as interactive television, virtual reality, faster Internet access, teleconferencing, medical imaging, advanced data services, and video-based applications (Sullivan, 1997). One thread of fiber optic cable, according to Sullivan (1997), could carry all calls made in America at the peak telephone usage moment on Mother's Day.

The cable industry is improving its customer service. On March 1, 1995, cable systems nationwide kicked off the On-Time Customer Service Guarantee (OTG), which promised on-time appointments for installation or the installation was free, and on-time service appointments or the customer received $20 (NCTA, 1998b). The industry relaunched OTG on March 1, 1997 to emphasize its continued dedication to customer service.

The cable industry is also offering Cable in the Classroom—free, nonviolent, commercial-free, educational cable programming. In addition, the industry is also offering cable modems to elementary and secondary schools for high-speed Internet access and is involved in developing distance learning programs and tutorials to show school teachers how to use the Internet in their classrooms.

Cable's Competitors

The cable industry is facing competition from DBS, MMDS, and SMATV systems (Albarran, 1996). These program providers take subscribers away from cable with promises of better reception and more viewing choices.

The cable industry is not too concerned about losing all of its customers to these services because of the inconvenience and expense. It does, however, need to worry about losing its *best* customers. The customers who tend to order premium and pay-per-view services—the customers that really make money for the cable system operators—are the ones who are switching to these other services. DBS, MMDS, and SMATV services may be more expensive than cable, but subscribers have shown that they are willing to pay more to get the channels they want.

Current Status

Today, cable is available to 97% of all television households in the United States. Sixty-five million households are cable subscribers, and the industry estimates that nearly 165 million people are reached by cable. The average subscriber pays around $27 per month for cable and gets more than 40 channels, with more than 45% of all subscribers receiving 54 channels or more (NCTA, 1998b). Cable systems get most of their revenue from subscribers, not advertisers. Advertising on local origination channels costs between $2 and $600 per 30-second spot, making up less than 5% of a cable system's gross revenue (Kizar, 1997).

The cable industry's operating revenues have increased slowly, but steadily, from nearly $25 billion in 1991 to $35.6 billion in 1995. Expenses have increased as well, from $20 billion in 1991 to more than $27 billion in 1995—most likely the result of cable plant upgrades (NCTA, 1998b).

Because cable channels receive subscription revenues (a set amount paid per subscriber) and advertising revenues, the audience of cable networks is measured two ways: reach and rating. Table 2.1 illustrates the reach of the top 20 cable networks in the United States, indicating the number of households that can receive each channel. (Note that the numbers in the table include other multichannel distribution services, including DBS and wireless cable.) Most channels charge local cable operators a fixed fee per month, per subscriber to carry a channel. These fees range from $.05 to $.25 per month for basic channels and up to $5 per month for premium channels.

As the amount of cable programming has grown, so have ratings. Cable's audience share has risen to more than 35% of all television viewing, with cable ad revenues reaching $7.9 billion in 1997. Total advertising revenues include $1.9 billion in local advertising revenue and $5.7 billion in cable network advertising revenue (NCTA, 1998a).

Factors to Watch

In the future, with the roll-out of high-speed cable modems, cable companies may need to address the issue of whether to regulate access to adult and pornographic material on the Internet, since they are required to do so for cable TV programming. Offering interactive multimedia services is also a possibility. Cable companies should be wary of DBS and MMDS as competitors. These industries threaten the much anticipated rate increases that cable companies have wanted to implement since the passage of the 1996 Telecommunications Act.

Table 2.1

Top 20 Cable Programming Services

Network	Number of Subscribers (in millions)	Year Founded
Discovery Channel	73	1985
ESPN	73	1979
TBS Superstation	73	1976
TNT	72.4	1988
C-SPAN	71.4	1979
TNN: The Nashville Network	71.4	1983
CNN	71	1980
USA Network	69.7	1980
Lifetime Television	69.5	1984
Headline News	68	1982
The Weather Channel	68	1982
AMC	67	1984
The Family Channel	66.9	1977
A&E Television Network	66.9	1984
MTV	66.7	1981
Nickelodeon/Nick at Nite	66	1979
CNBC	64	1989
QVC	63	1986
VH1	61.6	1985
The Learning Channel	60	1980

Source: NCTA, 1998a

Telephony is another potentially lucrative service cable companies are looking to offer. While each industry could easily become involved in the other as the result of improvements in HFC technology, it is more likely that cable will get into telephony than vice versa. It would be more economical for local and long distance telephone companies to become involved in each other's business than for either of them to try their hand at cable. Similarly, it would be far more beneficial for cable companies to try to enter the local telephone market.

What will happen to the cable industry when broadcast stations begin to transmit digital television signals in the near future? Will cable companies have to carry all of the new digital signals in addition to the analog signals? Look for cable companies and broadcasters to do battle once again over must-carry and retransmission consent of digital and high-definition television signals. If cable companies are forced to carry broadcasters' analog and digital signals, they will have to devote the equivalent of two channels to each broadcaster, perhaps forcing cable operators to drop a cable-programming channel.

Because many cable subscribers will still get their services through analog systems for a long time to come, the industry is poised to offer improved analog service to satisfy those customers who decide to wait to buy digital sets (Barthold, 1998b). The question, then, is when do they stop transmitting analog signals completely? Also, when high-definition television signals become a reality and more channels are available to programmers, will cable system providers have the bandwidth to broadcast the signals? And will they be able to justify offering more channels with less interest to the average viewer at the higher costs that will be necessary? Although these questions have no firm answers as of mid-1998, the next few years should see dramatic developments in this area.

The cable industry has seen a variety of changes in its short history. And it undoubtedly will see more in the future, as the federal government steps in once again to lower prices and increase competition, as more industries try to break into the business, and as improvements in hardware and software make cable more appealing to subscribers.

Bibliography

Albarran, A. (1996). *Media economics: Understanding markets, industries, and concepts*. Ames, IA: University Press.

Baldwin, T., & McVoy, S. (1988). *Cable communication*. Englewood Cliffs, NJ: Prentice-Hall.

Barthold, J. (1998a). The Internet access speed issue. *Cable World*. [Online]. Available: http://www.mediacentral.com/Magazines/CableWorld/News98/1998020208.htm/539128.

Barthold, J. (1998b). The next technology stop: Retail." *Cable World*. [Online]. Available: http://www.mediacentral.com/Magazines/CableWorld/News98/1998010513.htm/539128.

Barthold, J. (1998c). S-A takes field in digital battle thanks to 9 MSOs. *Cable World*. [Online]. Available: http://www.mediacentral.com/Magazines/CableWorld/News98/1998020206.htm/539128.

Bartlett, E. (1995). *Cable communications: Building the information infrastructure*. New York: McGraw-Hill.

Cole, Raywid & Braverman, LLP. (1996). *Summary of the Telecommunications Act of 1996*. [Online]. Available: http://www.crblaw.com/96summ.htm.

Crandall, R., & Furchtgott-Roth, H. (1996). *Cable TV: Regulation or competition*? Washington, DC: Brookings Institution Press.

Evans, S. (1996). Cable modems and the Internet. *New Telecom Quarterly, 4* (3), 36-41.

Johnson, L. (1994). *Toward competition in cable television*. Cambridge, MA: The MIT Press.

Kizar, E. A. (Ed.). (1997). *Broadcasting and cable yearbook, 1997*. New York: R. R. Bowker.

McConville, J. (1997). Cable wrestles with Internet free-speech issues. *New Telecom Quarterly, 5* (2), 63.

National Cable Television Association. (1998a). *Cable television developments, spring 1998*. Washington, DC: NCTA.

National Cable Television Association. (1998b). [Online]. Available: http://www.ncta.com/.

Sullivan, C. (1997). Fiber optics. *TeleTimes, 11* (3), 26, 30.

3

Pay Television Services

Jennifer H. Meadows, Ph.D.*

Today's television viewer is confronted with more choice than ever before. From premium channels to video on demand and direct broadcast satellites (DBS), the variety and number of pay television services has increased dramatically over the past few years. This chapter will discuss traditional pay television services, such as premium cable television channels and pay-per-view (PPV), as well as newer services such as video on demand (VOD) and near video on demand (NVOD).

The pay television services most of us know best are the premium channels generally associated with cable television. These services, such as Home Box Office and Showtime, offer a mix of popular movies, original programming, and sports without commercial interruption. Subscribers pay a monthly fee, usually around $10.00 per month per channel, above the basic cable fee. In addition to these premium channels, there is a specialized group of premium channels called mini-pays. These channels carry a much lower monthly charge, usually from $1.00 to $3.00 per month, and generally focus on specific kinds of programming such as sports (The Golf Channel), science fiction (the Sci-Fi Channel), and old movies (Turner Classic Movies). In some markets, these channels are bundled in specialized cable tiers, while in other markets they are offered on an à la carte basis.

Pay-per-view services have been offered to cable and DBS subscribers for a number of years. With PPV, a subscriber can order a single program for one set price. Programming on PPV ranges from popular movies and adult programming to sporting events and more specialized events such as rock concerts and professional wrestling. Depending on the cable or DBS system, the consumer can place a call to order the movie or event, or use the remote control to place an order. Movies are usually offered on several channels at staggered start times, while special events are typically offered on a one-time-only basis.

---

*Assistant Professor, Department of Communication Design, California State University, Chico (Chico, California).

With PPV, the events and movies are scheduled at set times. Viewers are subject to scheduled start times or must record the programming for later viewing. The advantage of near video on demand or enhanced pay-per-view is that the same movie can be offered on several different channels at varying start times. This gives the consumer a choice of multiple movies starting at closely staggered times. In the past, critics of PPV have complained that the restricted start times of movies on PPV have limited its success. NVOD resolves this issue by offering the viewer a greater number of start times, plus a larger selection of movies from which to choose.

Video on demand goes a step further by allowing the viewer to order from a wide variety of entertainment choices—whenever they want, with the press of a button. Additionally, the viewer can fast forward, rewind, and pause the program at any time. VOD puts control of the programming in the hands of the viewer instead of the video service provider, making the experience much more comparable to renting a home video than ordering PPV.

Services such as VOD and NVOD are made possible through such advances in technology as digital video compression, fiber optics, and new advanced set-top boxes. Further, alternative video service providers have introduced new opportunities for pay television services. Direct broadcast satellites, for example, presently offer NVOD services. New digital cable and MMDS services (also known as "wireless cable," see Chapter 6) are also offering NVOD and VOD services.

In order for all of these pay television services to be made available, the television household must be "addressable." Addressability means that the video service provider can communicate directly with the set-top box in each household, thus allowing the service provider to transmit a PPV movie to the consumer over a broadcast network without making it available to everyone. The set-top box decodes the blocking signal from the service provider and unscrambles or presents the requested programming.

In the past few years, several new technologies and competitors have emerged that will shape the future of pay television services. The passage of the Telecommunications Act of 1996 cleared the way for local telephone companies to offer video services. This drew telephone companies into the cable television business, but, to date, they've offered little in the way of services. The greatest competitive threat for cable companies has come from direct broadcast satellites. In addition, the arrival of broadcast digital television may herald the entry of the broadcast industry into pay television services, as networks and television stations decide whether to use one high-definition television (HDTV) channel or a multiplex of standard digital television (SDTV) channels (Dickson, 1997). In response to these changes and new developments in technology, cable companies are also going digital. These new digital cable services will allow a multitude of new and/or expanded pay television services (Coleman, 1997d). This chapter will discuss pertinent issues and highlight factors to watch in the fast-moving and quickly changing future of pay television services.

Background

Pay television has been around almost as long as television. As early as the 1940s, Zenith introduced Phonevision, a service which supplied movies via telephone lines. Customers could choose from three movies each day for $1.00 each (Gross, 1986). The Phonevision system never progressed beyond its trial period in 1951. In 1953, Paramount tested the Telemeter system, whereby customers inserted coins into a set-top decoding box in order to receive programming. Several other pay television systems were tested in subsequent years, but none of them ever took off. Part of this failure was the result of FCC regulations intended to protect threatened broadcasters and theater owners. Regulations included:

- Limitations on the rights of cable programmers to outbid broadcasters for programming.

- Requirements for local cable companies to lease decoder boxes to customers instead of selling them.

- A limitation of one pay television service per community—and only then in markets with at least four commercial stations.

These regulations were later rescinded, but not before they effectively stymied the growth of the fledgling pay TV services industry.

The 1970s and 1980s saw the proliferation of pay television services that used scrambled UHF signals and set-top decoder boxes. These services included SelecTV and ON-TV. However, signal stealing was a problem with these services because the decoder boxes that unscrambled the signals were easy to manufacture. Further deregulation in the early 1980s opened the subscription television market and made way for 24-hour pay services. However, competition to these over-the-air pay services arrived in the form of greater cable television penetration, premium cable channels, pay-per-view services, and home video. Moreover, UHF station owners realized that they could make more money broadcasting as independent stations rather than as subscription television services (Gross, 1986).

The next step for pay television services came in 1972 with the introduction of Home Box Office (HBO). As a video service delivered to cable companies via microwave technology, HBO was originally provided to cable customers in Wilkes-Barre, Pennsylvania. It quickly became a success despite a large subscriber turnover problem.

A commitment to new technology played a major role in HBO's future success. In 1975, HBO was beamed to a communications satellite, RCA's Satcom I, to become the first "national entertainment communications network" (Mair, 1988, p. 23). The world heavyweight boxing championship fight between Mohammed Ali and Joe Frazier, the *Thrilla from Manila*, was HBO's first big national sporting event, and it launched the pay service as the leader in premium pay television services, a spot it continues to hold today.

Pay-per-view was heralded as the future of pay television services in the early 1980s. Many industry analysts saw it as the real future of cable television because hit movies could be seen earlier than on network television, and VCR penetration was still quite low. The movie studios were very interested in PPV as a distribution channel for entertainment products and experimented with different release windows—the time it takes for a movie to go from theatrical release to pay cable, cable, home video, and broadcast television. Generally, a movie is released to home video before PPV. Service providers have argued that the windows need to be shorter because most viewers have already seen the movie on home video by the time it gets to PPV. This, they argue, is the reason why buy rates (the percentage of purchase out of the total number of addressable subscribers) for PPV movies have been disappointing.

Although buy rates for movies have been lackluster, PPV has been much more successful with adult services and event programming. Sporting events, such as boxing and professional wrestling, have consistently pulled in respectable buy rates.

Until recently, movie availability on PPV was constrained by:

- Limited channels.

- Limited number of start times.

- Complicated billing and ordering procedures.

- Lack of program control—you can't pause, fast forward, or rewind.

- Limited programming selection.

Operators attempted to market PPV as a convenient alternative to home video, yet, to this day, home video remains a multibillion dollar business, in contrast to PPV which just hit the billion dollar mark for the first time in 1997 (Umstead, 1997).

In the past several years, new video distribution services, as well as advances in communication technologies, have made new pay television services possible that, in many ways, overcome the limitations of services such as PPV and premium pay cable. Advances in compression and bandwidth capacity have made it possible for cable television operators to supply more channels on their cable systems. As a result, more premium channels are multiplexing—expanding their services from one channel to several channels. For example, HBO began to multiplex without an increase in subscription rates in the early 1990s. Showtime currently offers 18 multiplexed channels (New channel, 1997). Some multiplexed channels simply offer the same programming at different start times. Others offer different programming on each channel. One example is the Encore/ Starz Encoreplex which offers nine channels of genre-specific movies such as westerns, action, and true stories (Colman, 1997c). Figure 3.1 illustrates how the growth in cable system capacity since 1980 has resulted in an increase in the number of pay television channels and thus an increase in the number of pay television subscriptions or units.

Figure 3.1
Increase in Cable System Capacity

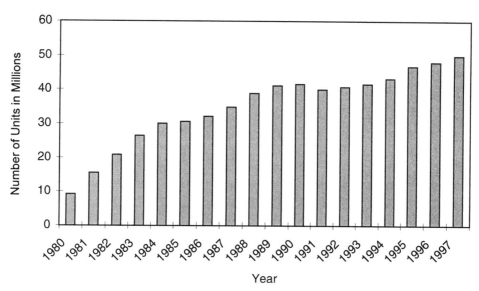

Source: NCTA

Additional channels also allow cable companies and alternative distribution services such as DBS and MMDS to devote more channels to NVOD services where hit movies can be scheduled every 30 minutes as opposed to every two hours.

Recent Developments

One of the most important recent developments in the pay TV services industry has resulted from the fallout from the Telecommunications Act of 1996. The act allowed the cable companies into the telephone business, and many cable companies rushed to upgrade their networks for new service offerings that took advantage of the increased bandwidth of their hybrid fiber/coax (HFC) networks. As a result, cable television customers in some parts of the United States can now receive Internet and telephony services in addition to traditional cable television. With the increased bandwidth also came increased channel capacity; therefore, more cable companies began to offer more PPV channels, more premium channels and their multiplexed versions, and more mini-pays.

Another important development is the evolution from analog to digital delivery services and programming. Digital video compression technologies allow more channels to be offered with superior picture and sound, and set-top boxes allow these digital signals to be shown on the subscriber's television. Direct broadcast satellite services have been offering digital delivery of television content for several years now (see Chapter 5). One DBS service provider, DirecTV, offers multiplexed versions of premium channels and features 61 NVOD PPV channels. Not surprisingly, DirecTV subscribers order more PPV movies than regular cable television subscribers. While only 7% of cable subscribers order a PPV movie per month, more than 30% of DBS households order PPV movies every month (Mitchell, 1998).

Large cable MSOs (multiple system operators) in the United States are currently experimenting with digital cable services. Cox Cable has Cox Digital TV, a system recently deployed in Orange County, California. This system uses a 750 MHz HFC network to deliver 150 entertainment channels including multiplexed versions of premium channels such as HBO (10) and Showtime (10), as well as 35 PPV channels which carry hit movies that start every half hour (Comcast Cable premieres, 1997). The service costs from $5.95 to $10.95 per month in addition to the regular cable rate.

For $10 per month, almost two million of Tele-Communications Incorporated's (TCI) 14.4 million cable customers can receive TCI's digital cable package. Other companies currently testing or offering some digital cable services include Jones Intercable, Comcast, MediaOne, and Century Communications. Most of these services use TCI's Headend in the Sky (HITS) digital satellite signal transport service (Coleman, 1997b). These digital services are made possible, in part, by a new wave of digital set-top boxes described as "a kind of magic box with substantial computing power, two-way capability, interactive programming guides, graphics accelerators, and in some cases cable modems" (Colman, 1997f, p. 51).

With the growth of digital cable and DBS services, pay television programming is also developing at a rapid rate. Premium channels such as Showtime and HBO are both rapidly increasing their channel capacity. Showtime has developed an 18-channel Premium Pak service which features channels much like the Encore package discussed earlier (New channel, 1997). Other cable networks are also multiplexing. For example, MTV is offering The Suite, a package of genre-specific music channels (Higgins, 1997).

While most digital cable systems available now offer NVOD, video on demand systems are being developed and tested. With these systems, subscribers can access programming immediately and have VCR functionality—they can pause, fast forward, and rewind the programming. One example is the OnSet system, which is currently being tested by Suburban Cable in the Philadelphia area. This system, designed by Diva Systems Inc., uses a scaleable video server that holds 1,000 films. Eight films can be streamed into a 6 MHz channel at one time and, using a compression ratio of 8 to 1, as few as two 6 MHz channels can offer up to 1,000 titles over a cable system designed to serve 500 homes per node (Ellis, 1998). Subscribers pay $5.95 per month to have access to the service, as well as a per-program charge.

One of the first VOD service offerings was Time Warner's now defunct Full Service Network (FSN). This service was deployed in Orlando, Florida from 1994 to 1997 and offered switched digital interactive multimedia services using an HFC network. Customers could order movies on demand, with full VCR functionality, from a library of 100 titles for about the same price as a video rental. Ultimately, FSN was a failure because customers did not use the interactive services. Consequently, the service could not generate enough revenue to pay for the cost of the infrastructure. Time Warner, though, is taking the information from FSN and applying it to their new digital video on demand service, Pegasus. They are planning to roll out this newer system in two phases, the first offering enhanced PPV (NVOD), multiplexing, expanded bandwidth, and an interactive program guide. The second phase will offer full video on demand and Internet services such as e-mail, chat, and WWW access (Full Service Network, 1997). This system is seen as competition for DBS's digital programming services. Advances in digital set-top boxes and the MPEG-2 transport standard for digital video are making services like this possible. Time Warner claims that 50% of their cable systems can already support the Pegasus architecture, and predicts it will be fully deployed throughout all its cable systems by 2000 (The Pegasus Program, 1997).

Although the involvement of local telephone companies in delivering television programming has not been as great as predicted when the Telecommunications Act of 1996 was enacted, some telcos are still working in the video service area. Bell Atlantic tested its Stargazer VOD system from 1993 until 1996. This system transmitted movies on demand to subscribers across regular telephone wires using asymmetrical digital subscriber line (ADSL) technology. Subscribers averaged 3.6 buys per month. Bell Atlantic planned to deploy a commercial VOD system, using data from the Stargazer trial and a switched broadband network in Philadelphia. However, while, in 1996, it looked like the telcos were going to become heavily involved in delivering video services, in 1997 and 1998, the telephone companies began to pull out of the video services business (Bell Atlantic, 1996).

Current Status

Things are looking good for pay television services. The number of addressable households in the United States grew from 32 million in 1996 to 37 million in 1997, an increase of 16%. The continued growth of DBS services and roll out of digital cable platforms is expected to push the number of addressable homes up even further.

PPV revenue, split between events and movies, continues to grow, and gross revenue in 1997 was $1.27 billion, a 32% increase over 1996. Showtime Event Television expects revenue in 1998 to rise to $1.7 billion (Showtime Event, 1998; Umstead, 1997; Colman, 1997a). (When analyzing PPV revenues, remember that the local cable system typically keeps about half of the gross revenue, with the other half going to the programmer or event promoter.)

PPV events earned $413 million in 1997 with boxing events, which continue to be one of the most popular events, contributing 61.2% of this revenue. The June 28, 1997 Tyson/Holyfield fight alone made $100 million. Professional wrestling is the next most popular PPV event, garnering 34.3% of the revenue. Request Television, however, expects boxing revenues to drop and wrestling revenues to rise due, in part, to the volatile and strange boxing events of 1997, highlighted by the now infamous ear-biting incident between Tyson and Holyfield.

PPV movies made $603 million in 1997, up from $246 million in 1995 (Mitchell, 1998; Cooper, 1995). This growth can be attributed to DBS, shorter video-to-PPV windows, and increased numbers of PPV channels. As mentioned earlier, DBS subscribers order more than three movies a month (Mitchell, 1998). The average video-to-PPV window was 48 days in 1997, and the PPV industry scored a big victory in 1998 over the Video Software Dealer Association which tried to expand PPV windows (Umstead, 1998). PPV services such as Viewers Choice and Request Television are adding channels to their services to increase buy rates. Viewers Choice research shows that increasing PPV movie channels from one or two—where buy rates are slightly less than 10%—to 10 translates into roughly 50% buy rates (Colman, 1997a). The average price for PPV movies in 1997 was $3.92 and is expected to drop in 1998.

Adult programming made $255 million in 1997. The two big players in adult programming, Playboy and Spice, merged in 1998, when Playboy bought The Spice Entertainment Company for $95 million. Playboy now runs four adult entertainment networks—Playboy TV, Playboy's AdulTVision, Spice, and the Adam and Eve Channel. The Playboy channels reached 17.9 million U.S. households in 1997, while the Spice channels reached 21.5 million households (Petrozzello, 1998).

Figure 3.2 shows the gross PPV revenue for 1997 for events, movies, and adult programming.

Figure 3.2
PPV Programming (in Millions)

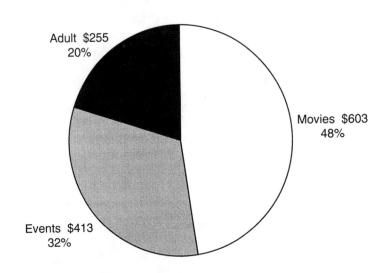

Source: Showtime Event (1998)

Factors to Watch

The future appears bright for pay television services. While it is still not a serious competitor to the home video industry, pay television service revenues continue to grow, while home video revenues have dropped (see Chapter 15). The audience will ultimately make the decision about the future success of pay television services. There are, however, some apparent trends:

- *Continued growth of PPV*—Expect PPV revenues to rise as more PPV channels are added to cable and DBS systems, and as DBS continues to add subscribers. Cable plant upgrades will lead to increased channel capacity and an increasing number of PPV channels and, thus, higher buy rates.

- *Continued multiplexing*—As bandwidth grows and video compression technologies expand, look for increased channel capacity leading premium and other cable networks to continue to multiplex. Theme multiplexes are growing in popularity, so look for more niche-oriented multiplexes of popular cable networks.

- *Expansion of pay television services*—The promise of digital broadcast television in a few years could lead to broadcast stations and networks offering pay television services. Broadcasters will decide if they are going to broadcast one HDTV channel or several Standard Definition Television (SDTV) channels which could include pay services.

- *Digital PPV services*—As cable companies upgrade their systems to digital transmission, look for increased digital PPV services offering NVOD and VOD.

- *Increased competition among video service providers*—Consumers can presently choose from a number of video programming providers including cable and DBS. Look for other providers to enter the market, including the broadcast stations and Internet service providers. In 1998, World Wide Web users can order special programming and events much like PPV. For example, Turner's World Championship Wrestling offered a pay-per-view Internet-only wrestling event using Real Audio (WCW Live, 1998).

Bibliography

Bell Atlantic video on demand efforts begin commercial transition. (1996, October 3). [Online]. Available: http://www.ba.com.

Colman, P. (1997a, December 8). PPV rides a roller coaster. *Broadcasting & Cable*, 103-104.

Colman, P. (1997b, December 8). Stand and deliver. *Broadcasting & Cable*, 42-52.

Colman, P. (1997c, November 24). John Sie: All the right movies. *Broadcasting & Cable*, 28-31.

Colman, P. (1997d, November 10). TVN pushes for digital PPV. *Broadcasting & Cable*, 60-62.

Colman, P. (1997e, November 3). Cox launches digital. *Broadcasting & Cable*, 55.

Coleman, P. (1997f, October 27). Making sense of set-tops. *Broadcasting & Cable*, 51.

Comcast Cable premieres new digital television technology. (1997, April 17). [Online]. Available: http:// www.comcast.com/about/index.htm.

Cooper, J. (1995, June 5). PPV's waiting game. *Cablevision*, 42.

Dickson, G. (1997, August 18). Low blows against high-def. *Broadcasting & Cable*, 46-50.

Ellis, L. (1998). *Valley Co. perks up VOD market.* [Online]. Available: http://www. mediacentral.com/ magazines/multichannelnews/news98.

Full Service Network. (1997, April 30). [Online]. Available: http://www.pathfinder.com/corp/.

Gross, L. S. (1986). *The new television technologies.* Dubuque, IA: Wm. C. Brown Publishers.

Higgins, J. M. (1997, December 1). Cable nets do digital. *Broadcasting & Cable*, 6.

Mair, G. (1988). *Inside HBO.* New York: Dodd, Mead & Co.

Mitchell, K. (1998). *PPV revenues to break 1.2 B mark in 1997.* [Online]. Available: http://www. mediacentral.com/magazines/cableworld/news97.

New channel for Showtime. (1997, December 22). *Broadcasting & Cable*, 28.

Onset of OnSet. (1997, December 8). *Broadcasting & Cable*, 97.

The Pegasus Program. (1998). [Online]. Available: http://www.pathfinder.com/corp/rfp/.

Petrozzello, D. (1998, February 9). Spicing up Playboy. *Broadcasting & Cable*, 44.

Showtime Event TV presented annual PPV industry overview. (1998). [Online]. Available: http://www. mrshowbiz.com/news/wire/971120/5-2-0.

Umstead, R. T. (1998, February 9). *VSDA is missing the target.* [Online]. Available: http://www. mediacentral.com/magazines/multichannelnews/news98.

Umstead, R. T. (1997, December 5). *Another view.* [Online]. Available: http://www. mediacentral.com/ magazines/multichannelnews/news98.

WCW Live events calendar. (1998). [Online]. Available: http://www.wcwwrestling.com/ppvevents.

4

Interactive Television

Paul Traudt, Ph.D.[*]

Interactive television (ITV) describes a range of two-way communication services between service providers (such as cable television systems) and end users. The most common technology includes a little black box that sits on top of the consumer's home television. This set-top box is connected to a local cable company, local telephone company, or both. Earlier set-top boxes included a corded remote control and allowed the interactive televiewer to send channel feedback in response to selected programming. Users could also shop or conduct financial transactions and business from home and order pay-per-view movies and sports.

Over the past two decades, capital investments in ITV have been enormous. Providers have deployed costly infrastructures, while offering less-than-compelling interactive services, resulting in limited consumer interest. Developers have continued to invest in ITV, hoping to strike it rich with the propriety hardware that defines the technology and programming for an entire industry. ACTV Entertainment, Comcast, Cox, MediaOne, Tele-Communications Incorporated, Time Warner, WebTV, Wink, and WorldGate are among ITV's major players.

Background

Early experiments in ITV ventures include Time Warner's two-way interactive cable experiment in the late 1970s, called QUBE, and on-screen videotext trials known as teletext in the early 1980s. Both met with indifference on the part of users. Time Warner launched the Full Service Network trial (known by insiders as QUBE II) in December 1994 in Orlando, Florida at a cost of $250 million. By May 1997, the media giant had

* Associate Professor and Coordinator of Telecommunications, Hank Greenspun School of Communication, University of Nevada, Las Vegas (Las Vegas, Nevada).

31

announced its intent to end the costly video on demand service. Time Warner cited the cost of per-household hardware and the failure to shift consumer interests away from home videocassettes as major reasons behind the unsuccessful venture. Only 4,000 subscribers took advantage of the residential phone service, movies, data banks, and electronic shopping offered by the system. Many industry observers suspected that the costly project represented the last full-scale effort at proprietary-based ITV because of the growth in popularity of Internet-based communications (Goldstein, 1997; Shiver, 1997; Marriage of, 1997).

Recent Developments

The failure of most interactive television ventures using cable television has prompted entreprenuers and proponents of interactive television to look for new ways to deliver interactivity to the television set. For many companies, the answer is an integration of computer and Internet technology with television technology.

The adoption of personal computers in U.S. homes has slowed, however, which has forced the PC industry to explore other ventures. Recent figures place PC penetration at 45%, up only 5% since 1996. In contrast, U.S. cable penetration now stands at 67%, so interest in converging personal computing and Internet networking with traditional televiewing was predictable. Microsoft led the charge to move personal computing and Web surfing out of the home office and into the dens and living rooms of America (Marriage of, 1997). The convergence of television and computer technology expanded the range of ITV services, including surfing the Web, shopping, e-mail, access to community information and services, pay-per-view movies and sports, and distance education.

Full-fledged entry by the personal computer industry repositioned ITV's technological and economic platforms from proprietary cable networks to proprietary computer hardware and software (Mossberg, 1997). Microsoft entered ITV in a big way in April 1997 with the acquisition of WebTV. Bill Gates was convinced that Microsoft's future depended on acquiring a larger share of the ITV market (Desmond, 1997). WebTV provided that market opportunity because it allowed subscribers to watch television programs while simultaneously surfing the Net (WebTV, 1998).

In response, other companies announced plans to provide competing services. Oracle Corporation announced plans in summer 1997 to integrate television programming with Website data, focusing on the problem of slow-loading pages (Clark, 1997). The Oracle system, as well as the majority of others, will transmit data via the television vertical blanking interval (the part of a television signal not displayed on television receivers). Intel announced its own competing version that will provide data using a Web-browser format (Freeman, 1997).

Internet service providers have also entered the fray. In fall 1997, America Online introduced Entertainment Asylum, a Website providing movie reviews, television program listings, celebrity interviews, and e-mail with online hosts (Kaplan, 1997).

In a related move, Bill Gates invested $1 billion in Comcast, a cable MSO (multiple system operator). This strategy was an endorsement of cable system bandwidth as a digital distribution medium, as well as an attempt to establish Microsoft's operating system as the standard for use in digital set-top boxes (Lesly, et al., 1997). Microsoft's alignment with major media such as MSNBC and the NBC Television Network further blurred traditionally distinct industries (Walker & Ferguson, 1998).

Internationally, British Interactive Broadcasting began producing interactive commercials which included icons for immediate consumer responses (Rogers, 1997).

Some competitors chose the technological middle ground, developing services that featured Websites linked to broadcast programs. Broadcast networks are attempting to capture a share of this market by producing interactive versions of such programs as game shows and courtroom dramas. NBC-TV produced virtual episodes of *The Pretender* and *Homicide*. As of mid-1998, such efforts have yet to generate profits (Tedesco, 1997).

Current Status

WebTV users related to the service as an extension of traditional televiewing experiences, rather than as an extension of computer usage. Among Internet services, e-mail communications and interactive chat were the most heavily-used features (see Figure 4.1). Users cite entertainment as the primary reason for using the service. Almost one-third of WebTV subscribers have personal computers, and nearly 15% already have access to online services (Tedesco, 1997). WebTV management claims that the service is not a marriage between television and the Internet—just better television. They predict sales of one million units by the end of 1998 and are counting on their electronic program guide to encourage adoption (Denton, 1997).

WebTV claimed 200,000 subscribers in November 1997, up 100% since Microsoft's acquisition the previous April. Factors contributing to this increase in subscriptions included the introduction of WebTV Plus, an improved set-top computer. Other improvements included a version of Hypertext Markup Language suitable for television, a 1.1 GB hard drive, a self-contained television tuner, and 128-bit encryption technology. There is a second proprietary modem called VideoModem that downloads Web pages, digital video, and audio at 1,000 Kb/s (kilobits per second) for immediate viewing. These downloaded sources can also be stored for later use or printed.

As of mid-1998, WebTV Plus hardware cost $199, plus an additional $50 for the wireless keyboard. There is a $19.95 monthly service charge (Caruso, 1997). Sony, Philips, and Mitsubishi produce and sell the Plus model. The Sony INT-W200 Internet terminal includes a 167 MHz RISC processor, an internal 1.08 GB hard drive, an internal 56 Kb/s modem, and 16-bit CD-quality stereo audio. Graphics processing enhances Web page appearance on home video monitors. The user interface is divided into a home page for television and one for the Web. The television home page provides a grid of television program listings. Users click on a program title and are switched directly to that program. The system is also designed to provide background information about television programs, including producer Websites. The WebTV Plus home page provides access to e-mail and other browser-based services. The system also allows users to monitor television programming while surfing the Web via a picture-in-picture function.

Figure 4.1
Internet Services

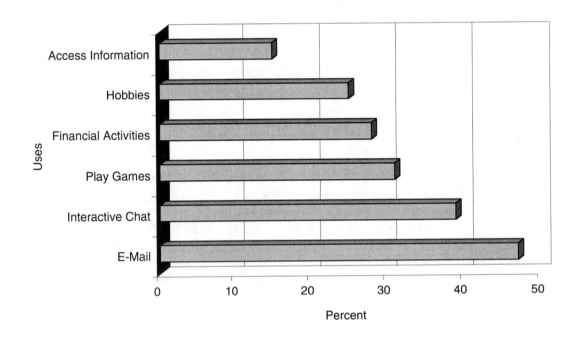

Source: P. Traudt

Critics cite the ease in setting up the system, but also lament that the latest version fails to live up to expectations. Most advertised features have yet to go online, and it may be fall 1998 or later before these additional services will be available. For example, crossover links will allow users to click on an icon superimposed over cooperating television programs. The result is a seamless production between program, Web-based content, and advertising content (Mossberg, 1998). The technology replaces the user's need to periodically download software upgrades necessary to support streaming media; however, WebTV does not fully support the latest releases of these applications.

Microsoft also teamed with *Moesha* producers (for UPN) in fall 1997. Customers with computers equipped with digital-receiver circuitry and Windows 98 could view episodes of this sitcom, type e-mail messages to producers and actors, and engage in chat room communications. Microsoft paid series producers $25,000 to $50,000 per episode (Hiltzik & Helm, 1997).

In January 1998, Microsoft agreed to provide ITV technologies to TCI for the production of Internet-ready set-top boxes. At the center of this agreement is an inexpensive computer chip that performs many of the functions offered by WebTV. The deal between Microsoft and TCI includes an order for five million set-top boxes installed with Microsoft's Windows CE operating system which will provide e-mail, Internet access, and video on demand to subscribers.

Comcast, Cox, MediaOne, and TCI have formed an ITV consortium with the @Home Network, and have ordered 15 million digital set-top boxes from General Instrument for $4.5 billion (Meyer & Stone, 1998). Digital set-top boxes sidestep problems in hardware becoming outdated because Internet, movies on demand, and other information processing will be performed at the cable headend (Silverthorne, 1998). Scientific-Atlanta, developer of PowerTV operating systems, plans to deliver up to 500,000 interactive digital set-top boxes during 1998 to nine cable operators. These operators include Adelphia Communications, Comcast Corporation, Time Warner Cable, and U S WEST Media Group's Media One. Scientific-Atlanta boasts four years of interactive field trials (Lambert, 1998). The corporation is talking with Microsoft about the possibility of merging WebTV technology with PowerTV. Flextech, a British subsidiary of U.S.-based Tele-Communications Inc., is in discussions with Microsoft about forming an interactive television alliance. A joint venture would represent a significant challenge to British Interactive Broadcasting, a venture backed by British Sky Broadcasting Group and British Telecommunications (White, 1998). Estimates vary, but digital set-top deployment should roll out before 2001. In the interim, the more successful current competitors are growing their subscriber base in an attempt to ensure their chances of long-term survival (Waltner, 1998).

Cable television recently established its new open cable standard, an initiative dictating technical requirements for ITV. The specifications include guidelines for writing software in both Windows CE and Java operating system languages for set-top boxes and on servers. The standard redefines the platform for ITV competitors, making proprietary advantages afforded by set-top boxes or software obsolete. Open standards will reduce the importance of hardware. TCI chose not to order WebTV boxes because it favors open cable standards (Thomas, 1998).

A key factor in the continuing evolution of interactive television is the continuing efforts by most of today's players to convert their analog systems to digital formats. Today's ITV systems work with high-end analog boxes, with many of these services coming online just as digital solutions are right around the corner. Cable TV's open standard will increase the importance of developing interactive content.

Ironically, the only broadly-successful form of interactive television uses comparatively simple technology. That application is television shopping, which uses an ordinary, one-way cable channel and the traditional telephone network. Although revenues have leveled somewhat since the dramatic growth experienced during the 1980s, television shopping remains strong, with sales totaling more than $3 billion in 1996 (Grant, 1997).

Factors to Watch

The convergence of computer networking and television has poured new life into ITV, creating a flurry of joint ventures, consortia, and company purchases. A number of factors will determine the degree of ITV's success in the next few years.

ITV can only grow if there is infrastructure in place to deliver interactive services to the end user. As of 1996, the top five multiple system operators in the United States had rewired less than 50% of their systems for two-way capability (Parsons & Frieden, 1998). Some estimate that the latest versions of ITV technology will find adoption in 5% of U.S. homes by 2001 (Thomas, 1998).

The interest of the potential user will ultimately determine the success or failure of ITV's latest evolution. Preliminary analyses suggest that home users continue to approach televiewing as a time for passive and relaxing entertainment, although they appear quite willing to catch up on e-mail and other computer-mediated forms of communication using the same system. Whether or not potential adopters see this type of technology as a substitute for other forms of home computing remains to be seen.

Truly seamless interactive programming services, the Achilles' heel of past efforts, will also be a key factor in the success of current ventures. The short-term prospects for ITV are good in that initial adopters of such technologies will be among those willing to experiment with, and pay for, such services. Successful ITV services will provide unique content enhancements not easily duplicated by competitors. These companies will also require solid partnerships with cable programmers, other video content producers, and cutting-edge Website developers in order to force ITV into America's televiewing mainstream.

Bibliography

Caruso, D. (1997, November 17). WebTV is Microsoft's linchpin in its drive for the interactive media market. *New York Times*, C5.

Clark, D. (1997, August 13). Oracle plans to integrate TV programs with data from the World Wide Web. *Wall Street Journal*, B7.

Denton, N. (1997, September 24). Making better TV. *Financial Times*, 15.

Desmond, E. (1997, November 10). Set-top boxing. *Fortune, 136*, 91-93.

Freeman, M. (1997, May). PCTV: "Must see" for geeks; Computer makers bet on data delivered along with the TV signal. *Mediaweek, 7*, 9-10.

Goldstein, S. (1997, May 17). Time Warner proves it's RIP for VOD; Direct mail, DVD tentative bedfellows. *Billboard, 109*, 58.

Grant, A. E. (1997). Television is the store: Direct response television. In R. A. Peterson (Ed.). *Electronic marketing and the consumer*. Thousand Oaks, CA: Sage.

Hiltzik, M., & Helm, L. (1997, July 28). The digital scramble. *Los Angeles Times*, D1.

Kaplan, K. (1997, October 27). AOL debuts Entertainment Asylum. *Los Angeles Times*, D1.

Lambert, P. (1998, January 23). One bonanza of a digital set-top deal. *Inter@ctive Week Online*. [Online]. Available: http://www.zdnet.com/sznn/content/inwo/0123/278043.html.

Lesly, E., Cortese, A., Reinhardt, A., & Hamm, S. (1997, November 24). Let the set-top wars begin. *Business Week, 3554,* 74-75.

Marriage of convenience. (1997, November). *Time Digital,* 60-64.

Meyer, M., & Stone, B. (1998, March 16). A post-PC future? *Newsweek,* 42-43.

Mossberg, W. (1997, August 7). The marriage of TV and home computer may last this time. *Wall Street Journal,* B1.

Mossberg, W. (1998). Internet television units show improvement after rocky start. *Las Vegas Review-Journal and Las Vegas Sun,* 3K.

Parsons, P., & Frieden, R. (1998). *The cable and satellite television industries.* Boston: Allyn & Bacon.

Rogers, D. (1997, May 29). Buying through the box. *Marketing,* 27.

Shiver, J. (1997, April 2). Time Warner's interactive TV project blinks. *Los Angeles Times,* D1.

Silverthorne, S. (1998). *Exec sees short run for WebTV.* [Online]. Available: http://www.zdnet.com/sznn/content/zdnn/0214/28524.html.

Tedesco, R. (1997, June 2). That's Internetainment: What's online to tempt the viewers. *Broadcasting & Cable, 127,* 54-55, 58, 60, 62, 64.

Thomas, E. (1998). *WebTV grabs lead but can it hold it?* [Online]. Available: http://www.zdnet/com/zdnn/content/msnb/0205/282667.html.

Walker, J., & Ferguson, D. (1998). *The broadcast television industry.* Boston: Allyn & Bacon.

Waltner, C. (1998, February 16). Will interactive TV players survive in digital era? *Inter@ctive Week.* [Online]. Available: http://www.dnet.com/intweek/print/980216/285886.html.

WebTV. (1998). [Online]. Available: http://www.webtv.com/tv/tune/homepage/index.html.

White, P. (1998). *Flextech, Microsoft in talks over TV venture.* [Online]. Available: http://cgi.zdnet.com/cgi-bin/printme.cgi?t=zdnn.

Direct Broadcast Satellites

Ted Carlin, Ph.D.*

A recent newspaper article used the following headline to introduce a story focusing on the direct broadcast satellite (DBS) industry: "Pizza-size satellite dish starts to deliver" (Ribbing, 1998, p. 1D). In 1998, after almost a decade of technology development, programming acquisitions, and legal maneuvering, the DBS industry has finally established itself as a stable and formidable competitor to cable television for multichannel television viewers. With over 6.6 million subscribers as of March 1998 (DTH subscriber, 1998), the DBS industry is aggressively pushing forward to expand its subscriber base before its window of opportunity closes.

This window of opportunity—to attract and retain as many DBS subscribers as possible—is slowly starting to close as the cable industry intensifies its conversion to digital technology. By using digital set-top converters, cable modems, and fiber optic cable infrastructure, cable operators are preparing to significantly upgrade the channel capacity and technical quality of their service over the next several years (Hogan, 1998c). Cable operators, through digital delivery of their services, are hoping to nullify two current DBS advantages:

(1)   Superior laserdisc-quality video and CD-quality sound.

(2)   More program channels due to greater transmission capacity.

The DBS industry, recognizing cable's digital progress, is using these advantages, and others to be discussed shortly, to establish itself as the choice for multichannel television in the United States.

---

*Assistant Professor of Radio/Television, Department of Communication and Journalism, Shippensburg University (Shippensburg, Pennsylvania).

Background

As originally conceived in 1962, satellite programming was never intended to be transmitted directly to individual households. After the Federal Communications Commission (FCC) implemented an "open skies policy" to encourage private industry to enter the satellite industry in 1972, satellite operators were content to distribute programming between television networks and stations, cable programmers and operators, and business and educational facilities (Frederick, 1993). The FCC assigned two portions of the fixed satellite service (FSS) frequency band to be used for these satellite relay services: the low-power C-Band (3.7 GHz to 4.2 GHz) and the medium-power Ku-band (11.7 GHz to 12.2 GHz).

In late 1975, Stanford University engineering professor Taylor Howard was able to intercept a low-power C-band transmission of the Home Box Office (HBO) cable network on a makeshift satellite system he designed (Parone, 1994). In 1978, Howard published a "low cost satellite-TV receiving system" how-to manual. Word spread rapidly among video enthusiasts and ham radio operators, and, by 1979, there were about 5,000 of these television receive-only (TVRO) satellite interception systems in use.

These 6- to 12-foot TVRO satellite dishes are commonplace throughout the United States, especially in rural areas not served by cable television services, and there are four million in use (Jessell, 1996). The large receiving dish is required to facilitate proper reception of the low-power C-band transmission signal. A number of factors prevented TVRO systems from becoming a realistic, national alternative to cable television for multichannel television service. These included the high cost of the TVRO system (around $2,000), the large size of the dish, city and county zoning laws, and the scrambling of C-band transmissions by program providers.

In the 1980s, a few entrepreneurs turned to the medium-power Ku-band to distribute satellite transmissions directly to consumers. By utilizing unused transmission space on existing Ku-band relay satellites, these companies would be the first to create a direct-to-the-home (DTH) satellite transmission service that would use a much smaller receiving dish than existing TVROs (Whitehouse, 1986). The initial advantages of DTH systems over TVRO systems included the higher frequencies and the higher power of the Ku-band, which resulted in less interference from other frequency transmissions, and stronger signals which could be received on the smaller three-foot dishes.

Many factors proved to be primary contributors to the failure of these medium-power Ku-band DTH services in the 1980s, including:

- High consumer entry costs ($1,000 to $1,500).

- Potential signal interference from heavy rain and snow.

- Limited channel capacity (compared with existing cable systems).

- Restricted access to programming (Johnson & Castleman, 1991).

DTH ventures by Comsat, United Satellite Communications, Skyband, and Crimson Satellite Associates failed to get off the ground during this period.

Also during this time between 1979 and 1989, the World Administrative Radio Conference (WARC) of the International Telecommunications Union (ITU) authorized and promoted the use of a different section of the FSS frequency band. The ITU, as the world's ultimate authority over the allocation and allotment of all radio transmission frequencies (including radio, TV, microwave, and satellite frequencies), allocated the high-power Ku-band (12.2 GHz to 12.7 GHz) for "multichannel, nationwide satellite-to-home video programming services in the Western Hemisphere" (Setzer, Franca & Cornell, 1980, p. 1). These high-power Ku-band services were to be called direct broadcast satellite (DBS) services. Specific DBS frequency assignments for each country, as well as satellite orbital positions to transmit these frequencies, were allocated at an ITU regional conference in 1983 (RARC, 1983).

A basic description of a DBS service, based on ITU specifications, was established by the FCC Office of Plans and Policy in 1980:

> A direct broadcast satellite would be located in the geostationary orbit, 22,300 miles above the equator. It would receive signals from earth and retransmit them for reception by small, inexpensive receiving antennas installed at individual residences. The receiver package for a DBS system will probably consist of a parabolic dish antenna, a down converter, and any auxiliary equipment necessary for encoding, channel selection, and the like (Setzer, Franca & Cornell, 1980, p. 7).

The FCC then established eight satellite orbital positions between 30°W and 175°W for DBS satellites. Only eight orbital positions are available for DBS because a minimum of 9° of spacing between each satellite is necessary to prevent the interference of signal transmissions. The FCC also assigned a total of 256 analog TV channels for DBS to use in this high-power Ku-band, with a maximum of 32 DBS channels per orbital position. Only three of these eight orbital positions (101°W, 110°W, and 119°W) can provide DBS service to the entire continental United States. Four orbital positions (148°W, 157°W, 166°W, and 175°W) can provide DBS service only to the western half of the country, while the orbital position at 61.5°W can only provide service to the eastern half.

The FCC received 15 applications for these DBS orbital positions and channels in 1983, accepted eight applications, and issued conditional construction permits to the eight applicants. The FCC granted the construction permits "conditioned upon the permitee's due diligence in the construction of its system" (see 27 C.F.R. Sect. 100.19b). DBS applicants had to do two things to satisfy this FCC "due diligence" requirement:

(1)   Begin construction or complete contracting for the construction of a satellite within one year of the granting of the permit.

(2)   Begin operation of the satellite within six years of the construction contract.

The original eight DBS applicants were CBS, Direct Broadcast Satellite Corporation (DBSC), Graphic Scanning Corporation, RCA, Satellite Television Corporation, United States Satellite Broadcasting Company (USSB), Video Satellite Systems, and Western Union. During the 1980s, some of these applicants failed to meet the FCC due diligence requirements and forfeited their construction permits. Other applicants pulled out, citing the failures of the medium-power Ku-band DTH systems, as well as the economic recession of the late 1980s (Johnson & Castleman, 1991).

In August 1989, citing the failures of the eight DBS applicants to launch successful services, the FCC revisited the DBS situation to establish a new group of DBS applicants (FCC, 1989). This new group of applicants included two of the original applicants, DBSC and USSB. These were joined by Advanced Communications, Continental Satellite Corporation, Direcsat Corporation, Dominion Satellite Video, EchoStar Communications Corporation, Hughes Communications, and Tempo Satellite Services.

From 1989 to 1992, not one of these DBS services was able to launch successfully. In addition to having problems raising capital, most were awaiting the availability of programming and the development of a reliable digital video compression standard. Investors were unwilling to invest monies into these new DBS services unless these two obstacles were overcome (Wold, 1996).

Cable operators, fearing the loss of their own subscribers and revenue, were placing enormous pressure on cable program networks to keep their programming off the new DBS services. Cable operators threatened to drop these program networks if they chose to license their programming to any DBS service (Hogan, 1995). These program networks were essential for DBS companies to launch their services because they had little money or expertise for program production of their own (Manasco, 1992).

In late 1992, DBS companies had this programming problem solved for them through the passage of the Cable Television Consumer Protection and Competition Act. The act guaranteed DBS companies access to cable program networks, and it "[forbade] cable television programmers from discriminating against DBS by refusing to sell services at terms comparable to those received by cable operators" (Lambert, 1992, p. 55). This provision, which has since been upheld in the Telecommunications Act of 1996, finally provided DBS companies with the program sources they needed to attract investors and future subscribers.

The other obstacle—establishment of a digital video compression standard—was solved by the engineering community in 1993 when MPEG-1 was chosen as the international standard. By using MPEG-1, DBS companies could digitally compress eight program channels into the space of one analog transmission channel, thus greatly increasing the total number of program channels available to consumers on the DBS service. (For example, the FCC has assigned DirecTV 27 analog channels. Using MPEG-1, DirecTV can actually provide their subscribers with 216 channels of programming.) In 1995, DBS companies upgraded their systems to MPEG-2, the improved broadcast-quality version.

With these obstacles behind them, two of the DBS applicants, Hughes Communications and USSB, were the first to launch their DBS services in June 1994. Utilizing the leadership and direction of Eddie Hartenstein (DirecTV) and Stanley Hubbard (USSB), Hughes established a subsidiary, DirecTV, to operate its DBS system, and then agreed to work with USSB to finance, build, deploy, and market their DBS systems together (Hogan, 1995). Hughes launched three satellites to the 101°W orbital position from 1993 to 1995. Both companies then signed a contractual agreement with Thomson Consumer Electronics to use Thomson's proprietary digital satellite system (DSS) to transmit and receive DirecTV and USSB programming (Howes, 1995).

DSS employs an 18-inch dish to receive the high-power Ku-band digital transmissions, a VCR-sized integrated receiver-decoder (IRD), and a multifunction remote control. Consumers must purchase the DSS receiving equipment from satellite retailers, consumer electronic stores, or department stores. They then have the choice of purchasing programming on a monthly or yearly basis from DirecTV, USSB, or both.

Recent Developments

After the successful launch of DirecTV and USSB in 1994, the FCC tried to force the other DBS applicants to bring their services to the marketplace. In late 1995 and early 1996, the FCC once again reevaluated the DBS applicants for adherence to its due diligence requirements. After several hearings, the FCC revoked the application of Advanced Communication Corporation, and stripped Dominion Satellite Video of some of its assigned channels for failing to meet these requirements.

Appeals by both companies were denied by the FCC, and these channels were then auctioned by the FCC in January 1996. MCI and News Corp., working together in a joint venture, obtained the Advanced DBS channels, while EchoStar obtained the Dominion channels to add to its previously-assigned channels. EchoStar also acquired the DBS channels from two other applicants, DirecSat and DBSC, through FCC-approved mergers in 1995 and 1996 (FCC, 1996).

In another merger, Loral-DBS, Inc., a subsidiary of Loral Aerospace Holdings, acquired the DBS channels of Continental Satellite Corporation. Continental was forced to turn over the channels to Loral-DBS after failing to meet previous contractual obligations with Loral Aerospace for launching Continental's proposed DBS satellite (FCC, 1995). Loral-DBS has yet to announce or launch its DBS service.

On March 4, 1996, EchoStar launched its high-power DBS service, the DISH Network, using the EchoStar-1 satellite at 119°W, becoming the third DBS applicant to successfully begin operations. Similar to DirecTV/USSB, EchoStar launched a second satellite in September 1996, in the 119°W orbital position to increase the number of DISH Network program channels to 170.

The DISH Network does not use the same DSS transmission format used by DirecTV/USSB. Instead, it uses the international satellite video transmission standard, digital video broadcasting (DVB), which was created after the DSS standard. Like DSS equipment, the DISH Network's DVB equipment utilizes MPEG-2 for digital video compression. What this means is that DISH Network subscribers can receive only DISH Network transmissions, and DirecTV/USSB subscribers can receive only DirecTV/USSB transmissions. The DVB system employs an 18-inch dish to receive its high-power Ku-band digital transmissions, a VCR-sized integrated receiver-decoder, and a multifunction remote control similar to DirecTV/USSB.

As with DirecTV/USSB, the DISH Network requires subscribers to purchase the DVB system, and then pay a separate amount for monthly or yearly programming packages. Also, like DirecTV/USSB, the DISH Network offers professional installation of the DVB equipment, or a do-it-yourself installation kit. Currently, both companies offer professional installation for $99. Because of competition, both companies have dropped prices by more than $100 since 1996. Both also offer various installation specials to attract new subscribers. The DISH Network, however, is the only company that sells its equipment directly from the factory via the Internet and an 800 phone number. Equipment can also be purchased from authorized satellite retailers.

The cable television industry did not ignore the implementation and growth of these DBS companies. In 1994, Continental Cablevision was intent on establishing a cable "Headend in the Sky" for consumers living in non-cable-access areas of the United States. It enlisted the support of five other cable operators (Comcast, Cox, Newhouse, Tele-Communications Incorporated, and Time Warner) and one satellite manufacturer (GE Americom) to launch a successful medium-power Ku-band DTH service (Wold, 1996). The service, named Primestar, transmitted 12 basic cable channels from GE Americom's medium-power K-1 satellite to larger three-foot dishes. Primestar offered far fewer channels than any of the cable operators' own local cable systems, so they believed that their cable subscribers would not be interested in Primestar as a replacement for cable service. (Primestar is not a true DBS service because it does not use FCC-assigned, high-power DBS channels, although most consumers are unaware of this discrepancy.)

As DirecTV and USSB began to prove that DBS was a viable service in late 1994, Primestar decided to change its focus and expand and enhance its offerings to compete directly with DBS. Primestar converted its 12-channel analog system to a proprietary DigiCipher-1 digitally-compressed service capable of delivering about 70 channels. In 1997, Primestar moved its service to GE Americom's medium-power GE-2 satellite, and increased its channel capacity to 160. To differentiate itself from the DBS companies, Primestar decided to market its service just like a local cable TV service by leasing the equipment *and* the programming packages together in one monthly fee. Subscribers were *not* required to purchase the Primestar dish, IRD, and remote, although equipment purchase was an option. Primestar continues to use this marketing approach.

Therefore, as of mid-1998, there are four companies operating DBS/DTH services in the continental United States. There are three DBS services (DirecTV, USSB, and the DISH Network) and one DTH service (Primestar). Table 5.1 summarizes the status of the DBS licensees.

Table 5.1

U.S. DBS Licensees

Orbital Position	61.5°W	101°W	110°W	119°W	148°W	157°W	166°W	175°W
Satellites in Orbit	EcStar-3	DBS-1, DBS-2, DBS-3		EcStar-1, EcStar-2, Tempo-1				
DirecTV Channels		27				27		
USSB Channels		5	3		8			
DISH Channels	11		1	21	24		1	32 (D)
MCI/NC Channels			28 (B)					
Loral Channels	11							
Tempo Channels				11 (C)			11	
DomSV Channels	8 (A)						8	
Channels Not Yet Assigned	2					5	1	

(A) EchoStar controls these channels through a channel-sharing agreement with Dominion
(B) Primestar is awaiting FCC approval to acquire these channels from MCI/News Corp.
(C) Primestar is awaiting FCC approval to acquire these channels from Tempo.
(D) EchoStar controls 10 channels, while its acquisitions—DBSC and DirecSat—control 11 apiece.

Source: T. Carlin

Current Status

U.S. DBS/DTH

The United States is the world's number one user of DBS/DTH services. As of March 1998, there were 6.6 million DBS/DTH subscribers. Table 5.2 summarizes subscribership figures for the industry since July 1994.

Table 5.2

U.S. DBS/DTH Subscribers

Date	Total DTH	DirecTV/USSB	DISH	Primestar
7/1/94	70,000	0	0	70,000
12/1/94	390,000	200,000	0	190,000
7/1/95	1.15 mil.	650,000	0	500,000
12/1/95	1.98 mil.	1.1 mil.	0	880,000
7/1/96	2.95 mil.	1.6 mil.	75,000	1.27 mil.
12/1/96	4.04 mil.	2.13 mil.	285,000	1.60 mil.
7/1/97	5.04 mil.	2.64 mil.	590,000	1.76 mil.
12/1/97	5.95 mil.	3.12 mil.	965,000	1.90 mil.
3/1/98	6.60 mil.	3.45 mil.	1.14 mil.	2.01 mil.

Source: SkyReports

Most of these 6.6 million subscribers are located in rural areas that are not served by a local cable system. According to the FCC, total multichannel television penetration in the continental United States, including 64.8 million cable television subscribers and 2.08 million C-band TVRO users, is just about 75%. This means that "the U.S. is running out of unserved homes to pitch, particularly in the boonies" (DBS knockin', 1998, p. 2).

It also means that DirecTV/USSB, the DISH Network, and Primestar are starting to aggressively seek out current cable customers for their services. Various new marketing campaigns by DirecTV/USSB and the DISH Network are being planned and implemented to attack the cable industry's most observable weaknesses:

- Rate hikes (due largely to increased programming costs).

- Lack of channel variety (due to limited analog systems).

- Customer service problems (due to past monopolistic practices).

While attacking these cable industry problems, DBS companies are also trying to solve two main issues impacting the industry at present: multiple television hook-ups in the home and subscriber access to local broadcast television stations and broadcast networks. Both issues are considered major impediments to the development and growth of DBS systems as true competitors to cable television (Hogan, 1998c).

When DBS companies began operations in the mid-1990s, the goal was to get the basic, one-TV system into as many rural subscriber homes as possible (Boyer, 1996). Due to declining costs, increased technology, and a new effort to attract cable customers, the focus has shifted to providing more user-friendly DBS services. According to Bill Casamo, executive vice president for DirecTV, the cost of a second receiver has always been a barrier to entry for some first-time subscribers. "As we go more into cabled markets, that becomes more of a factor," because cable customers are accustomed to seeing cable in multiple rooms of the home (Hogan, 1998e, p. 18).

As a result, DirecTV/USSB and the DISH Network have started marketing multiple TV setups for new subscribers. As of mid-1998, DirecTV/USSB is offering a basic "slave" unit for $99 with the purchase of a more expensive dual LNB DSS decoder (a dual LNB unit allows the receive dish to feed multiple receivers with different channels of programming simultaneously). The DISH Network is offering a scaled-down DVB decoder (Model 1000) for $129 to be used in an additional room. Prices can be expected to drop even further through increased competition and marketing. It is interesting to note that Primestar, owned and operated by cable companies, has always used dual LNB decoders since its inception. Primestar does charge subscribers $13 per month to lease an additional decoder, similar to local cable operators charging a monthly fee for additional cable boxes.

The second issue—access to local broadcast stations and networks—is much more difficult to overcome for DBS in most cabled communities. A provision in the Satellite Broadcasting Act of 1988 still prohibits DBS subscribers who live within the coverage area of local television stations from receiving any local TV stations or broadcast networks via their DBS system. Subscribers must connect a television antenna to their DBS system, or subscribe to local cable, to receive broadcast television stations. All of the DBS/DTH providers are allowed to provide broadcast stations that are available on satellite to those subscribers living outside of the coverage areas (i.e., rural, non-cable areas), and each offers various à la carte packages of stations and networks.

In a related issue—the DISH Network, because it has acquired a large number of DBS channels through the 1996 FCC DBS auction, mergers, and a channel-sharing agreement—has the channel capacity to deliver local stations back into their own markets. It is lobbying Congress and the U.S. Copyright Office to amend current U.S. satellite copyright regulations to allow it, and other interested DBS companies, to do so. The National Association of Broadcasters has supported this "local-into-local" approach, but issues of retransmission consent, must-carry, syndicated exclusivity, sports blackouts, and copyright protection—all issues currently being faced by cable operators—must still be resolved (Responses cool, 1998). The DISH Network sees local-into-local as necessary for its service to be able to compete on a level playing field with cable (EchoStar, 1998).

In terms of programming, DirecTV/USSB, the DISH Network, and Primestar have been able to acquire all of the top cable program networks, sports channels and events, and PPV events as envisioned by the 1992 Cable Act. What differentiates one DBS/DTH service from the other is how the program services are priced, packaged, and promoted. Each service has the following:

- On-screen program guides.

- Parental control features.

- Preset PPV spending limits.

- Instant PPV ordering using the remote control and a phone line hookup.

- Favorite channel lists.

- Equipment warranties.

- 800 phone numbers for customer service.

DirecTV

Programming on DirecTV consists of packages of basic cable channels and premium movie channels not found on USSB, as shown in Table 5.3. It also offers individual PPV movies, concerts, and sporting events through DirecTicket (i.e., movies for $2.99, boxing for $14.95). Using the remote control, subscribers can search the interactive program guide to access desired channels or to request PPV events. DirecTV, unlike cable, offers unique packages of college and professional sports (MLB Extra Innings, MLS Shootout, NBA League Pass, NFL Sunday Ticket, NHL Center Ice, and ESPN College Basketball and Football). It also offers CD-quality, commercial-free digital audio service, Music Choice, as part of its Total Choice package.

To see any of this programming, subscribers must purchase a DSS equipment package, available through a variety of retailers, and have the DSS system installed. DirecTV has authorized 17 different companies to manufacture the DSS equipment (including RCA, GE, Sony, Panasonic, and Sanyo), hoping to entice consumers with familiar, reliable brands. Prices vary according to individual retailers (including Best Buy, Circuit City, Sears, and Wal-Mart), the brand name chosen, and the complexity of the DSS system selected. Equipment prices can range from $149 to $499 for a one-TV system, plus $99 to $199 for installation.

Table 5.3
DirecTV Programming Packages

1   Total Choice PLATINUM $47.99 a month
    Includes over 85 Total Choice channels + over 25 specialty sports networks +14
    commercial-free movie channels.

2   Total Choice GOLD $39.99 a month
    Includes over 85 Total Choice channels + over 25 specialty sports networks.

3   Total Choice SILVER $39.99 a month
    Includes over 85 Total Choice channels + 14 commercial-free movie
    channels.

4   Total Choice PLUS ENCORE $33.99 a month
    Includes over 85 Total Choice channels + 8 ENCORE movie channels.

5   Total Choice $29.99 a month
    Over 85 channels of great entertainment, including 31 commercial-free Music
    Choice digital audio channels.

6   Select Choice $19.99 a month
    Over 40 popular channels of news, sports, and entertainment programming.

7   Plus DirecTV $14.99 a month
    A variety of news, sports, and entertainment programming for cable subscribers.

Source: DirecTV

USSB

Unlike DirecTV, USSB is built around premium movie channels. Because USSB was assigned just five DBS channels at 101°W, it can only offer up to 40 digitally-compressed channels. USSB decided to work with DirecTV and provide a complimentary movie-oriented channel to package with DirecTV, as shown in Table 5.4. As an enticement, USSB continues to offer one free month of Entertainment Unlimited to new subscribers. Each USSB programming package includes Big Events Channels, DSS Information Channel 999, and the DSS Edition of TV Guide for just $2.99 per month ($1.99 per month with the Entertainment Unlimited package).

Table 5.4

USSB Programming Packages

1 Entertainment Unlimited™ $32.99/month
*5 channels of HBO, 4 channels of Showtime plus Showtime Extreme, 3 channels
of Cinemax, 2 channels of the Movie Channel, Sundance Channel, 2 channels of
HBO Family, FLIX, fXM: Movies from Fox.*

2 Select Three $29.99/month
*Choose three of the following networks: Multichannel HBO, Multichannel Show-
time, Multichannel Cinemax, Multichannel The Movie Channel, Sundance Channel. Plus, you get HBO Family when you order HBO and fXM: Movies from Fox.*

3 Select Two $20.99/month
*Choose two of the following networks: Multichannel HBO, Multichannel Showtime,
Multichannel Cinemax, Multichannel The Movie Channel, Sundance Channel.
Plus, you get fXM: Movies from Fox.*

4 Select One $10.99/month
*Choose one of the following networks: Multichannel HBO, Multichannel Showtime,
Multichannel Cinemax, Multichannel The Movie Channel, Sundance Channel.*

5 Extras $4.99 each/month
*Take an extra movie channel for a small additional price. FLIX, fXM: Movies from
Fox.*

Source: USSB

The DISH Network

The newest and the most active of the DBS providers is EchoStar's Digital Sky High-
way (DISH) Network. Using aggressive pricing strategies for programming and DVB
equipment, the DISH Network reached one million subscribers faster than any other
DBS/DTH service by December 1997 (Hogan, 1998a). Marketing itself as the best value in
satellite television, the DISH Network offered its equipment for only $199 plus installa-
tion. DirecTV/USSB responded by lowering their DSS prices, resulting in a continuing
price war.

In January 1997, EchoStar's chairman Charlie Ergen, and News Corp.'s Rupert Murdoch, shocked the DBS industry by announcing a partnership to deliver a new DBS service, complete with local broadcast stations, to the continental United States. Called
ASkyB, this new DBS service would have combined DISH Network channels with Murdoch's MCI/News Corp.'s unused DBS channels to deliver over 500 channels to subscribers. Citing strategic management differences with Ergen, and EchoStar's unstable

financial picture, Murdoch unexpectedly pulled out of the project only a few months later to pursue other ventures, including the purchase of The Family Channel and a new DBS venture with Primestar described below.

In 1998, Ergen and the DISH network announced plans to deliver local-into-local broadcast television programming on its own, and eight new channels were added to its reorganized program line-up, as shown in Table 5.5. The DISH Network also offers a number of à la carte (i.e., The Golf Channel for $4.99 per month, Playboy TV for $14.99 per month, and international TV channels) and PPV movies (Dish-on-Demand for $2.99 per month).

Through its channel-sharing agreement with Dominion Satellite Video, EchoStar is marketing Dominion's Sky Angel religious programming to its subscribers. Subscribers must purchase a second dish antenna and point it at the EchoStar III satellite located at 61.5°W in order to receive the $9.99 per month service. This same second dish antenna would also be used to receive the DISH Network's local-into-local broadcast television programming and future Internet services.

Table 5.5
DISH Network Programming Packages

1	America's Top 60 CD	$28.99/month
	Includes all of the America's Top 40 channels, plus 16 more basic cable channels, one regional sports network, and the 30-channel DISH CD digital audio service.	
2	America's Top 40 CD	$19.99/month
	The basic DISH package which includes 40 basic cable channels, plus The Disney Channel.	
3	Movie Packages	$ per number of packages selected per month
	Subscribers can select up to four movie channel packages from HBO, Cinemax, Showtime Networks, and Starz/Encore. One package is $10.99, two packages are $19.99, three packages are $27.99, and four packages are $34.99.	
4	Multisport Package	$4.99/month
	Available only to America's Top 60 subscribers, this sports package includes all of the Fox Sportsnet affiliates and five other regional sports networks.	

Source: EchoStar

Primestar

After abruptly ending his proposed partnership with EchoStar to launch ASkyB, Rupert Murdoch sought an alliance with cable-owned Primestar to launch a DBS service in the United States using his MCI/News Corp. channels at 110°W. By distancing himself from the financial insecurities of EchoStar and the personal confrontations with Ergen, Murdoch attempted to establish a more secure financial and collaborative relationship with his fellow colleagues in cable TV.

Consequently, during late 1997 and early 1998, Primestar established agreements with Murdoch and another DBS applicant, Tempo Satellite, to acquire their high-power DBS channels at 110°W and 119°W to launch this new, unnamed, high-power DBS service. As of mid-1998, the FCC and the Justice Department had yet to approve the agreements because Primestar is owned by cable operators. Both agencies are concerned about potential monopolistic practices by Murdoch and Primestar's five cable-affiliated owners to eliminate cable TV's competition in the multichannel television industry through their combined ownership of local cable systems, cable programming networks, a DTH service, and a DBS service. The question being considered is, "Does the additional ownership of a DBS service give Murdoch and Primestar Partners too much control over the multichannel television industry?"

To complicate matters even further, Tempo is owned by TCI Satellite Entertainment (TSAT), a publicly traded spin-off of cable operator TCI, which is one of the six partners that own Primestar. TSAT must meet an FCC-mandated due diligence requirement in May 1998 or face losing the 11 channels at the 119°W orbital position. This may compel TSAT to seek a due diligence extension with the FCC, or to launch a separate high-power DBS service before the FCC deadline if the waiver is not approved. TSAT could then spin off the service to a non-cable owner if the FCC rules against the ownership transfer of Tempo to Primestar (Primestar presses, 1998).

In preparation for the merger, and in the hopes of becoming an independent company separate from its cable owners, Primestar is also restructuring its ownership to form a new corporation, Primestar, Inc. This new corporation would merge TSAT with Primestar Partners and would allow Primestar, Inc. to become the holder of all existing Primestar assets and all of Tempo's assets, including its DBS channels at 110°W and 166°W. TSAT's stockholders have approved the merger, subject to pending FCC approval (TSAT/Primestar restructuring, 1998).

In terms of programming its current medium-power DTH service, Primestar has settled on an "easy to get, easy to watch" marketing strategy that focuses on its cable-like plan of leasing its equipment and programming together for one monthly fee. With dual LNB receive dishes, free maintenance, and an 800 phone number, Primestar has effectively positioned itself as cable television via satellite (Hearn, 1998).

Like DirecTV/USSB, the DISH Network, and cable television services, Primestar uses various tiers of programming, as shown in Table 5.6. Primestar also offers à la carte selections like MLB Extra Innings, NBA League Pass, NHL Center Ice, The Golf Channel, and Playboy TV. Its PPV movie option, Prime Cinema, is available for $3.95 per movie.

Table 5.6
Primestar Programming Packages

1	**Prime Hits**	$59.99/month
	Includes all of the Prime Value tier, plus the Variety tier and Hollywood Hits.	
2	**Prime Entertainment**	$43.99/month
	Includes all of the Prime Value tier, plus the Variety tier, Starz!, and Encore networks.	
3	**Prime Value**	$32.99/month
	Includes 68 channels, with MSNBC Intellicast Regional Weather, a regional sports channel, the Disney Channel, and the 30-channel Prime Audio by DMX digital audio service.	
4	**Variety Tier**	$7.99/month
	Includes 16 basic cable channels not included in the Prime Value tier.	
5	**Hollywood Hits**	$27.99/month
	Includes multichannel HBO, Showtime, Starz!, and Encore.	
6	**Sports Tier**	$2.99/month
	Includes all of the regional sports networks carried by Primestar.	

Source: Primestar

International DBS/DTH

Although other countries have used satellites to transmit television signals to stations and cable systems, Japan was the first country to launch a DBS service in 1984 (Otsuka, 1995). Now, Japan and the rest of Asia are moving into digital DBS like the United States. In October 1996, Japan's largest satellite operator, JSAT, launched the country's first digital DBS system, PerfecTV. By March 1998, PerfecTV had over 300,000 subscribers for its 100-channel service and had decided to merge with a competitor, News Corp.'s JSkyB.

The new digital DBS service, SkyPerfecTV, will be able to deliver about 200 channels to its 500,000-plus subscribers (JSkyB, PerfecTV, 1998). And, DirecTV Japan, a digital DBS competitor launched by Hughes Corporation in December 1997, is currently offering its subscribers 85 digital channels, including exclusive coverage of Japanese baseball and soccer teams (DirecTV Japan, 1998).

STAR TV was launched in 1991 in Hong Kong, and is still the driving force for television in the rest of Asia. Within six months of STAR TV's launch, eight million viewers had tuned in. Today, STAR TV covers 53 countries, spanning an area from Egypt to Japan and the Commonwealth of Independent States to Indonesia, reaching an estimated audience of 260 million (STAR TV background, 1998). STAR TV, which was purchased by News Corp. in 1993, offers both subscription and free-to-air television services using AsiaSat 1 as its primary satellite platform, with additional services available on the AsiaSat 2 and Palapa C2 satellites.

In Europe, satellite consortiums SES Astra and Eutelsat continue to dominate the European DTH market. Using a number of satellites, each group has been able to provide over 100 channels to subscribers throughout the continent (Forrester, 1997). Luxembourg-based SES Astra has even announced plans to build and operate a "super-satellite" that would be capable of transmitting up to 1,000 digitally-compressed DBS channels.

Primary competition to SES Astra and Eutelsat has been from a number of recent national/regional DTH systems including News Corp.'s England-based BSkyB, France's CanalSatellite and TPS, Germany's DF-1, Italy's Telepiu, Norway's Canal Digital, and Spain's CSD and Via Digital. Digital DBS/DTH operators have done extremely well in France, Japan, and Malaysia, but have been under-performing in Germany and Italy (Francis, 1997).

Closer to the United States, in Latin America and Canada, DBS systems are also taking off. In Latin America, as deregulation and privatization of the telecom markets continues to spread through the region, the result has been fierce competition in satellite services. The leader in DBS in Latin America is Hughes Corporation's Galaxy Latin America, which provides a Latin America DirecTV service to 11 countries, including Brazil, Costa Rica, Mexico, and Panama. At the end of 1997, Galaxy totaled over $70 million in revenue from about 300,000 subscribers (Kessler, 1998).

In Canada, four companies are involved in the DBS/DTH industry. AlphaStar, the first company to launch a DBS service there in 1996, discontinued operations after a disappointing year. Its parent company, Tee-Comm Electronics, filed for bankruptcy in 1997 and is facing a lawsuit by stockholders for artificially inflating the value of AlphaStar and Tee-Comm's stock (Tee-Comm executives, 1998).

Also in 1997, Telesat Canada attempted to partner with a U.S. company, cable-owned TCI Satellite, to launch a U.S.-based service using its Canadian DBS channels. The FCC did not approve the merger, and now Telesat is preparing to launch its own DBS satellite operation. It is attempting to arrange additional financing and work out frequency conflicts at 110°W between Canada and the United States.

Star Choice Television is one of two Canadian firms that actually has satellite systems in operation. Star Choice had signed up only 5,000 subscribers through 1997, but it is optimistic that it can continue to attract consumers to its combination of local TV outlets and cable-like programming from the United States and Canada (Francis, 1997). The other company, ExpressVu, uses EchoStar's DISH Network equipment to operate a 180-channel DBS service. Subscribers must purchase the DVB equipment and a starter's tier of programming, and then they can add a wide range of specialty programming tiers including the Sports Bar, Kids Size, the Network Platter, and Film Feast. Both ExpressVu and Star Choice offer programming in English and French.

Factors to Watch

What was once an industry in search of reliable distribution technology and attractive programming is now an industry focused on brand awareness, marketing strategies, and strategic alliances. For example, DirecTV doubled its number of consumer promotions in 1998 to eight in the hopes of establishing its brand as the reliable market leader (DBS knockin', 1998). Key to these promotions are installation and programming price reductions and rebates.

DirecTV is establishing agreements directly with SMATV and MMDS services (see Chapter 6) to provide DirecTV to multiple-family dwelling unit (MDU) properties (apartments and townhomes). In addition, DirecTV has also formed a distribution alliance with SBC Communications and Bell Atlantic to allow these regional phone companies to offer DirecTV program packages through MDU and single-family home phone lines via digital set-top converter boxes.

SBC intends to market DirecTV packages with local phone service in a new Smart-Moves entertainment package, while Bell Atlantic might bundle it with its Bell Atlantic Plus program (Colman, 1998). SBC and Bell Atlantic will receive a sales commission for each home subscribing to DirecTV, while DirecTV will gain an additional revenue source and the ability to use these phone companies' local expertise in customer relations, installation and maintenance, and marketing.

GTE Corporation has also signed an agreement to market and distribute DirectTV to its customers. GTE will begin in a few markets, and then move to nationwide distribution.

USSB streamlined its service in 1998 by moving eight basic cable networks over to DirecTV to make room for more premium movie networks and PPV events. This should help sharpen USSB's focus on movies and big events, and provide DirecTV with more competitive programming to use in its battle for cable customers.

The rapidly-growing DISH Network, which has attacked cable television from its inception, is promoting its local-into-local broadcast television packages to consumers—and before Congress and the FCC—as the best way to challenge cable television's "monopoly" status in multichannel television. In March 1998, the DISH Network began introducing these $4.99 per month packages in Atlanta, Boston, Chicago, Dallas, New York, and Washington, D.C., with plans for more markets by 1999. Subscribers will have to purchase a second receive dish to get these packages off the EchoStar III satellite. And, until Congress amends copyright and satellite broadcasting laws, subscribers must live in ZIP codes not included in the local stations' coverage area. With cable rates continuing to rise since the implementation of deregulation in 1996, Congress has begun hearings concerning this and other cable competition issues.

Primestar is also awaiting congressional action on its acquisition of the MCI/News Corp. and Tempo DBS channels. Congress, with the support of the new FCC commissioners, the Justice Department, and the Commerce Department, is re-examining the role of cable operators in the DBS/DTH industry, cable's number one competitor. Fears of predatory pricing and other monopolistic practices have caused lawmakers to slow the pace of deregulation (Hearn, 1998).

These questions should be answered by late 1998 or early 1999. Regardless, the immediate future for the DBS/DTH industry will be filled with marketing campaigns attacking cable's pricing and service, while showcasing the virtues of digital quality and program variety. DBS companies will continue to use all avenues of marketing—network TV spots, infomercials, direct mail, in-store retail promotions and demonstrations, and sporting event sponsorships—to entice new subscribers. The big winner will be the consumer, as diversity of choice, a long-standing goal of U.S. communications policy, becomes a reality in the world of multichannel television.

Bibliography

Boyer, W. (1996, April). Across the Americas, 1996 is the year when DBS consumers benefit from more choices. *Satellite Communications*, 22-30.

Colman, P. (1998, March 9). DBS gets local help. *Broadcasting & Cable, 128* (10), 61-63.

DBS knockin' on cable's doors. (1998, March). [Online]. Available: http://www.mediacentral.com/magazines/cableworld.

DirecTV Japan announces new line-up. (1998, February 24). [Online]. Available: http://www.newsbytes.com.

DTH subscriber counts. (1998, March). [Online]. Available: http://www.skyreport.com/dthsubs.htm.

EchoStar. (1998, March). *Statement by Charlie Ergen in support of effective competition in video markets.* EchoStar press release.

Federal Communications Commission. (1989). *Memorandum opinion and order.* MM Docket No. 86-847, Washington, DC: FCC.

Federal Communications Commission. (1995). *Memorandum opinion and order.* MM Docket No. 95-1733, Washington, DC: FCC.

Federal Communications Commission. (1996, February 14). *Report No. SPB-37.* Washington, DC: FCC.

Forrester, C. (1997, November). Digital satellite in Europe: An expanding powerhouse. *Via Satellite*, 44-54.

Francis, G. (1997, September). Digital DBS: A look at the worldwide market. *Via Satellite*, 18-26.

Frederick, H. (1993). *Global communications & international relations.* Belmont, CA: Wadsworth.

Hearn, T. (1998, February 9). Primestar expects consent decree. *Multichannel News, 19* (6), 1, 62.

Hogan, M. (1995, September). US DBS: The competition heats up. *Via Satellite*, 28-34.

Hogan, M. (1998a, January 19). Demand remained strong for DBS in 1997. *Multichannel News, 19* (3), 33.

Hogan, M. (1998b, January 26). Primestar faces life with delays, uncertainty. *Multichannel News 19* (4), 1, 62.

Hogan, M. (1998c, February 2). Digital cable not immediate threat, says DBS. *Multichannel News, 19* (5), 12.

Hogan, M. (1998d, February 16). Primestar sets March roll-up, April launch. *Multichannel News, 19* (7), 1, 62.

Hogan, M. (1998e, March 2). DBS discounts 2nd receivers. *Multichannel News, 19* (9), 3, 18.

Howes, K. (1995, November). US satellite TV. *Via Satellite*, 28-34.

Jessel, H. (1996, February 5). The growing world of satellite TV. *Broadcasting & Cable, 126* (6), 59.

Johnson, L., & Castleman, D. (1991). *Direct broadcast satellites: A competitive alternative to cable television?* Santa Monica, CA: RAND.

JSkyB, PerfecTV to combine operations on May 1. (1998, March). *SkyReport.* [Online]. Available: http://www.skyreport.com/jskyb.htm

Kessler, K. (1998, March). The Latin American satellite market. *Via Satellite*, 17-26.

Lambert, P. (1992, July 27). Satellites: The next generation. *Broadcasting & Cable, 124* (31), 55-56.

Manasco, B. (1992, April). The U.S. multichannel marketplace in the year 2000. *Via Satellite*, 44-49.

Otsuka, N. (1995). Japan. In L. Gross (Ed.). *The international world of electronic media.* New York: McGraw-Hill.

Parone, M. (1994, February). Direct-to-home: Politics in a competitive marketplace. *Satellite Communications,* 28.

Primestar presses ahead with medium power. (1998, March). [Online]. Available: http://www.skyreport.com/ 213star.htm.

Regional Administrative Radio Conference. (1983). *Final report and order.* Geneva: ITU.

Responses cool to EchoStar. (1998, March 2). *Multichannel News, 19* (9), 55.

Ribbing, M. (1998, March 8). Pizza-size satellite dish starts to deliver. *The Baltimore Sun,* 1D, 3D.

Setzer, F., Franca, B., & Cornell, N. (1980, October 2). *Policies for regulation of direct broadcast satellites.* Washington, DC: FCC Office of Plans and Policy.

Setzer, F., & Levy, J. (1991, June). *Broadcast television in a multichannel marketplace.* Washington, DC: FCC Office of Plans and Policy, Working Paper No. 26.

STAR TV background. (1998, March). [Online]. Available: http://www.startv.com/startv/startv/sales/ back.html.

Tee-Com executives face shareholder lawsuit. (1998, March). *SkyReport.* [Online]. Available: http:// www.skyreport.com/tee-com.htm.

TSAT/Primestar restructuring inches along. (1998, March). [Online]. Available: http://www.skyreport.com/ 213star.htm.

Whitehouse, G. (1986). *Understanding the new technologies of the mass media.* Englewood Cliffs, NJ: Prentice-Hall.

Wold, R. N. (1996, September). U.S. DBS history: A long road to success. *Via Satellite,* 32-44.

6

Wireless Multipoint Distribution Services: MMDS and LMDS

Donald R. Martin, Ph.D.[*]

Multichannel multipoint distribution service (MMDS) technology, commonly called wireless cable, uses terrestrial microwave channels to distribute a varied range of telecommunications services to subscribers. Historically, this technology's primary use has been to transmit television programming services similar to those offered by wired cable television systems to residential dwellings that are equipped to receive microwave signals. However, the term wireless cable is a misnomer, as MMDS systems do not normally establish any physical connectivity between their headends and their subscribers, but they offer services which are similar to those offered by wired cable companies.

Local multipoint distribution service (LMDS) is a relatively new wireless technology that may become the successor to MMDS. The Federal Communications Commission (FCC) has allocated a substantial quantity of spectrum to LMDS, and licensees will be able to distribute many channels of "wireless cable," as well as offer wireless telephony, data, and other interactive services.

---

[*] Associate Professor of Communication, School of Communication, San Diego State University (San Diego, California).

Figure 6.1
Wireless Multiple-Channel Distribution System

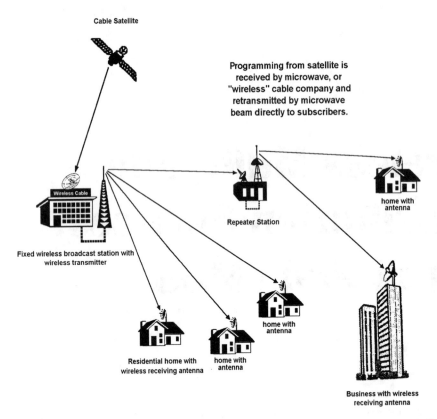

Cable Satellite

Programming from satellite is received by microwave, or "wireless" cable company and retransmitted by microwave beam directly to subscribers.

Wireless Cable

Repeater Station

home with antenna

Fixed wireless broadcast station with wireless transmitter

home with antenna

Residential home with wireless receiving antenna

home with antenna

Business with wireless receiving antenna

Source: R. B. Woodward

Background

In 1963, the FCC allocated a small portion of the microwave spectrum (2.150 GHz to 2.162 GHz) for the commercial distribution of television signals. This allocation consisted of one or two television channels in any given market and was called the multipoint distribution service (MDS). This was an attempt by the FCC to offer a means of delivering television signals to multiple subscribers using a relatively inexpensive technology. However, as the wired cable industry grew, it became apparent that a one- or two-channel MDS system would never be competitive in the video distribution marketplace.

As a result, the MDS industry began to seek more spectrum so that operators could offer more channels and provide meaningful competition to wired cable television systems. The portion of the spectrum that appeared to have the greatest potential for MDS expansion was the underutilized adjacent channels in the next microwave band (2.500 GHz to 2.686 GHz), which were allocated to the Instructional Television Fixed Service

(ITFS) and to the Operations Fixed Service (OFS). ITFS is a noncommercial educational service that is used by schools and colleges to distribute instructional programming to multiple sites for distant learning applications, while OFS is for governmental use.

In 1983, the FCC issued a *Report and Order* that reassigned 11 ITFS and OFS channels to MDS. The commission also authorized educational licensees to lease their remaining ITFS frequencies to MDS operators. The commission called this expanded service multi-channel MDS or MMDS. Through acquisitions and leases, an MMDS operator could then aggregate 33 channels into a system. Since a 33-channel system could be competitive with wired cable systems, the MMDS industry finally had the potential to become a viable competitor in the video distribution business.

During the 1980s, the MMDS industry experienced only modest gains. The slow growth of the industry was attributed to at least three factors:

- Undercapitalized speculators in MMDS channels.

- Educational institutions that were reluctant to lease ITFS channels.

- Financial markets that were cautious about investing in wireless systems that were limited to 33 channels.

In the early 1990s, several factors converged which enabled MMDS to begin to become a viable industry that could compete with cable. One of the more significant of these factors was the availability of programming. Since many of the more popular cable networks were partially owned by companies that had financial interests in the cable industry, MMDS operators sometimes had difficulty acquiring programming. This problem was resolved with the passing of the Cable Television Consumer Protection and Competition Act of 1992 which essentially stipulated that programming must be made available to all competitors at a reasonable price.

The promise of compressed digital technology was also instrumental in stimulating investment in the MMDS industry. Instead of 33 channels of analog television, new MMDS operators began to plan for digital systems. Digital compression allows more channels to be distributed within finite bandwidth. A compressed image is electronically sampled, and only those elements in the image that have changed since the last sampling are sent as updated information. Since all the information from each television frame is not retransmitted, less bandwidth is required to update the previous frame. Therefore, more signals can be "squeezed" into the space previously allocated for a single analog channel. For example, with a compression ratio of 4:1, four different channels of television can be distributed over a single 6 MHz television channel.

Using this compression ratio as an example, a wireless cable licensee can become competitive with a wired cable system by offering 132 channels of video over the 33 microwave channels used for MMDS. Digital compression also enables better signal encryption, thereby limiting non-subscriber piracy of the service.

During the early 1990s, schools and colleges also became more amenable to leasing their ITFS channels, as many of them had less use for the channels for instructional purposes. Therefore, MMDS systems finally had the programming availability and potential technical channel capacity to become viable competitors with wired cable systems.

The combination of adequate channel capacity and readily-available programming attracted a number of large companies to invest in MMDS in the early to mid-1990s. These companies were seeking a technology that would afford them a quick entry into the video distribution market previously dominated by the wired cable industry. Regional Bell operating companies (RBOCs), including Bell Atlantic, NYNEX (now Bell Atlantic), the former Pacific Telesis, and BellSouth, were among the largest companies to become interested in MMDS during this period.

Recent Developments

A number of recent developments have had a significant impact on MMDS in the United States. Many of these developments can be attributed to the "shake-out" of the residential video distribution market in the post Telecommunications Act of 1996 regulatory environment, as well as the perceived competitive need to offer services other than traditional video programming. One of the most important of these factors has been a decision by the RBOCs to scale down plans for developing MMDS systems. Bell Atlantic, of which NYNEX is now a part, has withdrawn from partnership with wireless cable operator CAI Wireless, Inc (Rising from, 1997). Similarly, SBC Communications, owner of Pacific Bell, terminated its negotiations to buy MMDS channels in San Francisco (Gibbons & Ellis, 1996). Pacific Bell has also apparently abandoned its plans to build new MMDS systems in San Jose and San Diego (Schlosser, 1997).

The decision by these RBOCs to reduce their presence in the MMDS business resulted in the dissolution of TeleTV, a company the telcos had created to realize economies of scale in purchasing and deploying MMDS equipment. The demise of TeleTV sent a message to the market that available capital and working capital for MMDS systems was on the decline (Rising from, 1997). Like CAI Wireless, other MMDS companies that had negotiated partnerships with the RBOCs now found themselves in significant financial difficulties (Orenstein, 1998). In general, stock and bond prices for wireless cable companies plummeted, as investors concluded that the future for MMDS would not be as lucrative as previously forecast (Higgins, 1997).

Despite this general decline, in May 1997, SBC-owned Pacific Bell deployed a new 150-channel MMDS system in the greater Los Angeles area (Schlosser, 1997). This system features 50 local television and cable channels, 30 digital quality audio channels, and a near video on demand system called Galaxy. It was being built before Pacific Bell owner SBC Communications began withdrawing from the MMDS market (Lynch, 1997). It should be noted that, as of early 1998, Pacific Bell has not aggressively marketed this system and may be seeking a buyer so it can withdraw from the MMDS business completely (Schlosser, 1997). BellSouth appears to be the only RBOC that is continuing to make a significant commitment to the operation of MMDS systems. They have purchased a number of wireless systems in Atlanta and other parts of Georgia and acquired MMDS in other communities (Rising from, 1997).

Despite the relatively slow development of MMDS in the United States, this technology appears to be flourishing in other countries. In mid-1997, there were over 5.5 million MMDS subscribers in 90 countries worldwide (Rising from, 1997). Many of the new systems are being built in less developed countries, which have little existing cable infrastructure, because MMDS can be deployed more rapidly than wired systems. MMDS is also being deployed in rural areas of developed countries such as the western Canadian provinces of Manitoba and Saskatchewan (Dickson, 1997).

A particularly interesting recent development is the shift in strategy by a number of MMDS companies from their sole focus on delivery of one-way video and audio services to the delivery of two-way data services. This expansion of services has been driven by the needs of Internet users for higher data speeds than can be delivered by conventional copper telephone lines. This is the same business sought by wired cable operators who have rebuilt their plants for two-way cable modem service.

In 1997, a number of wireless operators petitioned the FCC for permission to use the MMDS spectrum for two-way data services. While high-speed data services are not widely deployed in the wireless cable industry, there are systems that provide Internet services in Washington, D.C.; Las Vegas; Lakeland, Florida; Colorado Springs; Santa Rosa, California; and Nashua, New Hampshire (Cahoon, 1997). A number of industry and Wall Street investors consider MMDS voice and high-speed data services critical to the financial viability of the industry (Higgins, 1997). This data market may be emerging, as Spike Technologies has already developed the necessary hardware package to enable MMDS systems to convert some of their channel capacity for deployment of data services (Spike does, 1997).

One of the most promising developments in wireless distribution over the past several years has been the development of local multipoint distribution services (LMDS). This new service has been allocated 1.3 GHz of spectrum in the 28 GHz band. This spectrum allocation should be compared with the less than 0.2 GHz allocation available for traditional MMDS service in the 2.5 GHz band. In fact, the LMDS allocation is 17 times larger than the entire VHF television band and is large enough to enable LMDS licensees to offer subscribers a variety of sophisticated interactive digitized video, telephony, and data services (Smith, 1997). Therefore, LMDS has the potential to offer more video channels than any MMDS system. However, at the extremely high 28 GHz frequencies, more LMDS repeaters are needed to distribute a usable signal over a given coverage area. As a result, LMDS systems are being configured like cellular telephone systems, with multiple cells serving a cluster of subscribers within a three-mile radius. Subscribers receive the signals on six-inch square antennas (McConnell, 1997).

The extensive bandwidth will also allow LMDS operators to offer data services at speeds in excess of one gigabit per second (Gb/s) compared with 10 Megabits per second (Mb/s) for cable modems or 56 kilobits per second (Kb/s) for telephone modems (Leopold & Santo, 1998). Since wireless LMDS can easily reach subscribers' businesses and homes, it may solve the "last mile" problem and become a significant competitor to the existing wired networks provided by telephone and cable companies (Zeta, 1997).

CellularVision USA, Inc. has been offering "cable" television service via LMDS to over 12,500 Brooklyn, New York subscribers under a temporary experimental FCC authorization (Arnst & Gross, 1997). The FCC recently set rules for LMDS, and has auctioned licenses for systems to be deployed throughout the United States.

Current Status

There are approximately 250 operating MMDS systems in the United States serving an aggregate of over 1.1 million subscribers. However, in the United States, MMDS is still a small industry when compared with wired cable. Penetration levels are generally higher in areas where competition from existing wired cable and direct broadcast satellite (DBS) systems is less formidable. Currently, there are MMDS systems in 90 nations serving almost four million subscribers (Kreig, 1997).

While some U.S. MMDS systems are operating with healthy financial margins, much of the industry is facing difficult times. These problems have been created by the competition from wired cable and DBS and by the loss of investor confidence following the withdrawal of some of the RBOCs and others from the MMDS business.

On the other hand, the FCC has recently auctioned over 1,000 LMDS licenses. While few LMDS systems have actually been built, the number of issued licenses suggests that there is significant interest in developing this new area of the multipoint wireless distribution business (Arnst & Gross, 1997).

Factors to Watch

MMDS and LMDS have the potential to become viable competitors in the video distribution market. The future use of these technologies may be shaped by the following factors:

(1)   Most MMDS subscribers reside outside of the United States. They usually subscribe to wireless systems because a wired cable infrastructure is either small or non-existent in their region. MMDS companies will prosper in these areas only if they can continue to offer service at prices that make building wired cable systems or marketing DBS systems attractive for investors. However, if these operators invite competition by lagging behind technologically or charging unrealistic prices for service, they risk losing market share to wired cable, telephone, or DBS competitors.

(2)   In urban regions of the United States and other places where robust video distribution competition already exists, MMDS operators may survive to the extent

that they can offer high-speed data and other interactive services that are important to subscribers. It will be interesting to see if MMDS operators are able to respond to the need for high-speed Internet connectivity and attract new investment to their industry.

(3)   Future LMDS operators will have sufficient bandwidth to deliver an array of interactive digital services at incredible data rates. This distribution technology has the potential to become a significant competitor to all existing wired or wireless systems. The extent to which new licensees can attract the requisite capital and expeditiously deploy their systems will be crucial to the success of the LMDS industry.

MMDS is an older technology that may never achieve significant subscription levels in areas where there is existing wired cable competition. The future of the industry may be in offering unique, high-speed data and other interactive services instead of focusing solely on one-way "cable television" type services. Within the next few years, this interactive business will become a reality, or MMDS will pass into technological obsolescence. On the other hand, LMDS has more promise and may be the wireless distribution technology that captures a large share of the existing wired and wireless distribution markets.

Bibliography

Arnst, C., & Gross, N. (1997, April 14). A technology grows in Brooklyn. *Business Week*, 100.

Cahoon, J. (1997). How does wireless cable work? *Master Trading News*. [Online]. Available: http://www.mst.it/news.html.

Dickson, G. (1997, February 17). Broadband snags $30 million MMDS deal. *Broadcasting & Cable, 7*, 52.

Ernestine, D. (1998, February 5). CAI Wireless faces closure without new investors. *The Times Union* , E1.

Gibbons, K., & Ellis, L. (1996, November 18). PacTel scraps MMDS deal for Bay Area. *Multichannel News 47*, 1.

Higgins, J. (1997, May 19). PacTel delay another snag for wireless. *Broadcasting & Cable, 127*, 44.

Kreig, A. (1997, November 28). Take a second look at wireless cable. *Wireless Cable Association*. [Online]. Available http://www.wirelesscabl.com/Marktech.htm#article62.

Leopold, G., & Santo, B. (1998, February 16). FCC auction of local multipoint distribution service. *Electronic Engineering Times*, 8.

Lynch, S. (1997, May 30). Will new "wireless" cut cost of cable? *The Orange County Register*, A-14.

McConnell, C. (1997, March 17). Deciphering future of LMDS. *Broadcasting & Cable, 127*, 86.

Orenstein, D. (1998, January 3). CAI seeks outside aid to escape fiscal trouble. *The Times Union*, E1.

Rising from the ashes. (1997, July). *Cable & Satellite Europe*, 35.

Schlosser, J. (1997, October 6). PacBell's low-key digital company is quietly building California wireless system. *Broadcasting & Cable, 41*, 62.

Smith, D. (1997, April). LMDS at long last: FCC release rule. *Information Provider Newsletter*. [Online]. Available: http://www.dnai.com/~desmith/lmds_atlong.html.

Spike does two-way data over MMDS: An interview. (1997, April). *Information Provider Newsletter*. [Online]. Available http://www.vipconsult.com/spike2way.html.

Zeta, K. (1997). LMDS and broadband local networks for Asia. *Visual Institute of Information*. [Online]. Available: http://www.ctr.columbia.edu/vi/papers/ptc97.htm.

7

Advanced Television

Peter B. Seel, Ph.D. & Michel Dupagne, Ph.D.*

High-definition television (HDTV) is a specific type of advanced television (ATV) technology that represents the first significant change in this global communication medium since color images were added in the 1950s. First perfected in the 1970s by researchers at NHK (the Japan Broadcasting Corporation), the standardization of HDTV has become a contentious technological and political issue in Asia, North America, and Europe. The controversy has been fueled by debates over the forced obsolescence of existing television systems, the economic costs associated with making the change, and which nation or nations might dominate the development of such a widely-used technology. The related techno-political issues are important because they will influence almost all other forms of electronic media as these systems also complete the analog-to-digital conversion process. The shift to digital production and transmission will very likely have a greater effect on programming than whether or not the images are high-definition.

"Advanced television" is a generic term used by the Federal Communications Commission (FCC) in the United States to describe any technology that exceeds the present NTSC standard in audio and video quality (FCC, 1987a). Within this broad category, there are a number of subset acronyms:

- IDTV (improved-definition television).

- EDTV (extended- or enhanced-definition television).

- SDTV (standard-definition television).

- HDTV (high-definition television)

---

* Dr. Seel is Assistant Professor, Department of Journalism and Technical Communication, Colorado State University (Fort Collins, Colorado). Dr. Dupagne is Assistant Professor, School of Communication, University of Miami (Coral Gables, Florida).

While analog IDTV and EDTV have largely faded away in the United States in the face of emerging digital technologies, digital HDTV and SDTV have emerged as two key technologies in a new national ATV standard.

HDTV has been defined by the FCC as a system that provides image quality approaching that of 35mm film, has an image resolution of approximately twice that of conventional television, and has a picture aspect ratio of 16:9. At this aspect ratio of 1.78:1 (16 divided by 9), the television screen is wider in relation to height than the 1.33:1 (4 divided by 3) of NTSC. It is closer to the wide-screen images seen in movie theaters that are 1.85:1 or even wider. Figure 7.1 compares a 16:9 HDTV set with a 4:3 NTSC display—note that the higher resolution of the HDTV set permits the viewer to sit closer to the set, which results in a wider angle of view.

Figure 7.1
Wider Viewing Angle with HDTV

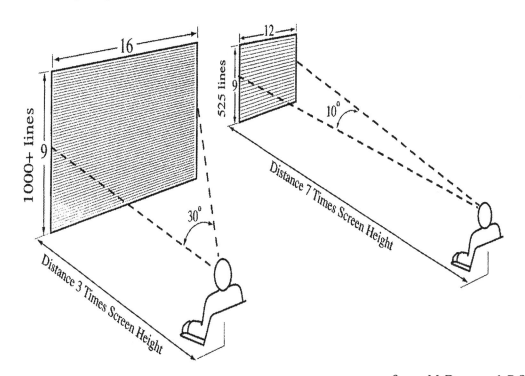

Source: M. Dupagne & P. Seel

SDTV—standard-definition television—is another type of ATV that can be broadcast instead of HDTV. Digital SDTV transmission will offer lower resolution than HDTV, but it will be available in both narrow- (1.33:1) and wide-screen (1.78:1) formats. Using digital video compression technology, it will be feasible for U.S. broadcasters to transmit four to six SDTV signals instead of one HDTV signal in the allocated 6 MHz digital channel. Thus, a television station would be able to retransmit a daytime soap opera while simultaneously broadcasting a local news channel, a sports channel, a business/financial channel, and perhaps a children's programming channel in SDTV. Some stations may reserve true HDTV single-channel programming for evening prime-time hours. The development of multi-channel SDTV broadcasting is an unintended consequence that was unforeseen by early HDTV researchers.

Background

In the 1970s and 1980s, Japanese researchers at NHK developed two related analog HDTV systems:

(1)   A "Hi-Vision" analog *production* standard with 1,125 scanning lines and 60 fields (30 frames) per second.

(2)   A "MUSE" *transmission* system with an original bandwidth of 9 MHz designed for direct broadcast satellite (DBS) distribution to the Japanese home islands (see Table 7.1).

Hi-Vision production equipment using the 1,125/60 HDTV format is presently in use throughout the world. Japanese HDTV transmission began in 1989 and now total 17 hours a day, from 7 A.M. to midnight (Nippon Hoso Kyokai, 1997).

In 1986, Japan and the United States attempted to have the Hi-Vision system adopted as a world HDTV production standard by the CCIR, a subgroup of the International Telecommunications Union (ITU), at a Plenary Assembly in Dubrovnik, Yugoslavia. However, European delegates lobbied for a postponement of this initiative that effectively resulted in a de facto rejection of the Japanese technology. European governments and their high-technology industries were still recovering from Japanese dominance of their VCR markets, and resolved to create their own distinctive HDTV standard that would be intentionally incompatible with Hi-Vision/MUSE (Dupagne & Seel, 1998).

Subsequently, a European R&D consortium, EUREKA EU-95, developed a distinctive system known as HD-MAC that featured 1,250 wide-screen scanning lines and 50 fields (25 frames) per second. This analog 1,250/50 system was used to transmit many European cultural and sporting events, such as the 1992 summer and winter Olympics in Barcelona, Spain and Albertville, France. However, since the 1990 development of digital HDTV technology in the United States, HD-MAC has been supplanted by a series of SDTV-type digital video broadcasting (DVB) standards for cable, satellite, and terrestrial broadcasting.

After it became apparent that there would be no single world HDTV standard in the wake of the rejection of the Japanese system at Dubrovnik, the FCC began a series of policy initiatives in 1987 that led to the creation of the Advisory Committee on Advanced Television Service (ACATS). This committee was comprised of 25 electronic media executives charged with investigating the policies, standards, and regulations that would facilitate the introduction of advanced television services in the United States (FCC, 1987b).

ACATS played a central role in the standardization of ATV services. Over 1,000 volunteers from television, cable, and other telecommunications organizations worked for eight years on three ACATS subcommittees to create a testing program to identify an ideal ATV broadcast transmission scheme.

Table 7.1
International ATV Standards

System	SMPTE 420 M	MUSE	DVB-T	ATSC DTV
Purpose	Production	Transmission	Transmission	Transmission
Region	North America, Japan (Hi-Vision)	Japan	Europe, adopted in other areas	North America
Image Coding	Analog	Analog	Digital	Digital
Aspect Ratio	1.78:1	1.78:1	1.33:1, 1.78:1, 2.21:1	1.33:1, 1.78:1
Lines	1,125	1,125	525 and 625, 1,080 and 1,152 (HDTV)	480, 720, 1,080*
Pixels/Line	n/a	n/a	Varies	640, 704, 1280, 1,920*
Scanning	2:1 Interlace	2:1 Interlace	1:1 Progressive, 2:1 Interlace	1:1 Progressive, 2:1 Interlace*
Bandwidth	30 MHz	6-9 MHz	6-8 MHz	6 MHz
Frame Rate	30 fps	30 fps	24, 25, 30 fps	24, 25, 30 fps*
Field Rate	60 Hz	60 Hz	50 Hz	60 Hz
Sound Coding	Digital	Digital	Digital	Digital

* As adopted by the FCC on December 24, 1996, the ATSC DTV image parameters, scanning options, and aspect ratios were not mandated, but were left to the discretion of display manufacturers and television broadcasters (FCC, 1996b). The ATSC lines are visible active lines, while others are scanning lines.

Source: P. Seel & M. Dupagne

Testing of a number of analog ATV systems was about to begin in 1990 when the General Instrument Corporation announced that it had perfected a method for digitally transmitting an HDTV signal. This announcement had a bombshell impact since many broadcast engineers were convinced that digital television transmission would not be technically possible until well into the 21st century (Brinkley, 1997). The other participants in the ACATS competition soon developed digital systems that were submitted for testing. At the end of the first round of tests, the Advisory Committee decided that none of the digital proponent systems could be identified as superior to the others and called for a second round of tests. Before that testing could take place, the three competitors in the testing process with digital systems—AT&T/Zenith, General Instrument/MIT, and Philips/Thomson/Sarnoff—decided to merge into a common consortium known as the Grand Alliance. With the active encouragement of the Advisory Committee, they combined elements of each of their ATV proponent systems in 1993 into a single digital Grand Alliance system for ACATS evaluation.

The FCC made a number of key decisions during the ATV testing process that defined a national transition process from NTSC to an advanced broadcast television system. One such ruling was the Commission's *First Report and Order* of August 24, 1990 which outlined a simulcast strategy for the transition to an ATV standard (FCC, 1990). This strategy required that U.S. broadcasters co-transmit both the new ATV signal and the existing NTSC signal for a period of time, at the end of which the NTSC transmitter would be turned off. Rather than try and deal with the inherent flaws of NTSC, the FCC decided to create a new television system that would be incompatible with the existing one. This was a decision with multibillion dollar consequences for broadcasters and consumers, since it meant that all existing production, transmission, and reception hardware would have to be replaced with new equipment capable of processing the ATV signal.

The FCC also proposed a transition window of 15 years from the adoption of a national ATV standard to the shutdown of NTSC broadcasting (FCC, 1992). This second ATV *Report and Order* caused consternation on the part of the television broadcast industry at what they perceived as too short a transition period. The nation was in the midst of a recession, and broadcasters were concerned about the financial cost of replacing their NTSC facilities. The transition cost for U.S. broadcasters has been estimated at $1.1 million to $2.2 million for ATV signal pass-through and up to $12 million for full station conversion (Dupagne & Seel, 1998).

Recent Developments

The Grand Alliance system was successfully tested during the summer of 1995, and a digital television standard based on that technology was recommended to the FCC by the Advisory Committee on November 28 (ACATS, 1995). In May 1996, the FCC proposed to adopt the *ATSC Digital Television (DTV) Standard* based upon the work accomplished by the Advanced Television Systems Committee (ATSC) in documenting the technology developed by the Grand Alliance consortium (FCC, 1996a). The proposed ATSC DTV standard specified 18 digital transmission variations as noted in Table 7.2. Stations would be able to choose whether to transmit one channel of high-resolution, wide-screen HDTV programming, or four to six channels of SDTV programs during various parts of the day.

The type of television owned by the viewer would dictate whether he or she was watching the program in high or low resolution on 16:9 wide-screen digital, 4:3 narrow-screen digital, or an older model 4:3 analog set.

Table 7.2
U.S. Advanced Television Systems Committee
(ATSC) DTV Formats

Format	Active Lines	Horizontal Pixels	Aspect Ratio	Picture Rate*
HDTV	1,080 lines	1,920 pixels/ line	16:9	60I, 30P, 24P
HDTV	720 lines	1,280 pixels/ line	16:9	60P, 30P, 24P
SDTV	480 lines	704 pixels/ line	16:9 or 4:3	60I, 60P, 30P, 24P
SDTV	480 lines	640 pixels/ line	4:3	60I, 60P, 30P, 24P

* In the picture rate column, "I" indicates interlace scan in *fields*/second and "P" means progressive scan in *frames*/second.

Source: ATSC

Before the FCC could adopt the new television standard, however, it had to resolve objections from several politically powerful media industry groups that disliked certain parts of it. A group representing the motion picture industry opposed the 16:9 aspect ratio, saying it was too narrow for televising most feature films, and sought to have it widened or to require broadcasters to transmit films in the same aspect ratio as they were photographed. Another group representing companies in the U.S. computer industry (e.g., Apple, Intel, and Microsoft) sought to have all the interlace scanning variations dropped from the proposed standard, contending that they would impede the convergence of television and computing technologies (Dupagne & Seel, 1998).

Interlace scanning is a form of signal compression that first scans the odd lines of a television image onto the screen, and then fills in the even lines to create a full video frame. The present NTSC standard uses interlace scanning to reduce the bandwidth needed for broadcast transmission. While interlace scanning is spectrum-efficient, it creates unwanted visual artifacts that can degrade image quality. *Progressive scanning*— where each complete frame is scanned on the screen in only one pass—is utilized in computer displays because it produces less eye-strain than interlace scanning. The computer industry group was willing to contest this fundamental standardization issue because the advent of digital transmission meant that every DTV set could function as a computer display and vice versa. DTV receivers will be capable of displaying small-font text (for example, a World Wide Web site) that would be illegible on a conventional NTSC television set.

In December 1996, the FCC finally approved a DTV standard that resolved the conflict in a remarkable way—it deleted any requirement to transmit any of the 18 transmission video formats listed in Table 7.2 (FCC, 1996b). Also, it did not mandate any specific aspect ratios or any requirement that broadcasters transmit true HDTV on their digital channels. Although the computer group did not succeed in forcing interlaced transmission out of the standard, it did leave the decision up to broadcasters. The FCC resolved the image aspect ratio controversy by leaving this basic decision up to broadcasters as well. They will be free to transmit digital programming in narrow- or wide-screen ratios as they wish, not an outcome that the motion picture industry wanted.

Other aspects of the ATSC standard remained unchanged, such as multichannel audio. The Grand Alliance adopted the Dolby AC-3 digital compression system after comparing other digital audio technologies such as Musicam. The AC-3 specifications call for a surround-sound, six-channel system that will approximate a motion picture theatrical configuration with speakers screen-left, screen-center, screen-right, rear-left, rear-right, and a subwoofer for bass effects. This audio system will enhance the diffusion of "home theater" television systems with multiple speakers and either a video projection screen, a direct-view CRT monitor, or a newly-developed flat-panel display that can be mounted on the wall like a painting.

Current Status

United States

By 1997, the U.S. HDTV debate had almost completely shifted from standardization matters to DTV implementation considerations, such as programming mix, station conversion, and consumer adoption. In April 1997, the FCC (1997a) issued a *Report and Order* that defined how the United States would make the transition to DTV broadcasting. The commission set December 31, 2006 as a target date for the phase-out of NTSC broadcasting. However, the U.S. Congress passed a bill in 1997 that would allow television stations to continue operating their analog transmitters as long as more than 15% of the households in a market cannot receive digital broadcasts through cable or DBS, or if 15% or more of local households lack a DTV set or a converter box to display digital images on their older television sets (Balanced Budget Act, 1997). This law gave broadcasters some breathing room if consumers do not adopt DTV technology as fast as set manufacturers would like.

To demonstrate their good faith in an expeditious DTV conversion, the four largest commercial television networks in the United States (ABC, CBS, NBC, and Fox) made a voluntary commitment to the FCC to have 24 of their affiliates in the top 10 markets on the air with a DTV signal by November 1, 1998 (Fedele, 1997). All of their affiliates in the top 30 markets will be broadcasting digital signals by November 1999. Table 7.3 outlines the rest of the roll-out.

Also in April 1997, consistent with the Telecommunications Act (1996), the FCC assigned each broadcaster a new channel for DTV operations (FCC, 1997b). In February 1998, it adopted a revised table of DTV allotments and established the final DTV core spectrum between channels 2 and 51 (FCC, 1998).

Table 7.3
U.S. Digital Television Broadcasting
Phase-In Schedule

Phase	Number of Stations	Market Size	Type of Station	DTV Transmission Deadline	Percentage of U.S. HHs*
1	24	Top 10	Voluntary	November 1, 1998	--
2	40	Top 10	Network Affiliates	May 1, 1999	30%
3	80	Top 30	Network Affiliates	November 1, 1999	53%
4	~ 1,037	All	All Commercial	May 1, 2002	~100%
5	~ 365	All	Noncommercial	May 1, 2003	--
6	~ 1,500	All	All	December 31, 2006	Planned NTSC Reversion Date

* Television households capable of receiving at least one local DTV broadcast signal.

Sources: FCC and TV Technology

Because over 65% of U.S. television households receive their signals from a cable television provider, digital conversion of cable facilities is essential to pass along DTV programming from broadcasters. Tele-Communications, Inc. (TCI), the largest cable operator in the United States, placed an initial order in February 1998 for 5.9 million digital set-top boxes for its 14.3 million customers (Desmond, 1998). However, the cable industry has been ambivalent about providing HDTV/SDTV service for two fundamental reasons:

(1)   They would have to provide an extra channel for every existing system channel under FCC simulcast rules, requiring a doubling of cable system capacity without increasing revenue.

(2)   SDTV technology would permit terrestrial broadcasters to transmit four to six of these channels simultaneously over the air, in effect transforming them into multi-channel providers in competition with cable news and sports channels.

Their hands may be forced by two important factors. Several key programming services, Home Box Office (HBO) and the Discovery Channel among them, have announced that they will provide true HDTV wide-screen programming to cable system operators in 1998 (Dickson, 1997). Cable customers with DTV sets will want to see these channels in high-definition, especially if they are early adopters who paid over $5,000 for their sets. Another factor for consideration by the cable industry is that their DBS competitors, specifically DirecTV, have disclosed plans to beam HDTV programming in the continental United States beginning in fall 1998 (Dickson, 1998a). The U.S. cable industry will be hard pressed to ignore HDTV transmission if their orbital competitors are actively promoting it as a viewer option.

What will DTV viewers be watching in the early stages? Movies are an obvious choice since they are already distributed on a high-definition medium—35mm motion picture film. They are also shot in a wide-screen aspect ratio that can be easily transferred to HDTV digital tape using a telecine machine. An estimated 70% of U.S. network prime-time programming is produced using motion picture film, making the conversion to HDTV largely a matter of image cropping (Stow, 1993). Warner Brothers Television has been photographing television series such as *Friends* and *ER* in wide-screen aspect ratios since 1994, although NTSC viewers cannot see 30% of the image that is on the film negative (Cookson, 1995). Hollywood studios such as Warner Brothers are shooting with wide-screen framing to protect their investment in these programs for syndication sales after widespread DTV broadcasting begins.

The first prototype DTV sets designed to meet the new U.S. standard made their debut at the Winter Consumer Electronics Show in January 1998. Priced between $6,000 and $12,000, the first sets will accept any of the 18 video formats listed in Table 7.2 and will go on sale in the fall of 1998 throughout the United States (Grotticelli, 1998). The set manufacturers are planning to target affluent early-adopter consumers who will not flinch at these initial high prices. Manufacturers are predicting sales of one million DTV sets by 2000, even at premium prices (Dickson, 1998b). However, it is expected that DTV sales will not really take off until prices drop to the $500 to $1,000 range, which is still more than consumers now pay for current NTSC sets.

According to FCC statistics, seven stations had already broadcast experimental DTV signals by February 1998, and 10 stations had received their DTV construction permits. After years of skepticism and reluctance, many U.S. broadcasters are now poised to embrace the DTV transition. A survey of 400 station executives conducted in December 1997 for Harris Corporation, a broadcast hardware manufacturer, revealed a higher level of confidence in DTV technology than in previous years. Of the respondents, 93% reported that they were either very likely or somewhat likely to convert their facilities to

DTV by 2002, and 66% claimed that they could now afford the cost of conversion, up from 42% in 1996. But broadcasters are still uncertain about their DTV programming strategy. Forty-four percent stated they still need to define the mix between HDTV and other digital services, 23% stated they would air their DTV programs primarily in HDTV, and the remaining 33% stated they would air multiple SDTV signals (Harris survey, 1998).

Drawn for BROADCASTING & CABLE by Jack Schmidt

"They're not sure of what to do with their new spectrum so they ordered an HDTV-SDTV-multiplexing combination antenna..."

Reprinted with permission.

In summary, the U.S. approach to HDTV and SDTV as part of a national conversion to digital broadcasting will eventually require the replacement of all broadcast facilities and most consumer sets with new equipment. It is a decision with multibillion-dollar consequences for all involved. There is no interim SDTV technology bridge to HDTV as the Europeans have planned, and the actual conversion will be swift, requiring less than a decade if the FCC's schedule is followed. By the year 2003, virtually 100% of U.S. television households will be able to receive at least one digital terrestrial signal. Similar plans for a conversion to digital television are underway in both Japan and Europe, but they have very different strategies for achieving this objective.

Europe

Formally established in September 1993, the Digital Video Broadcasting (DVB) Project includes 200 organizations (including Thomson and Philips) from over 30 countries. The group has defined European digital standards for satellite, cable, MMDS (wireless cable), and terrestrial broadcasting using MPEG-2 compression as a common denominator (Digital Video Broadcasting, 1996). The DVB standards are comparable to analog PAL and SECAM in terms of lines and frame rate (625 lines and 25 fps), and their transmission scheme uses a technology known as COFDM instead of the 8-VSB chosen by the Grand Alliance in the United States.

Until recently, DVB has focused most of its work on standardizing SDTV formats at the expense of HDTV. But this situation changed in July 1997 when the group issued a set of HDTV guidelines to ensure, among other things, interoperability between the decoder and encoder functionality for satellite, cable, and terrestrial transmission (DVB finally, 1997; http://www.dvb.org:80). This addendum will enable DVB standards to compete on a more equal footing with the U.S. ATSC DTV system, which is designed for both SDTV and HDTV. However, European satellite programmers and cable operators are in no rush to offer HDTV broadcasts in the short term and have instead opted for SDTV transmission. By January 1998, digital direct-to-home (DTH) service had developed slowly in Europe except in France where rival Canal+ and TPS had signed a total of 1,210,000 digital subscribers (Bulkley, 1998; Canal+, 1998).

Japan

Ever since Akimasa Egawa, then Director General of the Broadcasting Bureau at the Ministry of Posts and Telecommunications (MPT), declared in February 1994 that "The world trend is digital" (Choy, 1994, p. 5), Japan has begun reconsidering its commitment to MUSE broadcasting and contemplating the option of providing digital television. In fact, Japanese policy has taken the form of a dual-track approach. On the one hand, NHK and associated commercial broadcasters continue to offer a full-time HDTV schedule to Japanese viewers who own HDTV receivers (a total of 532,000 as of December 1997). Since October 1997, they have aired 17 hours of HDTV programs every day, with an emphasis on news, sports, and movies. The Nagano Winter Olympics Games represented yet another opportunity to showcase the quality and realism of HDTV production. From February 7 through February 22, 1998, there was nothing but Olympic coverage on the Hi-Vision channel (http://www.hpa.or.jp).

At the same time, Japan has launched a series of digital initiatives to catch up with the United States and Europe. In 1997, MPT proposed using the second BS-4 satellite, to be launched in 2000, for digital HDTV broadcasts. The satellite will have four transponders, each being capable of beaming either six digital SDTV channels or two HDTV channels. The proposed Japanese digital DBS system will include use of MPEG-2 encoding and multiple video formats, which will make it similar to the ATSC DTV standard adopted in the United States (Kumada, 1997).

Slower progress is being made in the terrestrial broadcasting area. In March 1997, the MPT announced that it would move up the introduction of digital terrestrial television (DTTV) from between 2000 and 2005 to before 2000 (Pollack, 1997). However, in March 1998, the Ministry revised the planned timeline to allow for two years of testing, followed by the introduction of DTTV in the three largest urban areas of the country in the years 2001 to 2002 (Ministry likely, 1998). Public and private broadcasters in Japan had criticized the accelerated pre-2000 timeline and had petitioned the MPT for more time to make the expensive transition to digital transmission.

Factors To Watch

The global diffusion of HDTV technology has taken much longer than its proponents had anticipated. Some delays were unforeseen, such as that caused by the development of digital transmission systems in the United States in 1990 and 1991. Other delays, such as the rejection of the Japanese Hi-Vision/MUSE system by the Europeans in 1986, were intentional. A case can be made that if the analog Hi-Vision system had been accepted as a global HDTV standard in 1986, the United States might never have conducted its ATV testing program—perhaps delaying the advent of digital broadcasting until well into the 21st century. The U.S. conversion to digital broadcasting between now and the year 2006 has spurred similar efforts in Japan and Europe. Proponents of both the U.S. ATSC system and the European DVB technology are now traveling the world seeking potential national adopters. In a situation that mirrors the PAL/SECAM standardization techno-politics of the 1960s, nations are being asked to stake their HDTV futures with one system or the other. The difference this time around is that both systems are digital, which will simplify transcoding programs from one global format to another.

The global development of digital television broadcasting is entering a crucial stage in the coming decade as terrestrial and satellite DTV transmitters are turned on, and consumers get their first look at HDTV and SDTV programs. Among the issues that are likely to emerge in 1998 and 1999:

- In the United States, over 65% of TV households receive their signals from cable. The FCC will make a decision in 1998 or possibly later concerning cable system requirements for carriage of over-the-air DTV channels. In light of the recent *Turner v. FCC* (1997) ruling, in which the Supreme Court affirmed the constitutionality of must-carry rules, this issue is likely to stimulate more passionate debate between the two industries (McConnell, 1998). Will the FCC decide to apply the must-carry rules to true HDTV channels only? If so, this action may dampen broadcaster interest in SDTV technology, because cable operators will not be required to carry multiple SDTV channels from each local broadcaster.

- In a related area, there is a controversy over whether cable giant TCI's digital set-top boxes will be able to reproduce high-definition images. It began in January 1998 when Gary Shapiro, president of the Consumer Electronics Manufacturers Association, criticized TCI for deploying new digital set-top boxes that "will down convert the HDTV broadcasts in either 1,080I or 720P formats to the lower-resolution 480P format" (Tedesco, 1998a, p. 17). TCI denied this, and claimed that its digital set-top boxes would have the capability to pass-through HDTV signals in the 1,080I format, or any other proposed HDTV format. At this point, however, it is not entirely clear whether these boxes will be configured to easily pass through HDTV signals to HDTV sets from either broadcasters or premium services, such as HBO (Brinkley, 1998; Colman, 1998). If cable providers do not give consumers easy access to HDTV programs, this factor may provide a competitive advantage to satellite-transmitted services such as DirecTV that plan to provide the highest-quality HDTV signals to their subscribers.

- In March 1997, President Clinton issued an executive order to establish the Advisory Committee on Public Interest Obligations of Digital Television Broadcasting, whose function is to study and recommend public interest responsibilities in return for use of digital television licenses (McConnell, 1997). When adopted sometime in 1998, what will those obligations involve? As a *quid pro quo* for the award of the spectrum for digital transmission, Vice President Albert Gore has asked the group to consider recommending that broadcasters provide a total of two hours of free airtime to candidates in the two months prior to an election (Stroud, 1998). Broadcasters are adamantly opposed to giving away political ad time that generates significant income during election periods. This promises to be a very contentious issue in the near future.

- ABC, CBS, NBC, and Fox announced their digital format decisions at the 1998 National Association of Broadcasters conference. CBS and NBC will transmit HDTV broadcasts as 1,080 interlaced lines (1,080I—see Table 7.2) in prime time, while ABC and Fox will go with 720-line progressive HDTV signals (720P) (McClellan & Dickson, 1998). ABC, NBC, and Fox will transmit 480-line progressive (480P) SDTV signals in various dayparts, while CBS will stick with 480-line interlaced digital SDTV transmissions. The 720P HDTV format is an interesting choice because it could permit the simultaneous broadcast of a second SDTV channel under certain circumstances. Some companies in the computer industry (e.g., Apple, Microsoft, and Intel) favor progressively scanned DTV transmissions that are computer-display-friendly and are encouraged by broadcaster acceptance of the 720P and 480P formats.

- Consumers will have the final say on the acceptance of DTV technology. How far will prices of DTV sets need to fall to make them acceptable to the majority of consumers? With prices in the $5,000+ range for large-screen displays, the initial market will be limited to the affluent technophiles with access to programming in a top 10 market or from DBS. Set manufacturers will need to amortize their significant research and development costs in DTV technology before set prices drop substantially, which probably means these reductions will not occur until after 2000. An unknown factor is going to be the influence of computer manufacturers who plan to use chips provided by Intel that can decode DTV signals for PC-based displays (Tedesco, 1998b). In January 1998, WETA-TV, one of the experimental DTV stations, successfully transmitted HDTV programming to a Pentium II PC equipped with an ATSC-compliant receiver card and an inexpensive "rabbit-ear" antenna placed near a window (WETA broadcasts, 1998). The prospect of millions of computer users having DTV display capability on the desktop may prod television set manufacturers to reduce prices more quickly.

In the coming decade, the merger of the television and the computer will create interesting programming options for viewers. Television programming will offer a level of potential viewer interactivity that has been just a pipe dream for the past two decades. DTV telecomputers with high-bandwidth connections to the Internet will have two-way transmission capabilities. The television of the 21st century will be a DTV model equally capable of telecasting the World Series in high resolution or delivering your Aunt Marge's personal Web page displaying her latest video production. As advanced television sets start to appear in stores in late 1998, it seems that we are going to need to redefine the meaning of the word "broadcasting."

Bibliography

Advisory Committee on Advanced Television Service. (1995). *Advisory Committee final report and recommendation*. Washington, DC: ACATS.

Balanced Budget Act of 1997. (1997). Pub. L. No. 105-33, § 3003, 111 Stat. 251, 265.

Brinkley, J. (1997). *Defining vision: The battle for the future of television*. New York: Harcourt Brace.

Brinkley, J. (1998, February 23). TV cable box software may blur digital signals. *New York Times*, C6.

Bulkley, K. (1998, January). Parlez-vous digital? *Cable and Satellite Europe*, 66.

Canal+ outpaces TPS. (February 16, 1998). *Broadcasting & Cable*, 38.

Choy, J. (1994, March 4). Future of Japan's advanced television system debated. *Japan Economic Institute Report*, 5-7.

Colman, P. (1998, March 9). Will cable be ready for HDTV? *Broadcasting & Cable*, 43.

Cookson, C. (1995). Introduction of wide-screen to television series production. *Symposium record of the 19th international television symposium and technical exhibition*. Montreux, Switzerland: The Symposium, 369-373.

Desmond, E. W. (1998, February 16). Malone again. *Fortune*, 66-69.

Dickson, G. (1997, July 14). Discovery pledges to go HDTV. *Broadcasting & Cable*, 82.

Dickson, G. (1998a, January 12). DirecTV ramps up for HDTV. *Broadcasting & Cable*, 1, 6.

Dickson, G. (1998b, January 12). HDTV sweeps CES floor. *Broadcasting & Cable*, 6-7.

Digital Video Broadcasting. (1996). *The new age of television: DVB*. Grand-Sarconnex, Switzerland: DVB.

Dupagne, M., & Seel, P. B. (1998). *High-definition television: A global perspective*. Ames, IA: Iowa State University Press.

DVB finally addresses HDTV. (1997, May). *Advanced Television Markets*, 1.

Fedele, J. (1997, September 25). DTV schedule breeds apprehension. *TV Technology*, 16.

Federal Communications Commission. (1987a). Advanced television systems and their impact on the existing television broadcast service. *Notice of Inquiry*. 2 FCC Rcd. 5125.

Federal Communications Commission. (1987b). *Formation of Advisory Committee on Advanced Television Service and announcement of first meeting*. 52 Fed. Reg. 38523.

Federal Communications Commission. (1990). Advanced television systems and their impact on the existing television broadcast service. *First Report and Order*. 5 FCC Rcd. 5627.

Federal Communications Commission. (1992). Advanced television systems and their impact on the existing television broadcast service. *Second Report and Order/Further Notice of Proposed Rule Making*. 7 FCC Rcd. 3340.

Federal Communications Commission. (1996a). Advanced television systems and their impact upon the existing television broadcast service. *Fifth Further Notice of Proposed Rule Making*, 11 FCC Rcd. 6235.

Federal Communications Commission. (1996b). Advanced television systems and their impact upon the existing television broadcast service. *Fourth Report and Order*. 11 FCC Rcd. 17771.

Federal Communications Commission. (1997a). Advanced television systems and their impact upon the existing television broadcast service. *Fifth Report and Order*. 12 FCC Rcd. 12809.

Federal Communications Commission. (1997b). Advanced television systems and their impact upon the existing television broadcast service. *Sixth Report and Order*. 12 FCC Rcd. 14588.

Federal Communications Commission. (1998, February 17). Advanced television systems and their impact upon the existing television broadcast service. *Memorandum Opinion and Order on Reconsideration of the Sixth Report and Order*. MM Docket No. 87-268.

Grotticelli, M. (1998, February 9). DTV dominates discussion at CES. *TV Technology*, 1, 24-25.

Harris survey shows broadcaster's optimism for DTV. (1998, February 9). *TV Technology*, 6.

Kumada, J. (1997, June). *The introduction of digital HDTV in Japan*. Paper presented at the "HDTV '97" workshop, Montreux, Switzerland.

McClellan, S., & Dickson, G. (1998, April 8). The lines are drawn. *Broadcasting & Cable*, 6-10.

McConnell, C. (1997, October 27). Gore's public interest panel meets, greets. *Broadcasting & Cable*, 17.

McConnell, C. (1998, March 9). DTV: Must carry, the sequel. *Broadcasting & Cable*, 44-45.

Ministry likely to delay ground-based digital TV. (1998, March 9). *The Nikkei Weekly*, 8.

Nippon Hoso Kyokai. (1997). *NHK factsheet '97*. Tokyo: NHK.

Pollack, A. (1997, March 11). Japan says it will move up introduction of digital television by a few years. *New York Times*, C6.

Stow, R. L. (1993). Market penetration of HDTV. In S. M. Weiss & R. L. Stow (Eds.). *NAB 1993 guide to HDTV implementation costs* (Appendix II). Washington, DC: National Association of Broadcasters.

Stroud, M. (1998, March 9). Gore group considers two hours of free airtime. *Broadcasting & Cable*, 50.

Tedesco, R. (1998a , January 26). CEMA sounds alarm over TCI DTV plans. *Broadcasting & Cable*, 17.

Tedesco, R. (1998b , March 9). Intel broke ranks but computer coalition holds. *Broadcasting & Cable*, 46.

Telecommunications Act of 1996. (1996). Pub. L. No. 104-104, § 336, 110 Stat. 56, 107.

Turner Broadcasting System, Inc. v. F.C.C., 117 S.Ct. 1174 (1997).

WETA broadcasts HDTV to a PC. (1998, February 23). *TV Technology*, 6.

8

Radio Broadcasting

David Sedman, Ph.D.[*]

> *The idea of digital revolution is implicitly an image of humankind stepping through a doorway into an unknown and fundamentally changed future. And it is a one-way journey, a doorway through which we can never step back to return to the comfortable media certainties of the past.*

> —Tony Feldman (1997) discussing the transformation of media to the digital age

One of the major developments in recent broadcast history was the Federal Communication Commission's action bringing television into the digital transmission age. Television stations are making the transformation to digital broadcasting with the purchase of new equipment while exploring their many options for offering new and improved services. As Feldman noted, there will be no turning back to analog for broadcasters moving into the digital age. The support for an advanced television system came not only from the broadcast industry and consumer electronics companies, but also from Hollywood and the computer industry.

The radio industry, on the other hand, is unlikely to attract comparable stimulus from outside forces as it struggles to decide when and how to enter the digital realm. While technological developments in the radio industry today are mundane, acting primarily to enhance the current AM-FM system, visionaries are reaching out to create a radio system with a more dynamic sound, greater reach, and a sweeping range of new features.

---

[*] Assistant Professor, Department of Communication Arts, Southern Methodist University (Dallas, Texas).

Background

The history of radio is rooted in 19th century wire transmission technologies—the telegraph (1820s) and the telephone (1870s). Guglielmo Marconi, generally considered the inventor of radio, first transmitted telegraphic dots and dashes without the use of wires in the 1890s. Wireless radio telegraphy was adopted by the maritime industry as a means of ship-to-shore and ship-to-ship communication.

In the early 20th century, tremendous advances led to radio telephony that allowed voice and music to be transmitted without wires. In 1920, the first radio station began operation. Radio networks were created later in the decade to provide local broadcasters with nationally-delivered programs. Innovative programming spurred receiver sales, and radio supplanted the phonograph as the most widely adopted consumer entertainment device. With the advent of television in the late 1940s, network entertainment programming gravitated to TV, and the radio industry realized that listening patterns had changed drastically because of television. With people listening for shorter periods, many stations switched to local programming with music formats accompanied by short bursts of news and information.

During the 20th century, two distinct radio transmission bands captured the largest audiences. The bands were named for the method used to superimpose an audio signal on a radio wave:

- Amplitude modulation (AM) which varies (modulates) the strength (amplitude) of a signal.

- Frequency modulation (FM) which varies the frequency of a signal.

The first radio stations transmitted AM signals only. The AM signal, although far reaching, is prone to electrical interference and limitations of fidelity. FM, which was developed in the 1930s and 1940s, provided a superior sounding service but was limited in range. The two bands continue to coexist in present-day radio. However, because FM is better suited to music format radio, it has overtaken AM to attract the largest portion of the radio audience. (Radio signals are also transmitted on the short wave and high frequency bands.)

A number of consumer audio devices have been introduced over the past 40 years that have spurred broadcast engineers into developing areas to augment and enhance a station's service. The increased use of stereophonic systems in the 1950s and 1960s helped lead to more FM stereo broadcasts and, during the 1980s and 1990s, AM broadcasters began making the transformation from monaural to stereo transmission. Four-channel, quadraphonic home systems of the 1960s led to experimentation with Quad-FM. Because home systems were not adopted by a critical mass of consumers, they disappeared from the consumer electronics marketplace, and experiments with Quad-FM were discontinued. When the digital compact disc replaced the analog vinyl record in the 1980s, many predicted that digital audio broadcasting transmission would replace analog systems by the early 1990s. This has not been the case in the United States or in most other countries.

Major changes in radio service are very difficult to institute. To be successful, a new service generally requires four levels of adoption:

(1) Approval by a governing body (such as the FCC in the United States).

(2) Acceptance by broadcast stations.

(3) Consent from the consumer electronics industry to design and market the new technology.

(4) Adoption by the public.

These factors, however, have not stopped engineers from continuing to design technology to improve upon radio's quality and scope.

Recent Developments

There are three basic categories that cover today's technological trends in the area of radio transmission. They are:

(1) Enhancements to improve the present-day, on-air transmission.

(2) Supplements to provide new services within the current radio system.

(3) New delivery modes to develop new transmission services incompatible to current radio.

Enhancements—Digital FM Exciters

Several North American transmitter companies, including Harris and Superior Broadcast Products, introduced digital FM exciters in the mid-1990s. The exciters enhanced stations' present transmission and, according to some stations, improved their effective broadcast radius. The technology provides the last link in the station's digital air chain. In a world in which digital communication technology is considered "good," radio stations adopting digital exciters could also launch marketing campaigns promoting digital transmission despite their position on the analog-transmitted FM band.

Enhancements—Extending the AM Band

Someday in the future, AM radio stations may gravitate to the digital domain and be able to compete more effectively with their present-day FM counterparts. Because this potential shift is years away in the United States, current AM broadcasters have been given some stopgap approaches for improving their service. One of these occurred in 1996 when the FCC (1996a) moved 83 stations from crowded frequencies on the AM dial (between 535 MHz and 1605 MHz) to the expanded portion of the band between 1605 MHz and 1705 MHz. This move provided stations with fewer interference problems because they were the first U.S. stations to occupy this portion of the spectrum.

Supplements—RDS

The notion of supplemental audio services is common to FM radio. FM broadcasters can utilize sidebands of their transmission signal to feed additional services to specially-equipped receivers. These subsidiary communications services (SCS) were instituted in the 1950s, and urrent uses include reading services for the visually impaired and background music heard in stores and office buildings. The latest use of sidebands is for the delivery of data known as radio data systems (RDS) or RBDS (radio broadcast data systems). RDS transmits information to specially-equipped receivers, sometimes called *smart radios* (Ammons, 1995). The information can include the "positioning statement" (slogan) of the radio station, the artist and song title of a musical selection currently playing, traffic and weather alerts, and advertiser information. Smart radios can also be programmed to turn on a receiver to transmit an emergency message and to automatically change to a station with the same format if the original station's signal becomes weak.

Although RDS has been used in Europe since the 1980s, it has diffused slowly in the United States. Because only a small percentage of radio stations are transmitting with RDS, there is little consumer knowledge of or demand for the service. Consequently, few consumer electronics makers are producing RDS hardware. In order to stimulate interest on the part of broadcasters, the Electronic Industry Association offered to share the cost of buying transmission equipment with stations in the top 25 U.S. markets if the station committed itself to the use and promotion of RDS services. About 300 stations made a commitment to use RDS by April 1996 (More than, 1996), but the number had fallen to roughly half that by 1998.

As is the case with all new broadcast technologies, the cost of the hardware starts out very high and, as the technology diffuses, the cost of equipment goes down. The first generation of receivers cost $25,000. In 1995, receivers cost $7,000 each; by early 1996, the receivers cost $2,700, and by 1998, the cost was about $1,500. For the technology to succeed with a critical mass of adopters, the price of the equipment must continue downward, and the number and range of services will have to increase.

Supplements—DARC

The data radio channel (DARC) is the high-speed data transmission equivalent to RDS. It is more than 10 times more powerful than RDS, and has sold well in Japan since its introduction in April 1995 (Barber, 1998). One DARC application, Digital DJ, provided service on WCBS-FM in New York, and the company hoped to expand to the top 10 U.S. markets by late 1998 or early 1999. Digital DJ is a subscription service offering stock updates and sports scores. A subscriber needs to purchase a specially-equipped, data screen-equipped AM/FM radio. Like RDS, DARC has yet to find its core audience in the United States. Successful demonstrations of radio-transmitted coupons, foreign language lessons, and providing directions to sponsors' places of business have, however, shown the bright potential of the system.

New Delivery Modes—The Internet

Radio stations have a presence on the Internet. The number of U.S. radio broadcasters that have home pages on the Internet has gone from a handful in 1994 to more than 4,300 by April 1998. The figure grows each week (BRS, 1998). Approximately 700 of these stations also provide audio samples of their broadcasts or simulcast their service on the Internet. Full-service Web-stations use applications such as Real Audio or Microsoft's Active Movie Streaming Format (ASF) to transmit their service 24 hours a day. The popularity of these services led to the publication of two books on Webcasting (Miles & Sakai, 1998; Miles, 1998). It also prompted the music licensing firms, Broadcast Music International (BMI) and the American Society of Composers and Publishers (ASCAP), to create a separate Internet-license for those wishing to play music on the Web.

The major technological development in Internet audio delivery was the transformation from "unicasting" to Internet protocol (IP) "multicasting." Unicasting—one-to-one communication—led to clogging of audio file servers. IP multicasting accommodates multiple users by treating them essentially as a single user. This more efficient delivery method promises to improve Internet service and audio quality.

New Delivery Modes—DAB

Probably the most anticipated transformation of radio technology will take place when digital audio broadcasting (DAB) becomes the predominant transmission technology. DAB is also referred to as DAR (digital audio radio) and digital sound broadcasting. Stations which use digital audio technologies such as CDs, minidiscs (MDs), digital audio workstations (DAWS), and digital cart players will be able to maintain the digital nature of their audio outputs. However, there are problems with adopting DAB. Stations must absorb the cost for new transmitters, and the buying public will have to purchase receivers capable of picking up DAB. If DAB were to be implemented in the United States just after the inauguration of advanced television system, as some industry analysts predict, the public might be hesitant to support yet another new technology. Also, there are questions as to which technical standard might best serve radio audiences of the 21st century.

One such question to be addressed is whether DAB should be satellite-fed or terrestrially broadcast. Satellite downlinks have provided stations with audio programming since the 1970s. More recently, satellite-delivered digital audio services have become widely available on cable television and direct broadcast satellite systems. In 1995, the FCC authorized spectrum for direct-satellite digital audio radio service in the United States. The FCC auctioned satellite DARS spectrum in 1997, in part to raise up to $3 billion to help balance the budget (Fleming, 1996). Results from the auction were disappointing, however, with bids totaling only $173 million.

Two companies scheduled to offer DARS in 1999 and 2000, respectively, are CD Radio and American Mobile Radio (Grant, 1998). The companies have proposed using a new breed of receiver that will deliver digital audio to silver-dollar-sized antennas installed on new or used cars. Drivers would pay a subscription fee to hear commercial-free and commercial-supported audio services. With the new system, a motorist could listen to the same station on a cross-country trip.

Terrestrial DAB is another imminent possibility for radio broadcasters. Three types of terrestrial DAB have attracted industry scrutiny:

- In-band on-channel (IBOC).

- In-band adjacent-channel (IBAC).

- Out-of-band, which requires a new band of spectrum (Jurgen, 1996).

Out-of-band service would lead to the obsolescence of current radio sets if broadcasters discontinued analog service. The advantage of out-of-band is that engineers are not trying to retrofit an old technology. Instead, a newer system generally allows for greater freedom and a wider range of services. The out-of-band service that has become a de facto standard in Europe, Canada, Mexico, and many other countries is known as Eureka 147. Standardizing Eureka 147 in the United States would lead to the end of IBOC and IBAC work. It would also foster compatibility with neighboring countries and the nations which have already adopted the system. However, U.S. broadcasters prefer an in-band digital system and question the availability of spectrum for the out-of-band approach (Stimson, 1998).

The easier system to inaugurate would be an IBOC system. IBOC would leave all stations on the same frequencies that they presently occupy. The in-band approach would allow users of current analog radio sets to continue using them even after instituting IBOC service. Two IBOC systems under consideration in the United States are Digital Radio Express (DRE) and USA Digital Radio (USADR).

IBAC systems also leave FM broadcasters in their current broadcast slots. They allow broadcasters to transmit digitally in adjacent sidebands. However, no IBAC systems have been developed which would allow for digital AM broadcasts, and there are interference concerns with simultaneous FM analog transmissions. Given the uncertainties with a conversion to digital audio broadcast, U.S. radio should remain a predominately analog-transmitted AM and FM medium at the turn of the century.

Factors to Watch

A New Economy of Scale

With more than 12,000 radio stations in the United States, a concern about widespread adoption of new radio broadcasting technology is that individual stations might not have the financial clout to institute changes. However, the economy of scale in radio has continued to evolve over past years. Radio's unprofitable years in the late 1980s and early 1990s led to FCC ownership rule modifications which allowed competing stations to combine personnel and broker programming time through LMAs (local marketing agreements). In 1992, the FCC (1992) also eased its duopoly rule that had long prevented an owner from holding more than one AM or one FM station in a given market.

The FCC rules allowed stations to streamline their operations, and many stations began to rely heavily on network-delivered programming instead of creating their own locally originated shows. Profits soared following the passage of the Telecommunications Act of 1996 which repealed limits on the number of radio stations a single licensee could own. Furthermore, the FCC relaxed limitations on the number of stations an entity could own in a single market (FCC, 1996b). (In radio markets with more than 45 stations, an owner can hold licenses of up to eight commercial radio stations, of which not more than five can be in the same service [AM or FM]. In markets with 30 to 44 stations, they can own seven stations, but not more than four in the same service. In markets with 15 to 29 stations, they may own six stations, but not more than four in the same service. In markets with fewer than 15 stations, they may own five stations, but not more than three in the same service and not more than 50% of the stations in that market.) The economics were so robust that the market value of radio-related stocks increased an average of 400% between 1993 and 1998 and 110% during 1997 alone (Rathburn, 1998).

As a result, group owners continued to buy more stations, and consolidation within the industry continued. In the first quarter of 1997, radio station transactions (not including mergers) amounted to more than $4 billion compared with less than $800 million in the same period during 1995 and $2 billion in 1996 (Changing hands, 1998). By contrast, the total radio transactions for the entire year of 1991 totaled less than $1 billion (Ditingo, 1995). Total advertising revenues for U.S. radio stations increased 10% to $13.6 billion from $12.4 billion in 1996. The 1997 ad revenue total represented a 50% increase when compared with 1991's total of $8.0 billion (Radio revenues, 1998).

The ever-expanding group owners should have more financial clout when it comes to buying innovative technology; look for them to make "group buys" for their stations. In addition, some group owners already have direct ties to the innovations being tested and, therefore, might be more interested in adopting the technology. Equipment makers began reporting increased sales of solid-state radio transmitters thanks to the new mega-radio chains (Petrozzello, 1996). The adoption of solid-state transmitters is a signal of purchasing power and a move away from tube-based models. Perhaps more important, however, is the fact that radio transmitters are built to last for 10 to 20 years or longer. These major transmitter purchases may thus signify a corporate vision among some group owners that digital audio broadcasting is years away.

Rolling Out New Radio Technology

The years 1999 and 2000 are key roll out periods for radio technology. Three subcarrier services tested in the mid-1990s—the NHK FM Subcarrier Information System (FMSS) known as the Digital DJ, the Seiko High Speed Data System (HSDS), and Mitre's Subcarrier Traffic Information Channel (STIC)—will try to make it in the U.S. marketplace. RDS service providers have some reasons to be optimistic. In part, because of decreasing receiver costs, Ford announced it would ship 600,000 RDS-equipped cars in

1999 and, by early 2000, there will be one million RDS-equipped Fords on the road. Thus, RDS might begin to attain the critical mass it needs to thrive in America.

As noted, CD Radio and American Mobile Radio are scheduled to begin DARS service in 1999 and 2000, respectively. USADR's terrestrial DAB plans call for the first receivers to be available by December 2000. With station conversion costs estimated at between $70,000 and $150,000, small and medium market stations will have to make some difficult decisions on equipment purchasing plans (McClane & Stimson, 1998). When these new services have their chance in the marketplace, the industry should have a clearer vision of what radio service will be like in the 21st century.

Bibliography

Ammons, B. (1995, September). *RBDS for your station*. Tempe, AZ: Circuit Research Labs. [Online]. Available: http://www.aloha.com/~cpiengrs/rbds.htm.

Barber, D. A. (1998, March 18). Digital DJ has big hopes for DARC. *Radio World*, 12, 14.

BRS Radio Consultants. (1998, April 5). *Commercial radio stations on the Web*. [Online]. Available: http://brsradio.com/.

Changing hands. (1998, March 30). *Broadcasting & Cable*, 46.

Ditingo, V. (1995). *The remaking of radio*. Boston: Focal Press.

Federal Communications Commission. (1992). *Revision of radio rules and policies*. Washington, DC: FCC.

Federal Communications Commission. (1996a). *Mass Media Bureau announces revised expanded AM broadcast band improvement factors and allotment plan*. Washington, DC: FCC.

Federal Communications Commission. (1996b). *FCC revises national multiple radio ownership rule* [NM 96-12]. Washington, DC: FCC.

Feldman, T. (1997). *An introduction to digital media*. London: Routledge.

Fleming, H. (1996, October 7). DARS auctions will help underwrite federal agencies. *Broadcasting*, 20.

Grant, L. (1998, February 9). 3,000 miles, 1 radio station. *USA Today*, 6B.

Jurgen, R. (1996, March). Broadcasting with digital audio. *IEEE Spectrum*, 52-59.

McClane, J., & Stimson, L. (1998, March 18). Walden: "This is the future of radio." *Radio World*, 1, 17.

Miles, P. (1998). *Mecklermedia Internet world guide to Webcasting*. New York: John Wiley & Sons.

Miles, P., & Sakai, D. (1998) *Internet age broadcaster*. Washington, DC: National Association of Broadcasters.

More than 200 U.S. radio stations have agreed to use Radio Data System. (1996, March 28). *Communications Daily*.

Petrozzello, D. (1996, June 3). Radio posts solid sales of solid-state. *Broadcasting & Cable*, 74-75.

Radio revenues increase 10% in 1997. (1998, February 2). *Radio online*. [Online]. Available: http://www.radio-online.com/.

Rathburn, E. (1998, April 8). Wall Street tuned to radio. *Broadcasting & Cable*, 58.

Stimson, L. (1998, March 4). NRSC group meets new DAB player. *Radio World*, 1, 8.

Viner, A. (1996, February). *Reinventing radio*. Presentation at Communications: The New Media Conference, Toronto.

COMPUTERS & CONSUMER ELECTRONICS

I f there is one theme underlying the developments discussed in this book, it is "the impact of digital technology." Nowhere is that impact more profound than in the computer industry. This year's computer technology will unquestionably be replaced in less than two years by technology that has up to twice the performance at almost half the cost (a phenomenon known as "Moore's Law"). These advances in computer technology in turn lead to advances in almost every other technology, especially those consumer products incorporating microprocessors or other computer components. The next eight chapters illustrate the speed, direction, and impact of the continuous innovation in computer technologies across a wide range of computing and consumer electronics technology.

The next chapter explores the manner in which computers have moved beyond text to incorporate the melange of video, audio, text, and data known as "multimedia." The following chapter addresses the most significant emerging application of personal computers, explaining the Internet and the World Wide Web as an information resource. E-mail has emerged as one of the most ubiquitous applications of computer technology in business, earning a chapter of its own, Chapter 11. The application of computers and other related communication technologies in the office setting (Chapter 12) then provides an illustration of the manner in which these technologies are revolutionizing commerce around the world. At one point, prognosticators predicted that networked computer technology would lead to a "paperless office." Chapter 13's discussion of document printing technology provides a clear indication of how and why we are using more paper than ever.

The longest chapter in this section, Chapter 14, discusses a set of technologies that may have the greatest long-term potential to revolutionize the way we live and work—virtual reality. The production and distribution of video and audio programming is the subject of the last two technologies in this section. The home video chapter reports on the incredible popularity of existing analog video formats, and on new, digital technology that is beginning to challenge the analog incumbents. Finally, the digital audio chapter reports on the early outcome of the battle between competing analog and digital audio technologies, with digital casualties as well as victors.

In reading these chapters, the most common theme is the systematic obsolescence of the technologies discussed. The manufacturers of computers, video games, etc. continue to develop newer and more powerful hardware with new applications that prompt consumers to continually discard two- and three-year old devices that work as well as they did they day they were purchased, but not as well as this year's model. Most software distribution, from movies and music to television and video games, has been based upon the continual introduction of new "messages." The adoption of this marketing technique by hardware manufacturers assures these companies of a continuing outlet for their products, even when the number of users remains nearly static.

An important consideration in comparing these technologies is how long the cycle of planned obsolescence can continue. Is there a computer or piece of consumer electronics so good that it will never be replaced by a "better" one? Will technology continue to advance at the same rate it has over the past two decades? How important is the equipment (hardware) versus the message communicated over that equipment (software)?

Finally, each of these chapters provides some important statistics, including penetration and market size, which can be used to compare the technologies to each other. For example, there is far more attention paid today to the Internet than to almost any other technology, but less than one in five U.S. households has access to the Internet (with even smaller penetration levels in other countries). On the other hand, the VCR is now found in about nine out of 10 U.S. homes. In making these comparisons, it is also important to distinguish between projections of sales and penetration, and *actual* sales and penetration. There is no shortage of hyperbole for any new technology, as each new product fights for its share of consumer attention.

9

Multimedia Computers and Video Games

Tina M. Hudson*

Multimedia is the integration of text, audio, graphics, animation, and full-motion video which can be disseminated on multiple platforms. A direct result of the convergence of the communications, entertainment, and computer industries, multimedia uses every type of media that currently exists. It promises to continue shaping the technology that supports it, and it will alter the way we live as we enter the new millennium.

There are two basic categories of multimedia: networked and stand-alone. Networked multimedia includes Internet-based applications and most forms of videoconferencing. Non-networked multimedia includes CD-ROMs, video games, and interactive television. Speed, reach, availability, and number of users are some of the defining limitations and capabilities of each category (Grant, 1998). This chapter will detail the background, recent developments, current status, and future prospects for multimedia and the two most popular platforms—CD-ROMs and video games.

Multimedia programs, or titles, are typically geared toward entertainment, education, and training. They are distributed by means of multimedia personal computers (MPCs), video game consoles, multimedia networks, virtual reality systems, the Internet, and public access kiosks. While MPCs and video game consoles enjoy prominence in distributing multimedia, other available platforms are gaining in popularity, and there is potential for the number of platforms to increase.

* Graduate Student, College of Journalism & Mass Communications, University of South Carolina (Columbia, South Carolina).

The key defining element in multimedia is its interactivity, which is defined as the dialogue between the user and the computer. The term can also be described as the ability of the user to control or interact with the computer or the multimedia title. Other media, such as the television, also combine text, graphics, audio, and video; however, the user cannot interact with or control what is happening on the television screen (Holsinger, 1995).

The development of the first audio compact disc by Philips and Sony in 1982 was the beginning of multimedia technology. CD-ROM drives have evolved since 1985 to meet the increasingly high storage demands of multimedia titles. Specifications for CD formats are published in "colored books," which provide details of size, amount, and type of data that can be stored on the optical disk. See Table 9.1 for a complete list of the formats and specifications. More recently, the desire to deliver high-quality video has led to the development and recent introduction into the marketplace of the digital versatile (or video) disc (DVD). DVDs have the potential to replace CDs for storage and mass distribution, given that they can hold seven times more data than a CD (DVD: Delayed no more, 1997).

Table 9.1
Colored Book Specifications and Formats

Disc Format	Disc Type	Book Specification
Audio CDs (CD-DA)	CD-DA	Red Book
Data CDs (CD-ROM)	CD	Yellow Book
Data CDs (professional use)	CD-i	Green Book
Recordable CDs	CD-MO, CD-WO, CD-RW	Orange Book
LaserDisc	CD Extra	Blue Book
VideoCD	DVD	White Book

Source: A. McFadden

Industry formats for developing technology are becoming increasingly more difficult to standardize. In particular, competing format standards for the next-generation DVD, the DVD-RAM, could have a considerable effect on mass adoption of DVD as an eventual replacement for the CD-ROM. This confusion over standardization will profoundly affect adoption of this new technology (Andrews, 1998).

While CD-ROM is the preeminent platform for delivering multimedia titles, another popular platform is the video game console. Video game consoles are essentially computers without monitors and keyboards, and they are designed to play quick animation on a television screen. These systems are small enough to sit on top of the TV or VCR, and they can handle animation playback more efficiently than the most powerful PC. Two of the largest manufacturers of video game consoles, Nintendo Entertainment Systems and Sony, are currently engaged in heavy competition for the multibillion dollar video game market.

Background

Multimedia

Multimedia has been around for nearly 15 years, when sound, video, graphics, and text were combined with the introduction of the first CD-ROM. Multimedia is distributed over a number of platforms, including MPCs and video game consoles. The evolution of these platforms was parallel, as improvements to CD-ROM technology for use in MPCs aided in the development of the consoles, which are basically computers without keyboards (Grant, 1998).

Multimedia has benefited both the private and public sectors, with many titles being developed for educational, training, and business purposes. Application titles, such as encyclopedias, have been developed exclusively for use in the classroom. Businesses, large and small, are able to cut costs by increasing employee productivity through the use of desktop publishing and presentation software to produce their own programs. The private sector also enjoys a wide variety and number of multimedia titles, including ones for entertainment, education, and productivity.

The computer-based multimedia platform is either for playback or authoring of titles. A playback system must be certified by the Multimedia PC Working Group, which defines the minimum requirements for MPC components and peripherals. MPC specifications were developed through industry consensus and have been central to the rapid adoption of the personal computer. They have also spurred the development of a growing and established base of standardized, multimedia-ready computers, which, in turn, precipitated the ongoing development of MPC software and peripherals (New multimedia standard, 1995).

The latest specifications or industry standard for MPCs is MPC-3, which was introduced in June 1995 and updated in February 1996. MPC-3 is an upgrade to two previous specifications that set minimum requirements for the multimedia components of the PC, including, among others, the processor, RAM, hard drive, audio, and CD-ROM. Other peripherals such as the monitor, speakers, and graphics capabilities also have minimum requirements.

Authoring systems consist of the computers and hardware that multimedia developers use in designing multimedia titles. Authoring tools are used by programmers and consist of a computer language, such as ScriptX, that communicates with the binary code of the computer so the programmer can tell the computer what to do. The early days of programming multimedia presentations revolved around Apple Computer's HyperCard and its HyperTalk programming language. In 1990, HyperCard's technology was brought to the IBM PC platform (Miastkowski, 1990).

Due to the complexities with the computer language involved in writing early multimedia programs, only professional programmers were able to create them, requiring companies to invest heavily in custom programs. The development of sophisticated software and hardware to assist with multimedia design has lowered the cost of producing multimedia titles and made it possible for more individuals to develop their own programs (Holsinger, 1995).

Creating a multimedia title is similar to the process of making a movie, and it involves many people in the design, production, and distribution phases. In the design phase, the production team determines the applications, platforms, user experiences, and overall budget. An application that is to be directed toward home entertainment systems, for example, will carry a platform on NES, Sony, or another manufacturer of video game consoles. Applications geared toward MPCs will be designed to run on Windows or Macintosh. The production budget is typically much less expensive than making a movie, usually between $275,000 and $400,000. It is rare for a multimedia title to cost more than a million dollars (Holsinger, 1995).

Multimedia titles are produced on large, expensive hard disk drives and transferred to more durable and less expensive media for distribution, such as a CD-ROM disc or video game cartridge. CD-ROMs are especially cost effective. It costs approximately $2,000 to create a master disc and about one dollar per disc to duplicate (Holsinger, 1995).

Graphics and animated graphics are also an integral feature of the multimedia title, and video games tend to rely heavily upon them. According to Holsinger (1995), "one very consistent lesson in multimedia production is that people don't like reading huge amounts of text on a screen" (p. 59). As a result, programmers use graphics more frequently to explain a concept, provide information, or increase enjoyment of the title.

Standardization of multimedia platforms is of extreme concern to multimedia publishers. Because of the rapid and ongoing development of technology, the need for standards continues to complicate the development of software titles. Delays in developing titles often result from controversy concerning developing standards for a new technology. Delays in standards are also compounded by the emergence of new or improved technologies.

CD-ROMs

The development by Philips and Sony in 1982 of the audio compact disc (CD-DA), a digital optical storage media, led to CD-ROM, the most utilized optical storage media for multimedia today. The specifications, referred to as the Red Book standards, include a 12-cm diameter optical disc capable of storing up to 72 minutes of high-quality sound encoded using 16-bit linear PCM at a 44.1 kHz sampling rate (Jeffcoate, 1995). The development of disc standards for this storage medium opened the door for a multitude of CD players to enter the market, as long as they were capable of playing a CD-DA disc. Standardization contributed to the rapid adoption of the technology, which resulted in audio compact discs and CD players outselling cassette tapes and players by 1992.

In 1985, the technology of the CD-DA was expanded to store prerecorded digital data in addition to audio. The specification for this expanded format, known as the *Yellow Book*, supported up to 650 MB of prerecorded digital data and was the original CD-ROM. This first-generation CD-ROM stored primarily text and was capable of supporting still images, graphics, and audio. However, it was not possible to have all three together at once. This format was also operating system dependent, requiring different versions for each platform. Macintosh required the Hierarchical Filing System (HFS) format, while MS-DOS file systems used the IS 9660, a directory format covered by an ISO standard (Holsinger, 1995).

Alternative formats to Philips' and Sony's CD-ROM standards were developed, such as Panasonic's 3DO Multiplayer system. While the 3DO Multiplayer was capable of playing CD-DA discs, the different characteristics and features of this alternate format rendered it incompatible with all other platforms. This incompatibility led to the eventual demise of 3DO and other alternate formats (Jeffcoate, 1995).

These earlier CD-ROMs continued to be in the form of read-only memory formats and primarily involved the mass distribution of information, prerecorded and stored in analog or digital form. While CD-ROMs have high storage capacities, flexibility, and longevity, and are very proficient at playing back audio and graphics, the quality of video playback was limited. Recent technology, discussed later in this chapter, has since been developed to address these limitations.

Video Game Consoles

Video games for home entertainment were introduced in the early 1970s with the release of Pong—the first game machine—and the video arcade. In 1971, Magnavox developed Odyssey, the first home video game system that connected to a television. Several years later, Atari developed an Odyssey-like machine that played *Pong* which experienced decent sales (Kent, et al., 1997).

The first "programmable" or cartridge-based home game console was Fairchild's Channel F in 1976. Cartridges could be inserted into these home consoles, allowing, for the first time, multiple-game playing capabilities. These early home consoles used several animation processors in conjunction with a fairly low-power microprocessor and were capable of transferring as much as several megabytes of data per second to the system's CPU (Holsinger, 1995). Channel F was the first of many console games to take advantage of dedicated hardware that increased the speed of animation and actually surpassed MPCs at delivering animated graphics. The actual workings of these proprietary consoles are trade secrets, and only the manufacturers know the hardware construction.

The first generation home game consoles used 8-bit systems, a format standard that set minimum requirements for the processor's speed and graphics capabilities. Fairchild and other manufacturers, including Mattel and Coleco, did not gain widespread popularity, and sales were unimpressive.

This early failure for home game consoles resulted from the manufacturers' inability to deliver the arcade experience it claimed players would have. Early home game titles were translations from successful arcade titles, mixed with original work. Manufacturers of home systems typically developed games for their own machines, until designers left and formed third-party publishing firms. While both Atari and Sega experienced tremendous success with such titles as *Asteroids*, *Pac Man*, and *Frogger*, video arcade games did not transfer well to the home console.

These translated games were supposed to deliver the same gaming quality as arcades; however, they failed miserably. Additionally, the tremendous success the arcade industry was experiencing at this time, coupled with relatively low cost, did not add to the home video console's appeal. In 1977, Atari released their first programmable video computer system (VCS), the Atari 2600, which experienced early success with such titles as *Atari Football* and *Space Invaders*. It wasn't until 1979, however, with the release of the title *Asteroids*, that year-round sales got strong. Atari continued to rely on translations of popular arcade video games and, in 1980, with their release of the home game version of *Space Invaders*, sales for the Atari 2600 sky rocketed.

Although Atari experienced tremendous success with the Atari 2600, the success was short-lived. Video games dropped in popularity for about five years. Then, in 1986, Nintendo released the NES, a 16-bit cartridge-based home video system, which sold for $199. NES outsold its competitors 10 to one (Kent, et al., 1997). Both Sega and Atari released comparable systems—the Sega Master and Atari 7800—but both failed miserably. In 1989, TurboGrafx, the first PC-based system, was developed by NEC, and sold for $400. Competitors, however, declined to develop similar systems. Sega had some success with Genesis, another cartridge-based console that sold for $249.95. Nintendo continued to remain a strong leader. The first portables were also released in 1989, Nintendo's Game Boy ($149.95) and Atari's Lynx ($179.95) (Kent, et al., 1997).

The first half of the 1990s saw continued improvements in the hardware for video game consoles and the graphic capabilities of the software titles designed for these consoles. By 1992, Sega was a major competitor, with a strong showing of their *Sonic the Hedgehog* titles. Sega's first CD-based console, the Sega CD, however, was not successful and may have enjoyed more success had they provided easier access to third-party software developers. Consequently, the system was abandoned within three years. Sony, a new player in the home video console market, began working on a 32-bit CD-based console after negotiations with Nintendo to develop a joint system fell through.

The mid-1990s saw Atari and Sega's share in the market continue to decline primarily because of new competition from Sony. In 1993, Sega announced a 32/64-bit console, the Saturn, and the third edition of its popular *Sonic* title. At this time, Sega controlled 50% of the video game market. Nintendo also announced in 1993 its plans for a 64-bit system, Project Reality. In 1995, Sega released the Saturn to unimpressive sales, perhaps because it was released months before it was publicized, catching software developers off guard, with only a few titles available initially. Nintendo released *Donkey Kong* and began to regain previous lost market share as its sales began to catch up with those of Sega.

In 1995, Sony released the PlayStation, a 64-bit CD-based console, for $299—$100 less than expected. This machine experienced strong sales and received praise from retailers and consumers. It appeared that CD-based technology was the wave of the future. However, it was another bad year for overall sales (Kent, et al., 1997).

A full year after Sony released its CD-based system, Nintendo finally came out in late 1996 with Nintendo64, a 64-bit system not CD-based. Nintendo64 was a strong seller, but Sony continued to lead the industry in share. Sales were strong for the first time in several years. It would appear that the new 64-bit systems were the impetus for a growing industry, as Nintendo and Sony virtually control the entire market, leaving Sega to lag far behind in third place.

In 1993, Senators Joseph Lieberman (D-CT) and Herbert Kohl (D-WI) investigated violence in multimedia. This investigation led to the joint development of voluntary and involuntary ratings (About the Entertainment, 1998). These ratings were designed to provide information to consumers regarding video game content. Currently, two classification systems are used in the United States: the Entertainment Software Rating Board (ESRB) and the Recreational Software Advisory Council (RSAC). While many say the ratings have led to the development of more violent games, none of these games appear to be making best-seller lists.

ESRB ratings are age-based categories which were implemented in September 1994. These ratings are voluntarily used by publishers for all platforms, including personal computers, CD-ROMS, and video game consoles. Publishers submit their games to the ESRB, and the games are categorized into five age groups based upon the amount of violence, language, and sexual content in the game. Along with the age category designation, the degrees of the material content, or "descriptors," are indicated on the packaging (About the Entertainment, 1998).

RSAC ratings, also established in late 1994, are based on providing content information instead of using age as a factor. These ratings, which have been criticized as being confusing to the consumer, use an "All" rating for those games suitable for all audiences. For those games not suitable to all audiences, a ranking scale of one to four is used, which differentiates between the levels of violence, nudity/sex, and language found in the game (About RSACi, 1998).

Recent Developments

Multimedia

The latest chip technology addressing the delivery of multimedia is the MMX. MMX has instruction sets designed specifically to accelerate multimedia functions (New chips speed, 1997) and are extensions to a PC's processor. MMX chips were developed by Intel and are in all Pentium II computers.

Competition remains fierce between the developers of processing chips. Other chip vendors, including AMD and Cyrix, went to court to acquire rights to manufacture MMX-extensions for their processor chips. Different types of instruction sets have also been designed by chip vendors such as Digital Equipment Corporation and Sun Microsystems. DEC's latest processor is the 533 MHz Alpha 21164, which includes such features as floating-point performance for 3-D and MMX-like extensions called Motion Video Instruction (MVI) for enhancing digital video (New chips speed, 1997).

New copy protection technologies have recently been developed for existing CD-ROM and DVD-ROM drives by Hide and Seek Technologies, which, when added to the structure of a disc during manufacturing, allows for one-time software installation or limited use. Variations on this technology also provide for protection of video CD or DVD movies, allowing for one-time or limited viewings. This new technology could result in opening new distribution channels for movies, such as vending machines, and alter the way movies are rented. This technology can also be extended to CD-R discs using a different protection method, requiring the user to remove an adhesive "ripcord" once installation is complete (Starrett, 1997).

Copyright protection concerns delaying the introduction of the DVD player centered around implementation of "video watermarking," a new encryption technology that prevents disc copying and software piracy in DVD systems. The content scrambling system (CSS) had been previously defined as the licensable specification for DVD-video systems. However, CSS protection technology is not approved by the DVD Forum, a 10-member group of manufacturers and publishers who jointly defined the specifications for the new technology. To complicate matters, many of the early multimedia upgrade kits, including some from DVD Forum members, contained CSS descrambling chips (Kroeker, 1997).

CD-ROMS

Recent developments in CD-ROM technology arose from playback limitations and increasing demands from software producers and consumers alike to provide more CD-player flexibility and better video technology. These technological advances dealt with:

- The way a disc is read by the player.

- The player's capability of writing, erasing, and rewriting on discs.

- Improved video playback.

Until recently, data had been recorded using *constant linear velocity* (CLV) mode—each sector on a disc holds the same amount of data whether the sector is located on the longer, outer tracks or the shorter, inner tracks. To read those sectors at the discs' center at the same rate as those at the outer edge of the disc, the motor has to speed up and spin the disc faster on the inner tracks (What does, 1997). This speeding up and slowing down adds stress on the motor, or torque, and slows down playback (The evolution, 1997).

To reduce torque and speed up playback, compact disc recordables (CD-Rs) were developed utilizing *constant angular velocity* (CAV) mode and write-once, read-many (WORM) technology. The specifications for CD-R can be found in two parts of the *Orange Book*.

In the CAV mode, a mode also used in hard disk drives, the drive spins the disc at a constant rate, instead of adjusting for the outer or inner disc location, as with the earlier CD-ROM drives. The drive *head* reads the data at varying rates depending upon what part of the disc is being read. For outer edges, the drive head reads the data at higher speeds than when reading data at the inner tracks. These speeds range from 12x (1,800K per second) to 24x (3,600K per second) (The evolution, 1997). Hybrid CD-ROMs combine both CLV and CAV modes, spinning the disc at a constant rate or reading data at the same rate (The evolution, 1997).

Additionally, CD-R allows a user to copy or record information from another drive, such as a hard disk, onto a blank CD-R disc. The disc is being altered, and information cannot be erased from the CD-R disc once it has been written to (WORM technology). Software programs have been developed that allow for multisessions, or to add more information to a disc after the CD-R has been "burned" (or recorded). Multisessions are typically used now with linked multisession discs that enable multimedia programs to expand to more than one disc without losing any valuable storage space.

In 1996, compact disk rewritable (CD-RW) technology was introduced, providing for recording, erasing, and rewriting data on blank CD-RW discs. CD-RW drives use magneto-optical technology. This phase-change technology differs from WORM in that the infrared laser does not make deformations in the recording dye layer. Instead, the state of material on the disc's surface is changed from a crystalline or transparent layer to an amorphous or nondescript form with a visible laser (Andy McFadden, 1997). The first generation of CD-RW drives only allowed for playback of CD-RW discs and were criticized in the industry for violating the tradition of compatibility. Presently, CD-RW drives capable of playing both CD-R and CD-RW technologies are becoming available.

DVD-ROM

The emergence of a new *high-density* rewritable optical storage media led to the development of the DVD, with format specifications found in the *White Book*. DVD was designed to play back broadcast-quality video, and one disc can store up to 4.7 GB of data per side (Andrews, 1998)—seven times the storage capacity of CDs. Future DVD-ROM media, capable of storing data on both sides of the disc and using dual-layer media, will provide 17 GB of storage, 27 times the amount of today's CD-ROMs. DVD-ROM utilizes MPEG-2 compressed video playback, which is visibly superior to MPEG-1. MPEG-2 can display a resolution of 720 by 480 pixels, four times that of MPEG-1. While MPEG-1 is designed to play on the earliest CD-ROMs, MPEG-2 requires a significant hardware upgrade.

In addition, DVD-ROM will not degrade from copy to copy, nor after continuous use, and it has a better range of colors containing the exact information as the original. Perhaps more important, this medium can be compressed into many sizes depending upon how you use it, indicating it can be easily developed for video game consoles, MPCs, and can even adapt to the various platform operating systems.

DVD-ROM systems arrived in stores in March 1997, after delays by manufacturers in developing final specifications as a result of copyright concerns by the movie studios (Parker, 1997), and the continuing industry split over incompatible, competing standards (Kay, 1998). This delay has certainly had an effect on adoption, but the fact is that the confusion surrounding DVD-ROM and DVD-RAM (random access memory) may have a greater effect on adoption.

There are presently two versions of DVD-ROM that have been introduced to the market, DVD-1 and DVD-2. Limitations with DVD-1, including inability to read CD-R or CD-RW drives, led to the development of DVD-2 (Andrews, 1998). DVD-ROM is the official format approved by the DVD Forum (Parker, 1998).

DVD-ROM players lag behind the fastest CD-ROM drives in overall performance, and, while DVD-ROMs can read DVD movies intended for consumer DVD players, DVD-ROM discs will not be readable by TV-based players (Andrews, 1998). Additionally, there are only limited quantities of DVD-ROM discs available, as multimedia developers scramble to produce titles that will work in these new machines.

The second-generation DVD will have similar capabilities as the CD-RW in that it will be capable of writing, erasing, and rewriting data. DVD-RAM was adopted as the standard format in 1997 by the DVD Forum, and it is expected to be released in 1998. This standard format, however, requires a protective covering, or cartridge, making it incompatible with existing DVD-ROM drives. This development led several Forum members to split from the group and develop a format that was not only compatible, but backward compatible with existing DVD-ROM and CD-ROM drives.

This alternate format, DVD+RW (rewritable) uses phase change, grove-only recording similar to the CD-RW. It will be capable of reading CD-ROM, CD-R, and CD-RW media, as well as DVD-Video and DVD-ROM. This format does not require a cartridge, and it can hold up to 3 GB of data per side. This format is backed by Sony, Philips, and Hewlett-Packard and is expected to be released in summer 1998. Additionally, a third format, Multimedia-Video File, which can hold up to 5.2 GB of data per side, has been developed and is being backed by NEC. A release date had not been given as of spring 1998 (Parker, 1998).

Video Games

During 1997, Sony dropped the price for the PlayStation to $149, initiating a price war. Nintendo64, which originally sold for $249 when introduced in 1996, followed suit. Sega did not lower the $299 price of its Saturn. Although Nintendo does not have the greater market presence, it was clearly the winner of the price war as it gained in market share, perhaps at Sega's cost. Sega clearly lost this battle, as it fell to third in market share.

Many retailers say the development of the 64-bit industry has resulted in a rejuvenation of the video game industry. Video stores are also benefiting from rentals of games. Retailers have seen sales triple and quadruple since the introduction of the new generation systems (Atwood, 1997).

Multimedia titles continue to be developed for 64-bit machines, while the 32-bit machines have lost popularity. Of the top 20 game titles for 1997, 14 were first introduced in 1997. In 1996, approximately 66% of next-generation (32-bit and higher) software sold; that figure climbed to 90% in 1997. An estimated 48 million next-generation software game titles sold in retail stores last year according to the NPD Group (Kenedy, 1998b).

Only about 40 games have been released for the Nintendo64, but hundreds of titles are available for the Sony PlayStation (Thomas, 1998). Sega Saturn has only about one-third the number of successful titles as the Nintendo64. The disparity in the number of titles available for both Nintendo and Sony are direct reflections of the software licensing procedures of each console manufacturer. Licensing fees to third-party software developers for Sony and Sega machines simply require a request for hardware specifications to either manufacturer and payment of a one-time licensing fee. Nintendo maintains substantially more quality control over games played on their machines and requires designers to submit game concepts. Nintendo works with the developer in creating a title that Nintendo believes will be successful. Publishers can create up to five titles a year for Nintendo and are required to pay a licensing fee for each title.

Current Status

Multimedia

High-tech multimedia titles have begun to replace the ho-hum, as technological improvements to hardware and peripherals introduced "feel," improved graphics, and increased data processing. While computer sales lagged during the end of 1996, unit sales were up due, in part, to the MMX-enhanced computer systems, Pentium II computers, and the decrease in price, to under $1,000, for lower-speed computers. Strong computer and peripheral sales spurred software retail sales to $4.9 billion in 1997, up 19% over 1996 (Sales surge, 1998).

The number of computers shipped worldwide in 1997 saw the biggest gain for Hewlett-Packard, Dell Computer, and Compaq Computer (PC shipments, 1998). Compaq, the worldwide and U.S. leader in PC shipments, saw an increase of 51% over 1996 shipments in the United States and maintained a 16.6% share of total shipments. Dell Computer came in second with 9.5% of 1997 shipments, a 67.1% increase over 1996 shipments. Packard Bell's shipments, which declined 4.4% over 1996, still came in third with 9.2% of total shipments (see Table 9.2).

By the end of 1997, computer superstores, such as Computer City and CompUSA, were the dominant retail outlets for software sales, achieving 41% of software revenue for that year. Consumer electronic superstores garnered 12.9%, down from 13.3% in 1996, while office superstores increased their share to 12.5%. Software-specialty retailers fell from 13.6% in 1996 to 10.8% in 1997 (Retail sales, 1998).

The competition for shelf space continues to plague new and old developers of multimedia software titles. Packaging of titles and associations with big-name developers can determine if a title is placed on a lower shelf or not. It has also become more difficult to find software titles because large retail chains have started charging software companies for shelf space. Some software titles can have up-front costs of from $10,000 to $20,000 per title per chain (Rothman, 1998).

Table 9.2
Top Five 1997 U.S. PC Shipments

Manufacturer	Total Shipments (Units/ Millions)	% of Total Shipments (Units/ Percent)	% Change from 1996 (Units/ Percent)
Compaq	5.14	16.6%	51.0%
Dell	2.94	9.5	67.1
Packard Bell	2.84	9.2	-4.4
IBM	2.74	8.9	23.6
Gateway	2.17	7.0	35.2

Source: Computer Retail Week

Productivity and education software titles are not selling as well as video game titles. One impact of this decline was that stock in one of the top producers of multimedia titles, Broderbund, was downgraded by analysts when sales from their entertainment line barely offset their losses from education and productivity titles.

One of the biggest software events of recent times centered on the all-time best selling CD-ROM title to date, *Myst*, and how its sequel *Riven* would fare (Atwood, 1997). Riven was ranked first in the November 1997 top-selling PC games list published by *PC Data*. (*Myst* was in the number two spot.)

Toy makers such as Mattel and Hasbro enjoyed surprising success in the multimedia software market in 1997, while titles backed by major movie studios, such as LucasArts, did not perform as well as expected. Of the top 20 titles as of November 1997, nine were published by toymakers. In contrast, only one major studio had a top 20 title (Thomas, 1998).

CD-ROMS

Prices for blank CD-R discs have declined rapidly since their introduction and now cost between $2 and $3 each. Prices for these discs are determined by market demand and royalties paid to the recording industry (Oakes, 1998). The royalties result from the 1992 Audio Home Recording Act which requires a payment by disc manufacturers for each blank consumer disc sold to be made to the Recording Industry Association of America. Prices for the CD-R player range from $399 to $650, and prices for internal players are expected to drop below $250 in 1998 (Sengstack, 1997).

DVD-ROM kits and upgrades are available in a wide range of prices, from $350 to $799. These kits contain the software and additional hardware to bring the MPC up to

DVD-ROM speed. These kits include, in addition to the drive, an internal IDE drive cable, external video wires, internal audio cables, a decoder card, a VGA loop-back cable, and DVD-ROM discs. Blank DVD-ROM discs are sold separately for about $50 (Waring, 1998).

Video Games

Major players in the video game market have seen their number decrease by half in the past three years, and analysts who predicted in 1995 that the market wasn't strong enough to support four players were off the mark. While the number of leaders did indeed decrease, the fall from grace by Sega and 3DO was largely due to technology and missed opportunities, not a weak market for video games. The two current leaders, Nintendo and Sony, simply had machines that were superior in performance and provided multimedia titles that were more appealing to the consumer. Early sales figures reported by the NPD Group, an independent market research firm, reflect a 51% increase over 1996 in the retail sales of video games to $5.5 billion (Kenedy, 1998a).

Sony maintains 48.5% of the market share for next-generation systems (64-bit) and 54.3% of the market share for systems, software, and peripherals sales combined. According to Sony's executive vice president and chief operating officer, Kaz Hirai, since the introduction of the PlayStation in late September 1995, 8.7 million systems, 47.3 million software units, and 19.8 million peripherals have been sold worldwide (Kenedy, 1998b, 1998a). There are hundreds of titles available at almost half the cost of Nintendo64 titles. Sony titles rounded out the bottom half of the top 10 with four titles, averaging in price at $36.00. Sony's top title, *Final Fantasy*, sold for $51.00.

Nintendo introduced Nintendo64, its next-generation system (although still cartridge-based) a year after Sony, and although it has a 43.3% market share, it has not released any figures regarding total sales to date (Kenedy, 1998b). Nintendo did introduce, in 1997, the first-ever device that responds through vibrations to action occurring on screen. The Nintendo64 Rumble Pak is sold separately at a suggested retail price of $19.95 or with the title *StarFox64*, for $69.95. Nintendo has released 45 titles to date, with 18 additional titles expected in 1998 at an average cost of $61. Nintendo64 sold the top five video games in 1997, with two of the top three titles in the *Mario Brothers* family, the game that started it all for Nintendo (Nintendo64 game list, 1998).

Sega has not experienced the same success with its next-generation system, the Saturn, even though it was introduced several months before the Sony PlayStation. The Saturn maintains an 8.3% market share since its debut in March 1995 (see Figure 9.1).

Figure 9.1
Video Game Marketshare

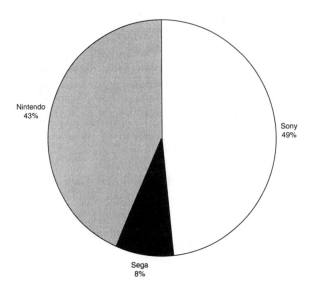

Source: Kenedy (1998a)

Factors to Watch

The recent breakthrough developments in DVD-ROM and DVD-RAM technology will almost certainly change the direction of multimedia, CD-ROMs, and video games. With Sony experiencing such tremendous success with their CD-based system over the past several years, it would appear that, if competitors want to increase their market share or even simply maintain their footing, they will have to develop consoles using DVD-based technology. Nintendo64 has announced its plans, including specifications, for its CD-based console, but has not set a release date. Sega is also developing a new system, with a 1999 release date, described as their "super" console that has "advanced technical capabilities," provides business opportunities to third-party publishers, and provides "revolutionary game play" for consumers. No other technical specifications have been provided; however, Sega did reveal that its platform would deliver gaming experiences "never before possible" (Sega confirms, 1998).

Networked multimedia will continue to gain in popularity and will be a major competitor for the end-users' valuable discretionary time. As the Internet improves security in delivering such applications as electronic cash for shopping and other sensitive monetary transactions, as well as providing more online gaming capabilities, this platform will certainly see a dramatic increase in use.

Technology is moving at such a rapid pace that leaders in the video game industry as well as the PC-based industry will be skipping levels. Rather than going up the ladder, rung by rung, the manufacturers of hardware, peripherals, and possibly even platforms, will be leaping two steps at a time. The ability of DVD to play on any platform is a powerful feature, and could very well be the "killer app" that the world has been waiting for. While early figures indicate slow adoption, these numbers do not necessarily reflect overall diffusion into the market.

Video games and PC entertainment software will continue to be highly competitive with each other; however, the trend will be toward developing a title to play on both platform types. With the introduction of DVD into the PC, we will see titles that perform like never before. The process of publishing titles will become more complicated as the fight for shelf space continues, and factors such as packaging and peripherals will play important roles in determining what gets prominent shelf position and what does not.

Another trend that could have great impact as far as increases in sales is a focus of multimedia titles toward young girls. While multimedia entertainment titles and video games have notoriously been directed toward the eight- to 16-year-old boy, present research indicates that girls can also benefit from computer gaming. Recent successes with Purple Moon's *Rockett Adventures* and Mattel's line of Barbie CD-ROM titles are signs that times are changing. We should see the number of titles published specifically for young girls increase dramatically over the next several years.

With the introduction of DVD technology, MPC specifications will be updated in the near future. MPC-3 level specifications originally were adopted in July 1995, and were updated February 1996. Minimum specifications will have to include DVD technology, even though CD-ROMs are still prevalent and the adoption rate of the DVD is slower than anticipated. While DVD+RW will have the capability of playing on CD-ROM drives, if multimedia titles are to be developed using this new technology, PCs must be certified with DVD-related specifications.

Growing concern over copyright protection will increase and continue to hamper adoption, except among the early adopters. Until hardware or software is developed to satisfy those who have the most to lose—the movie and recording industries—this issue will not go away. Manufacturers, who developed the latest technology causing such uproar, will be looked upon to address and remedy the problem.

Conclusion

The past several years have seen tremendous technological advances directly impacting many facets of multimedia. These advances have addressed the increasing need for greater storage capacities, flexibility, and longevity, in addition to providing improved interactivity to the end user. While the introduction of CD-ROM has led to these advances, future developments, led by DVD-ROM, will continue to change the face of multimedia and launch it into the next century.

This chapter has explored computers as a new medium. Computers are also playing a critical role in virtually all other media. The following article provides a great set of examples of how computers are changing not only the content, but the basic manner in which media organizations function.

Bibliography

About RSACi. (1998). *Recreational Software Advisory Council Home Page*. [Online]. Available: http://www.rsac.org/fra_content.asp?onIndex=1.

About the Entertainment Software Rating Board. (1998). *Entertainment Software Rating Board Home Page*. [Online]. Available: http://www.esrb.org/news.html.

Andrews, D. (1998, January). DVD finally! *PCWorld*, 195-207.

Andy McFadden's CD-Recordable. (1998, January 19, 1998). [Online]. Available: http://www.cd-info.com/CDIC/Technology/CD-R/FAQ.html.

Atwood, B. (1997, August 23). Enter*Active Christmas: Multimedia offerings include *Myst* sequel, handheld pets & rap gaming. *Billboard*, 67.

Bennett, H. (1997, September). Running OPC: The best thing for CD-R, but what about DVD? *Emedia Professional*, 52-53.

Block, D. (1997, May). Packaging your multimedia product: Where are the resources? *Emedia Professional*, 100.

DVD: Delayed no more. (1997, July). *PC World*. [Online]. Available: http://www.pcworld.com/hardware/cd-rom_drives/articles/jul97/1507p143g.html.

The evolution of the CD-ROM drive. (1997, July). *PC World*. [Online]. Available: http://www.pcworld.com/hardware/cd-rom_drives/articles/jul97/1507p143f.html.

Fitzpatrick, E. (1997, August 23). Retail finds windfall in new games. *Billboard*, 87.

Grant, A. E. (1998). *Interactive multimedia*. Unpublished manuscript.

Hardware hit hardest by price erosion. (1998, January 5). *Computer Retail Week*. [Online]. Available: http://www.techweb.com/se/directlink.cgi?CRW19980105S0023.

Holsinger, E. (1995). *How multimedia works*. Emeryville, CA: Ziff-Davis.

An introduction to the Multimedia PC Working Group. (1998). *Software Professionals Association Home Page*. [Online]. Available: http://www.spa.org/mpc/over.htm.

Jeffcoate, J. (1995). *Multimedia in practice: Technology and application*. New York: Prentice Hall.

Kay, R. (1998, January). DVD stands for DiVideD. *Byte*. [Online]. Available: http://www.byte.com/9801/sec5/art15.htm.

Kenedy, K. (1998a, January 23). U.S. video game sales hit $5.5 billion. *Computer Retail Week*. [Online]. Available: http://www.techweb.com/se/techsearch.cg?action-View&doc_id=INV19980123S0008.

Kenedy, K. (1998b, January 30). Sony says software will drive console sales. *Computer Retail Week*. [Online]. Available: http://techweb.cmp.com/crw/news98/sony0130.html.

Kent, S., & Horwitz, J., & Fielder, J. (1997). *History of video games*. [Online]. Available: http://www.videogamespot.com/features/universal/hov/index.html.

Kroeker, K. (1997, August). More copy-protection delays for DVD-video on PCS? *Emedia Professional*, 14.

Lanctot, R. (1997, February 3). Retail software revenue grew 12.7% in '96—PC games, productivity titles fuel record December sales. *Computer Retail Week*. [Online]. Available: http://www.techweb.com/se/directlink.cgi?CRW19970203S0013.

Lewis, M. (1998, February 3). Sugar, spice, and everything nice computer games girls play. *MSNBC*. [Online]. Available: http://www.slate.com/Millionerds/98-02-03/Millionerds.asp.

McMakin, M., & Parry, J. (1998, January 13). Rewriting: Simply a drag. *New Media*. [Online]. Available: http://newmedia/com/newmedia/98/01/feature/CD_Rewritable.html.

Miastkowski, S. (1990, July). Plus gets hyper across platforms. *Byte*, 111-112.

New chips speed multimedia. (1997, June 3). *New Media*. [Online]. Available: http://newmedia.com/Today/97/06/03/New_Chips_Speed.html.

New multimedia standard for personal computers delivers full-screen, full-motion video and wavetable sound. (1995, June 12). *Multimedia PC Working Group*. [Online]. Available: http://www.spa.org/mpc/mpc_pr.htm.

Nintendo64 disk drive. (1998). *Nintendo Home Page. Nintendo64 hardware*. [Online]. Available: http://www.nintendo.com/n64/64dd.html.

Nintendo64 Rumble Pak. (1998) *Nintendo Home Page. Nintendo64 hardware*. [Online]. Available: http://www.nintendo.com/n64/rumblepak.html.

Nintendo64 game list. (1998) *Nintendo Home Page. Nintendo64 game list*. [Online]. Available: http://www.nintendo.com/n64/gamelist.html.

NPD reports video games had banner year in 1997. (1998, January 26). *NPD Home Page*. [Online]. Available: http://www.npd.com/corp/press/press_videogame1.htm.

Oakes, C. (1998, January 30). Cheaper CD recorders, pricier discs. *Wired*. [Online]. Available: http://www.wired.com/news/news/business/story/9963.html.

Parker, D. (1997, November). DVD completes half of a vicious circle—With a twist. *Emedia Professional*, 84.

Parker, D. (1998, January). The many faces of high-density. *Emedia Professional*, 60.

PC Data lists top-selling software for December 1997. (1998, February 5). *USA Today*. [Online]. Available: http://www.usatoday.com/life/cyber/tech/review/cdrom.htm.

PC shipments increase for HP, Dell, Compaq. (1998, January 27). *Computer Retail Week*. [Online]. Available: http://techweb.cmp.com/crw/news98/mkts0127.html.

Retail sales hit $29 billion despite PC price plunge. (1998, January 5). *Computer Retail Week*. [Online]. Available: http://www.techweb.com/se/directlink.cgi?CRW19980105S0021.

Rosenbush, S. (1998, January 6). DVD's dismal premier. *USA Today*. [Online]. Available: http://www.usatoday.com/life/cyber/tech/ctb884.htm.

Rothman, D. (1998, March). The software search. Where has all the software gone? *Home Family PC*. [Online]. Available: http://www.zdnet.com/familypc/content/9802/columns/savvy.html.

Sales surge despite lack of "killer app." (1998, January 5). *Computer Retail Week*. [Online]. Available: http://www.techweb.com/se/directlink.cgi?CRW19980105S0024.

Sega confirms new console in the works. (1998, January). *Sega Home Page, Sega Central*. [Online]. Available: http://www.sega.com/central/press_releases/jan98/newplat.html.

Sengstack, J. (1997, June). Easier, cheaper recordable CDs finally break the mold. *PC World*, 68-72.

Starrett, B. (1997, August). Hide and Seek announces non-intrusive CD copy protection. *Emedia Professional*, 13.

Thomas, Jr., E. (1998, January). Interactive game sales light up holidays for desperate retailers. *MSNBC*. [Online]. Available: http://www.msnbc.com/news/133598.asp.

Waring, B. (1998, January 13). Stop making sense. *New Media*. [Online]. Available: http://newmedia.com/NewMedia/98/01/testsuite/Cool_Tools.html.

What does 12X-20X mean? (1997, September). *Emedia Professional*, 58.

Computers in Media: Media Asset Management

Joan Van Tassel, Ph.D. *

One of the latest and most interesting applications of computer technology is variously called "media asset management," "digital content management," the "server-based newsroom (or station)," or the "automated newsroom (or station)." In film studios, media asset management is the favored term. For Internet and Intranet content creation companies, computer industry firms, and organizations that are not in the entertainment industry, digital content management may be more descriptive.

No matter what it is called, media asset management is not a single technology. Rather, it is a complex combination of hardware, software, and business practices that allows companies that produce, market, and manipulate information products to perform more work, more efficiently.

A completely different terminology has grown up in the broadcast and cable industries. Video servers were first adopted to insert commercials into the play-to-air or cable program stream. Now, local stations are putting servers in their newsrooms to edit and produce their local news shows, as NBC affiliate KHNL-TV in Honolulu did, one of the first U.S. television stations to adopt this technology. Local broadcast executives predict that the server-based station is only a few years away.

At the network level, CBS News is installing the IBM Digital Library and asset management system. Fox Broadcasting converted its Network Center and now plays out all its programming, commercials, promotions, and long-form programs from networked video servers. And NBC is in the process of building the Genesis digital server system. Cable networks also employ video servers for playback, including Viacom channels Showtime and M2. Probably the most sophisticated media asset management system in operation as of early 1998 is the one used by Discovery Communications, which makes all of its owned material available to versions of the Discovery Channel, now available in 22 foreign territories.

In radio, automated stations operate with almost no human intervention at all, except a part-time engineer who makes sure the systems keep running. A computer draws down a syndicated signal, stores it, inserts local spots, and plays it out, reprocessing the signal to make it conform to the station's FCC-assigned channel. A salesperson works outside the station, an accounting service handles the business administration, and an owner writes and cashes the checks.

---

* Associate, Technology Futures, Inc. (Malibu, California).

Like all organizations, media companies, such as film studios, post-production facilities, and television and radio stations, have used computers for business administration—human resources, payroll, accounting, billing, and sometimes traffic. At the same time, digital equipment has become more important in the arena of program creation and post-production. Digital machines provide nearly all special effects and titles. Finally, much of the signal processing for final distribution, including modulation for broadcast and multiplexing for cable, is accomplished digitally.

However, these digital islands—business, content creation, post-production, and distribution—have remained as separate built-up areas of activity. Media asset management can tie together some or all of these islands into a single, computerized domain, bringing the same efficiencies computers brought within each area to activities across departments and even the entire organization.

The Technology Platform

Four components make up a media asset management system: a hardware platform, software, a network, and a set of procedures for the acquisition, indexing, storing, retrieving, and tracking of material. Each element requires a thorough understanding of the implementing organization's current operation and needs, as well as how the needs will change over time.

The hardware includes a central computer or a distributed system of networked computers that function in parallel. Depending on how large the organization is, there may be any number of workstations attached to the network, accommodating upwards of several thousand simultaneous users. There are multiple layers of software to run the network and to provide security, communication, and user interfaces. Then, there is shared application software, such as asset management programs. Finally, local machines are loaded with the programs users need to do their particular jobs.

The network can be a local area network (LAN), an organizational wide area network (WAN), a within-enterprise Intranet that uses HTTP, a telephone company provided virtual private network (VPN), or even the public Internet. Many organizations use a combination of networks to link a wide variety of users—employees, suppliers, vendors, and customers.

The procedures and practices of an implementing organization define how the media asset management system must be customized to fit the organization. Few companies can adopt a turnkey system and expect a smooth transition to an all-digital operation. Automated radio stations, and to some extent, television stations may be exceptions, however, since they tend to have similar operational procedures even before bringing in a media asset management system.

Benefits of Adoption

An end-to-end media asset management system brings business efficiencies wherever it is implemented. Organizations usually put in a system in one area of their business. For example, a TV station could start with just the newsroom and a film studio might begin by putting video, still images, drawings, and text needed just for marketing, advertising, and publicity.

In addition to efficiency, a media asset management system allows for collaboration among employees, suppliers, vendors, and customers. And more than one worker can access the same asset to perform work, allowing people to work simultaneously on the same project or to complete multiple projects at the same time without waiting for the original material to be available.

Collaboration and multiple access permit staff to accomplish many tasks in less time than they did before the system was implemented. For example, the contractual rights and obligations that are attached to a video or audio clip or a photograph can be attached to the material, allowing rapid determination of what the company owns, how it is allowed to use the media, and how much should be paid for it. Thus, the system can simplify appropriate royalty payments and executives can approve work quickly. All these functions reduce the cycle time to market when time is important, as it often is with information products such as films, television programs, magazines, and books.

Media asset management also lets rights-holders exploit the material they own more effectively. They have immediate access to rights and availabilities and can recycle and distribute material quickly. Further, rights-holders may find that they can market component elements of the product in addition to the finished product, providing another revenue stream.

Finally, media asset management can save considerable money by reducing the number of times material is digitized and the considerable costs of transporting negatives, tapes, and reels from one location to another. Many companies send images and audio clips, sometimes even video clips, across the Internet or another computer network. Over the course of a project, saving transport costs can add up to a substantial sum.

Barriers to Adoption

The cost of large-scale media asset management systems can be daunting, in both money and time. Implementing the system requires much planning and a careful, thorough introduction. There are also some important issues that are not entirely solved, including methods of indexing and cataloguing, search and retrieval techniques, and system security. In spite of these barriers, media asset management systems are rapidly becoming an integral application of computer technology in mass media.

10

The Internet and the World Wide Web

Philip J. Auter, Ph.D.*

When this chapter was first published in 1996, the Internet and the World Wide Web were just beginning to enter the global consumer arena. The Internet had advanced from an experimental to an elite stage of communication—and was on its way to becoming a form of mass communication.

Today, just two short years later, the "Net" has become a major influence for people young and old across the globe. It is used so regularly, for so many diverse purposes, that it is safe to say that the Internet has truly become a new channel for all forms of human communication.

The Internet is a way of linking many small local area networks (LANs) into a global database of amazing proportions. The interconnections allow instant access to a world of information. Each individual network is administered, maintained, and paid for separately by individual commercial, government, private, and educational institutions or individuals (Eddings, 1996; Gaffin, 1996). Although Internet access has been predominantly free in the past, commercialization and growth of this new channel of communication have driven many sites to charge, or at least include advertising. And most "onramps" to the information superhighway—Internet service providers—charge at least $19.95 per month for the service.

Although terms like Internet and Web have become synonymous for all forms of computer-mediated communication, a more accurate way to define the Internet is as the channel for a variety of new forms of communication. Some of the most important ser-

---

* Assistant Professor, Department of Communication, University of West Florida (Pensacola, Florida).

vices provided by the Internet include e-mail, file transfer, access to remote computer systems via telnet and gopher, newsgroups and mailing lists, and the World Wide Web.

Electronic Mail

Electronic mail, or e-mail, is a way of sending almost instantaneously—and nearly free—an electronic letter between anyone in the world that is linked via a LAN, computer, modem and phone, or even fax machine. Although basic concepts behind e-mail parallel those of regular mail, e-mail offers the speed of a telephone call with the detail of a letter. It can be one-to-one, like regular mail, or can be used as a method of mass distribution (Eddings, 1996; Gaffin, 1996). E-mail is discussed in more detail in Chapter 11.

File Transfer

File transfer protocol (FTP) allows computer users to download and upload shareware programs, documents, and pictures to and from databases that are stored in archives at hundreds of sites around the world. FTP allows remote log-in to distant computers to access and transfer files. Files are typically compressed to save space and allow faster transfer. Compressed files must be decompressed before they can be used (Eddings, 1996; Gaffin, 1996).

Telnet and Gopher

Telnet allows a user to have remote access to databases, library catalogs, and other information sources around the world. For instance, the University of Michigan's Department of Atmospheric, Oceanographic, and Space Sciences supplies weather forecasts to U.S. and foreign cities, along with skiing and hurricane reports (Gaffin, 1996). Type the following into a Telnet program (not a Web browser):

telnet madlab.sprl.umich.edu 3000

and you will find yourself at their main page. From here, you can access a variety of current and historical weather information.

A Gopher is client/server software that uses Telnet and other Internet applications to send information back and forth between the gopher server and a distant client. Gopher servers exist on almost every large, publicly-accessible computer on the Internet, and they allow access to a system's resources including FTP, search programs, newsgroups, and much more (Eddings, 1996; Gaffin, 1996). Gopher was originally created at the University of Minnesota (the "Golden Gophers")—hence the name (Gaffin, 1996). The name also refers to the fact that the software "burrows" through the Internet to find information.

Mailing Lists and Newsgroups

A mailing list is an Internet database of people interested in a particular topic. Anyone can post to a list, but only subscribers will receive the posts. When a message is sent to the list, it automatically goes to all list members.

A moderated list is one that is screened by an administrator for duplicate messages or unacceptable content. Unmoderated lists have an administrator, but she or he does no censoring of messages whatsoever. List servers are even more automated and can subscribe, unsubscribe, and perform several other commands without the aid of an administrator (Eddings, 1996; Gaffin, 1996).

Newsgroups work much like mailing lists and may be moderated or unmoderated. Unlike mailing lists, however, newsgroup data is stored on all local sites that subscribe to the newsgroup. Because of this, only the most recent portion (thread) of the discussion may be available. Some sites do archive data, however (Eddings, 1996; Gaffin, 1996).

The World Wide Web

The World Wide Web is the most recent and exciting development on the Internet. Also a client/server distributed system, the Web creates a rich graphical environment on more powerful personal computers, incorporating enhanced text, graphics, sound, and moving visuals. Creative use of newer technologies has allowed the Web to evolve into a powerfully interactive communication environment. For individuals whose computers or servers cannot handle the information transfer load of the Web, a text-based equivalent known as "Lynx" is an option.

A Web browser, such as Netscape or Microsoft Internet Explorer, creates a unique, hypermedia-based menu on your computer screen. "Hypermedia" is the foundation of the Web and is a different, nonlinear way of linking data. One can jump around in hypermedia documents, clicking on highlighted topics, and switching to new documents—often at entirely different locations on the globe. Web servers maintain pointers or links to data that is spread out over the entire Internet (Eddings, 1996; Gaffin, 1996).

The *best* Web pages are true hypertext documents, taking advantage of the concept of nonlinear information linking by allowing users to jump from concept to concept at many logical points. The *worst* hypertext documents fall into two extremes—traditional text works simply "dumped" into the Net with few or no links, and documents where the author has gone "hyperlink crazy." The proliferation of this new form of text has resulted in new, nonlinear instruction environments that appear to enhance student learning as measured by exam scores and student perceptions. (For an excellent list of manuscripts that discuss the effects of nonlinear instruction, see University of Queensland psychology professor, Dr. Simon Dennis' bibliography page [http://www.psy.uq.edu.au:8080/~mav/ bibliography.html].)

The Web browser handles all connections and switches. Netscape was once the dominant Internet browser because of its easy-to-use interface, and because Netscape Communications Corporation [http://home.netscape.com/] offered the software free as a download from their Website. But Microsoft's Internet Explorer [http://www.microsoft.com/ie/] has stolen much of Netscape's thunder. Although some services, such as America Online, have proprietary browsers, on closer inspection, many of them turn out to be legally modified versions of Explorer or Netscape.

Bundled with the Windows 95 operating system—a standard on most PCs purchased in America beginning in the mid-1990s—Explorer quickly landed on more desktops than Netscape could ever hope for. In response, the U.S. Department of Justice has looked into claims of monopolistic tactics by Microsoft, even as Netscape's market share and stock value dropped. As of mid-1998, no decision had been reached on this issue.

The Web can be used for FTP, "global" Net searches, access to newsgroups, database searches, etc. Using browsers, you can link to "home pages" of businesses, government organizations, educational institutions, and individuals.

Figure 10.1
Philip J. Auter Home Page

Source: P. J. Auter

Hypermedia documents are written in hypertext markup language (HTML)—a series of codes that define the graphical nature and links on a page (see *A Beginner's Guide to HTML* at http://www.ncsa.uiuc.edu/General/Internet/WWW/HTMLPrimer.html). HTML is the key to the hypermedia format of the Web, but it is an unfriendly scripting language to work in. WYSIWYG (what-you-see-is-what-you-get) software has changed that, however.

Figure 10.2
Partial HTML Code for Phil Auter's Home Page

```
<HTML><HEAD>
<TITLE>Philip J. Auter, Ph.D.</TITLE>
<BODY BGCOLOR="#FFFFFF">
<CENTER><IMG SRC="Auter/images/uwf/uwfsmall1.jpg"></CENTER>

<IMG BORDER=4 Align=Right SRC="Auter/images/AuterP.jpg" ALT="Philip J. Auter
photo"><CENTER>

<IMG SRC="Auter/images/vidcam01.gif"> <B><FONT SIZE=6><A NAME="auter"> Philip
J. Auter </A></FONT></B><IMG  SRC="Auter/images/animations/macani.gif"><BR>
<FONT SIZE=5><I>Assistant Professor of Communication</I></FONT>
<P>
<IMG WIDTH=300 SRC="Auter/images/shortblueline.jpg"></CENTER>
<IMG  SRC="Auter/images/animations/emailani.gif">
<A HREF = "Auter/autervita2.html#address"><FONT SIZE=3>Address</A>
<P>
<B>Phil Auter</B> <A HREF = "Auter/autervita2.html#teaching experience">teaches</A>
courses in journalism, television production, media history, basic interpersonal communi-
cation, and new communication systems -- such as the Internet and World Wide Web.  He
conducts <A HREF = "Auter/autervita2.html#publications">research</A> in audience uses
of and gratifications from television and the Internet.  He also helps to coordinate the UWF
<A HREF = "communication.html">Department of Communication Arts</A> website.
<P>
Phil received his Ph.D. from the <A HREF = "http://www.uky.edu/CommInfoStudies/">Col-
lege of Communication & Information Studies</A> at the <A HREF = "http://www.uky.edu/
">University of Kentucky</A> in 1992.  Having experience in both print and broadcast
media, Phil taught at the <A HREF = "http://www.evansville.edu/">University of Evans-
ville</A>, in Indiana, and the <A HREF = "http://www.usouthal.edu/">University of South
Alabama</A>, in Mobile,  before coming to <A HREF = "http://www.uwf.edu/">UWF</A> in
1998.
<P>
```

Source: P. J. Auter

WYSIWYG Page Development Software

Until the mid-1990s, Web page authors were forced to work directly with arcane HTML code. Rather than focus on creativity, interactivity, and graphic design, home page managers had to learn to decipher the less-than-user-friendly HTML tag system. Finally, in the mid-1990s, a number of software producers began producing WYSIWYG software that allows home page producers to focus on the look of their pages—laying out text and graphics in a manner similar to traditional desktop publishing applications. One popular

WYSIWYG Web authoring program is Adobe PageMill [http://www.adobe.com/prodindex/PageMill/main.html].

Similar to other WYSIWYG Web authoring software, PageMill was designed for the non-technical user. Pages are written and designed in a word processor-style environment. Results appear as they will when accessed by Web browsers such as Netscape and Internet Explorer. Styles can be applied and images resized, and PageMill checks and corrects URL (uniform resource locator) links as they are copied and pasted throughout Web site documents (Adobe Systems Incorporated, 1998). And newer WYSIWYG Web page software allows advanced features (such as forms, CGI scripts, and Java applets) to be previewed.

The Domain Name System

All these applications rely on the assumption that each computer in the "network of networks" can find any other computer linked to cyberspace. The domain name system (DNS) establishes an Internet address for every single computer account with Internet access. Major domains maintain lists and addresses of other domains at the next level down—and so on—to the end computer user. U.S. domains use three-letter identifiers and are divided by application or theme. The primary U.S. domains are gov (government), org (other organizations on the Internet), edu (educational institutions), com (commercial companies connected to the net), mil (military installations on-line), and net (companies and groups concerned with Internet administration). The rest of the world uses two-letter country codes as their top domains (Eddings, 1996). In reading an address, domain hierarchy goes from right to left. For example, in pauter@uwf.edu, the major domain is "edu" (for educational institution). The next domain is "uwf" which represents the University of West Florida in Pensacola; "pauter" is the address on that computer for Professor Phil Auter's e-mail account. With this information, anyone in the world with Internet access can e-mail Phil Auter a letter. Web pages have similar addresses, known as URLs. They are based on the same system, but can sometimes be much longer. Take, for example, Phil Auter's home Web page address:

http://www.uwf.edu/~commarts/faculty/homepages/auter.html

"http://" is the standard command to any Web browser that it will be searching for (and retrieving) a document using hypertext transfer protocol. It is always used when searching for Web pages. In fact, since this is the standard beginning of a Web page URL, many browsers are configured so that you do not need to type it in. "www.uwf.edu" is the domain of the computer at the University of West Florida on which campus web pages exist. "commarts," "faculty," and "homepages" are three levels of subdirectories on the "uwf" system. Finally, "auter.html" is the actual page. The ".html" suffix is a common, but not mandatory, appendage to many Web page files.

Although some computer literacy is required, the Internet and Web continue to become more user-friendly, more visually appealing, more globally accessible, and more necessary in our daily lives. Much has changed about this new medium in its relatively short history.

Background

The history of the Internet and World Wide Web dates back to Cold War tensions between the United States and the former Soviet Union. The U.S. government formed the Advanced Research Projects Agency (ARPA) to establish a U.S. lead in science and technology applicable to the military in the 1960s. ARPA worked with the RAND Corporation to create a successful way to communicate after a nuclear war. RAND came up with the idea of a "network" that would have no central authority and that could operate even if a number of its major nodes lay in ruin (Gaffin, 1996).

During the 1960s, researchers experimented with linking computers to each other and to people through telephone hook-ups, using funds from ARPA. They developed a packet switching technology that allowed multiple users to share the same data lines—unlike any telephone lines to that date (Gaffin, 1996).

The first node of the new network was at UCLA. Stanford, the University of California at San Bernardino, and the University of Utah completed the original four-node system known as ARPANET by December 1969 (Zakon, 1997). Shortly thereafter, individuals with access to the network developed electronic mail, remote log-in to distant computers, and the ability to transfer files via data lines. The world now had detailed communication (and data transfer) at the speed of a phone call. ARPANET began online conferences with the help of several college (and one high school) students. Elite scientific discussions later changed to more general, mass appeal topics (Gaffin, 1996).

During the 1970s, ARPA supported the development of internetworking protocols for transferring data between different types of computer networks (Gaffin, 1996). Many LANs being developed in both academia and commercial industries began establishing gateways to the Internet to allow for electronic mail transfer. Links soon developed between ARPANET and counterparts in other countries (Gaffin, 1996).

In 1983, this network of networks became known collectively as the Internet. At about this time, ARPANET was split into ARPANET and MILNET (the Military Network). The Defense Communication Association required the two subgroups to use specific data formatting, called protocols, designed for Internet transmission known as TCP/IP (which stands for "transmission control protocol" and "Internet protocol"), allowing internetwork communication. This change also made it possible for additional nodes to come online. The domain name server addressing system was established in 1984 (Zakon, 1997).

The Internet expanded at a phenomenal rate. Hundreds, then thousands, of colleges, research companies, and government agencies began to connect their computers to this worldwide network. The National Science Foundation (NSFNET) began providing backbone service to U.S. supercomputers by 1986 (Zakon, 1997). This action led to an even faster dissemination of the Net from the government into university and commercial arenas. NSFNET linked mid-level nets that, in turn, connected universities, LANs, etc. Also in 1986, the first freenet [telnet://kanga.ins.cwru.edu]—a network designed to provide an Internet link to individuals with limited access to traditional avenues—was formed in Cleveland, Ohio and administered by Case Western Reserve University (Eddings, 1996).

In 1989, Tim Berners-Lee created the World Wide Web while working at CERN, the European Particle Physics Laboratory (Berners-Lee, 1998). However, it was not released to the public until 1991 (Cailliau, 1995; Zakon, 1997). This text-only Web, although a powerful research tool, did not take the public by storm. It would take the more graphical and user-friendly Mosaic-enhanced Web to begin doing that in 1993 (Zakon, 1997).

In 1990, the Electronic Frontier Foundation (EFF) was formed by the founders of Lotus Development Corporation, Mitchell Kapor and John Perry Barlow, to address social and legal issues arising from the impact on society of the increasingly pervasive use of computers as a means of communication and information distribution. EFF [http://www.eff.org/] began its mission of encouraging public debate on telecommunications and societal issues as well as supporting litigation that extends First Amendment rights to Internet-published work (Electronic Frontier Foundation, 1990). The same year, ARPA-NET was decommissioned by the Defense Communications Agency because NSFNET and mid-level nets had superseded it (Zakon, 1997).

Until the early 1990s, the Internet community was strongly against going commercial. Because government and educational institutions had primarily supported the Net, its primary purpose was considered research and education, although it was becoming more and more of an entertainment medium. Attempts to post advertisements to list-servs and discussion groups were violently opposed by members. Later, some Usenet newsgroups were specifically developed as "classified ad" services, where individuals could post information about second-hand items for sale.

Commercial companies did not really begin to participate in the Internet until the emergence of the World Wide Web. With the introduction of the graphically-oriented Web, the growing number of commercial online services, and the dissemination of the Net to wider areas of society, the definition of the Internet changed dramatically. Such alien concepts as subscriptions, ads, leasing arrangements, and product sales were not only becoming accepted, it was becoming more and more apparent that they would shoulder much of the burden of financially supporting the rapidly growing medium (Meroz, 1995).

By 1992, the number of host computers topped one million. In 1993, the White House [http://www.whitehouse.gov/WH/Welcome.html] and the United Nations [http://www.un.org/] were on the Net, and Internet Talk Radio (transcripts available at http://www.es.net/hypertext/talk-radio.html) began broadcasting and uploading compressed audio files of complete radio broadcasts for distribution on the Net. Other radio stations followed their lead (Zakon, 1997). In 1994, First Virtual [http://www.firstvirtual.com/], the first "cyberbank," came into existence. (The company now specializes in electronic messaging.).

Yahoo! [http://www.yahoo.com/], one of the first WWW search engines, was released to the Internet public in 1994. More recently, Digital Equipment Corporation introduced Alta Vista [http://www.altavista.digital.com/]—a service that makes heavy use of multi-threading 64-bit addressing capabilities to search for any word in any document published on the Web (including "hidden" text) or in Usenet discussion groups (Digital Equipment Corporation, 1997). Led by pornography sites, many Web page developers soon realized that more people would find their sites if dozens of key search words were hidden within the

HTML source code. Quite a few search engines have sprung up since then, many of which also offer "people finders," mapping programs, and even Web page language translation. (An excellent database of search engines can be found at http://hawking.NHGS.Tec.VA.U.S./internet_resources_index.html).

The G7 meeting in February 1995, hosted by the European Commission at Brussels, focused primarily on the effects of the World Wide Web on global society (Cailliau, 1995). NSFNET reverted to a research network, and U.S. backbone traffic was now routed through interconnected network providers (Zakon, 1997).

According to Zakon (1997), Radio HK, the first 24-hour Internet-only radio station, began offering programming online in 1995. (It ceased operation in 1996.) Leaders in audio streaming since 1995, Real Networks [http://www.real.com/] announced RealVideo software for real-time streaming of video across the Net in 1997 (Real Networks, April 1997). Although many people still do not have computers or Internet connections powerful enough to take full advantage of video streaming, the merger of the TV and the PC has begun.

The Web became the service with the greatest traffic on NSFNET in 1995 as well. Finally, traditional online dial-up services such as CompuServ and America Online began providing Internet and WWW access. The commercialization of the Net continued with a number of Net-related companies going public. Netscape, creator of the most popular Web browser, had the third largest ever NASDAQ IPO share value as of August 9, 1995. And the Vatican [http://www.vatican.va/] went online in 1995 as well (Zakon, 1997).

Web pages, originally useful because of their hypertext links, began evolving into rich, textured multimedia documents in the mid-1990s. CGI scripts and Java applets (mini-applications) allowed the Net to become interactive with forms, hit-counters, calculators, and other scripts—making the Web an even more powerful tool for business, education, and entertainment (Sun Microsystems, Inc., 1998). Image maps—pictures that include several imbedded hyperlinks—offer users a more visual method of navigating through a database of information. Animated icons, streamed audio and video, and plugins such as Macromedia's Shockwave Flash also now enhance the graphical interface with sound and motion. (For an excellent example, check out the Fox Kids Cyberstation website [http://www.foxkids.com/launch.htm]. You'll need Macromedia's Shockwave Flash to get the full effect [http://www.macromedia.com/].)

In 1996, America Online initiated a revolution in Internet service pricing by offering unlimited access for only $19.95 per month. Demand swelled as users switched from other Internet service providers, and many new users entered the market. Even though AOL spent millions of dollars upgrading their service, and other providers lowered their prices to match AOL, there was still a brief bottleneck as demand far-outstripped access.

Commercialization of the Net allowed for exponential growth in for-profit and subsidized sites—which generated greater demand from a larger user-base. Once the Internet went commercial, profitability became the big question. One attempt to stimulate online purchasing was "cybercash" accounts at Net banks like First Virtual (First Virtual, 1995). The need for these "middlemen" in online purchasing has almost vanished, however, due to the relative safety of a secure socket layer (SSL) of communication developed by Netscape and now widely adopted (Netscape Communications Inc., 1998). Data encrypted by SSL is considered to be so secure that some companies such as Wal-Mart Online are willing to reimburse you for any fraudulent charges made to your credit card that are not covered by your credit card issuer if your information is stolen while at Wal-Mart Online.

Still, much like broadcast TV, many Web sites don't "sell" anything, but rather offer information and links to additional services. More and more banner advertising is springing up on these pages, but to date, few are profiting from Net advertising. One organization that has bucked the trend by earning profit from Internet advertising is DoubleClick [http://www.doubleclick.net/].

Recent Developments

Over the last few years, the Net and the Web have grown at exponential rates. In response, the government and other users have tried to expand the Net itself and increase the speed of data transfer. All of the events have led to a marked increase in use by the corporate world and average individuals.

Access to Internet service now comes in three flavors: the original education/government institution route, online services such as AOL, and newer Internet service providers (ISPs). All three forms offer some type of basic Internet access—including e-mail, file transfer, news and chat groups, and Web browsing. Most also allow users—for an additional, nominal fee—a limited amount of storage space for establishing their own presence on the Web. But despite their similarities, these services do have their differences.

Today, institutional access—once free to all employees (and, in the case of universities, students)—is often limited to on-site services. Away from work, even employees and students will frequently have access only to text-based e-mail and Web services. Many universities have begun contracting out graphical Web service to agencies like Campus MCI [http://www.campus.mci.net/] because their internal computer services group cannot handle the usage load, and the fiscal managers can no longer afford to foot the bill.

Despite what elite "Netizens" may think of the company, America Online [http://www.aol.com/] may have done as much as any other organization to link up the large, Net-phobic masses. Once AOL evolved from a closed-system service, it became the epitome of the ultra-simplified online service experience. It offers simple access, as well as a wealth of online and phone-assisted help. Most important for the inexperienced user, through its service and its browser software, AOL prepackages much of the Internet into easy-to-access categories.

Whether they are large corporate entities such as AT&T Worldnet [http://www.att.net/] or "mom-and-pop" operations like Dibbs Internet Services [http://www.dibbs.net], ISPs differ from online services primarily by their lack of additional "layers of protection" between the user and the Net. More experienced users often gravitate to ISPs because they consider these "newbie" features to be cumbersome.

A number of important mergers and acquisitions have changed the landscape of Internet service providers. Many of the smaller outlets have consolidated. Even bigger concerns have been bought out. For example, CompuServe was purchased by AOL in 1997. Meanwhile, other corporations, such as Sprint, have sold off their Internet service divisions in an attempt to divest themselves from what can still be called a risky business. With all the mergers, acquisitions, and closings in this field, one truly needs a scorecard to identify the players.

Companies represented on the Net can be divided into "real-world" businesses with an Internet presence, such as Wal-Mart Online [http://www.wal-mart.com/] and Web-only cyber-companies, such as the highly successful online bookstore, Amazon.com [http://www.amazon.com/], that cannot be accessed any other way. Net-based profits are still limited however.

Although few companies have yet to make profits from the Net, many have discovered that they can better serve their current customer's needs for information. College loan clearinghouse, Sallie Mae [http://www.salliemae.com/]—like many other lenders—allows borrowers to track their account status online. So do major credit card companies (e.g., AT&T Universal Cards [http://www.att.com/ucs/]). Main Street can track events on Wall Street with free services offered by most of the brokerage houses (e.g., Stockmaster at [http://www.stockmaster.com/]), and the list of corporate service sites continue to grow.

Regardless of what they offer the consumer, almost all companies online are easier to find—thanks to a major change in the URL addressing system. InterNIC [http://rs.internic.net/]—established in 1993 by AT&T, General Atomics, and Network Solutions, Inc., and supported by the National Science Foundation—was designed, among other things, to clean up and simplify registration of Internet "addresses" (the URLs we type to go places on the Web) (InterNIC, 1998). For a fee of $100, almost any unused domain name can be registered. Corporations and individuals alike snapped up these easier-to-find domain names, making them more accessible for the average user.

The Net has also seen an explosive growth in the presence of the "regular citizen," not just to utilize its services, but to actually establish their own sites. With a limited amount of money and some knowledge of Web page design, anyone now can be a "publisher" of information. Dramatically faster computers and modems, along with WYSIWYG publishing software and $20 per month ISP access can all be credited for this evolution.

Current Status

Determining the number of users and user demographics of the Internet and WWW as a growing medium is challenging at best. Surveys almost always consist of convenience samples and frequently cannot screen out multiple responses made by the same individual. It is almost impossible to differentiate between households, individuals, and even complete strangers because, despite potential security issues, people often let others use their accounts. Additionally, there has seldom been a way to corroborate responses given to Internet-based surveys: A respondent claiming to be a 45-year-old female may actually be a 12-year-old male. Finally, even quasi-accurate results become quickly outdated as the Net community continues to grow at exponential rates.

A number of online surveys have been performed under a variety of conditions and with varying results—many are unscientific and seldom can they claim to be representative. With that caveat in mind, what follows are excerpts from the executive summaries of two Internet surveys performed by respected institutions: the Graphics, Visualization, and Usability (GVU) Center at Georgia Tech University in Atlanta and CommerceNet in conjunction with A. C. Nielsen.

The GVU Eighth Survey

The Graphics, Visualization, and Usability Center was established at Georgia Tech in 1991. The eighth survey, run from October 10, 1997 through November 16, 1997, collected over 10,000 responses (Georgia Tech, 1997).

Survey results showed that female usage had increased to nearly 40% of the sample. The fact that older, more experienced users tended to be male suggested that the influx of female users is a new trend. Although most of the users connected to the Net via modem, 27% were using 33.6K modems rather than slower devices. Thirty-nine percent of the respondents had upgraded their connection speed in the previous year. Most people spent 15 minutes or less searching before they began finding useful information (64%) (Georgia Tech, 1997).

Most of the respondents considered the Internet indispensable for e-mail (84%) and Web browsing (82%). The next most popular service—chat—fell a distant third (22%). Fully 56% of those surveyed had more than one e-mail account (all surveyed had at least one), and 46% had created a Web page. Almost half of all respondents (45%) used the Web one to four times per day. As with the previous survey, females use it slightly less frequently than males (Georgia Tech, 1997).

Privacy (30.49%) and censorship (24.18%) were the Internet issues that respondents considered most important. Despite privacy concerns, fully 25% did not know what cookies (i.e., unique, persistent, session identifiers) were (Georgia Tech, 1997).

Although many individuals still do not shop over the Internet due to security concerns, Internet commerce has blossomed. The most often-cited reason for using the Web for shopping was convenience (65%), followed by availability of vendor information (60%), no pressure from salespeople (55%), and time savings (53%). Men and women both ranked convenience first, but females valued no pressure from salespeople (54%) slightly higher than vendor information (51%) (Georgia Tech, 1997).

The CommerceNet/Nielsen Survey

Unlike most surveys of Internet/WWW usage, results of the CommerceNet/Nielsen Internet Demographics Surveys are based on completed telephone-based interviews of U.S. and Canadian citizens. A. C. Nielsen attempted to employ rigorous scientific methodology in order to obtain a sample that accurately and proportionally reflected the population of Internet users and non-users in both countries (CommerceNet/Nielsen, 1997).

Although their surveys have been widely debated, CommerceNet and A. C. Nielsen claim that their surveys can be generalized to the population with an acceptable margin of error.

Some of the key findings of the CommerceNet/Nielsen survey were:

- In 1997, there were approximately 58 million Internet and 48 million Web users in the United States and Canada that were 16 years of age or older.

- Twenty-six percent of the North American population—59 million individuals—use electronic mail.

- An important new statistic is electronic commerce activity. Almost 10 million people have made at least one purchase on the Web—a growth of 50% over previous studies. One inhibiting factor to Net shopping was a lack of knowledge about SSL. Only 14% of those surveyed were aware of this new encryption technology (CommerceNet/Nielsen, 1997).

CommerceNet and A. C. Nielsen claim that their survey is rigorous, but it is important to note that no standardized method of accurately measuring Web demographics and usage has yet to be accepted. It will be important to watch for further surveys, and always identify their source and methodology.

Factors to Watch

Many of the predictions made in the last edition of this text—and by other pundits—about the Internet have been about as accurate as psychic hotline information. Technological and cultural changes occur so rapidly and unexpectedly in this domain that it is almost futile to attempt to guess what will happen next—but it is fun. So without any further ado, let's look into some future concerns.

How fast can the Internet go? Currently, most people are limited to using 56K modems that, due to line limitations and ISP incompatibility, seldom connect faster than 28.8K. But the government is beginning to fund "Internet-2"—the next generation Internet (NGI)—a new backbone that could be up to 1,000 times faster than the existing system (Tan, 1997). This innovation would even speed up existing Ethernet connections—people who, usually through their office, are "directly wired" to the Net. Meanwhile, although an official standard for 56K modems was finally adopted (3Com, 1998), look for future competition from cable, digital satellite, and even broadcast options. Although WebTV [http://www.webtv.com/]—now a subsidiary of Microsoft—has made only limited inroads into the online user-base, some television stations are now actually using part of their bandwidth to broadcast a condensed version of Internet services to computers with special radio signal receivers. Experimental delivery services—promising more, better, and faster—will continue to proliferate, even if they are short-lived.

Expect online profit-making to take a big jump over the next few years, as more and more companies learn how to make their products and services appealing to the growing online community. As Internet market analysis improves, Net advertising will become more carefully targeted, and revenues should increase. But the big question about Internet dollars is "are they new money, or simply shifting of dollars out of traditional shopping and advertising outlets?"

Traditional news agencies, such as CNN [http://www.cnn.com/] and the *New York Times* [http://www.nyt.com/] as well as online publications such as *The Drudge Report* [http://www.drudgereport.com/] have increased the speed of news delivery, but also the attempt to "scoop" the competition. Over the course of several months in late 1997 and early 1998, many of these publications had to retract stories about President Bill Clinton that were perhaps published a little too hastily. Is irresponsible journalism going to be an occasional occurrence or a new trend?

One current and future concern is Internet taxation and regulation. Although President Clinton denounced a proposed state sales tax on Internet commerce in 1998, in almost the same breath, he proposed a national Net service tax to fund the wiring of public schools. Although the Communication Decency Act was found to be unconstitutional, many organizations and politicians are poised to restrict content in one form or another (Reardon, 1997).

Netizens should also be concerned that Microsoft may once again force itself into a dominant software position, this time in the Web browser arena. Only the government can stop this from happening—if they choose to do so. And we should also be on the lookout for new computer operating systems that may be built entirely around Web browser software. Microsoft's early versions of Windows 98 worked that way.

Will the whole world become wired? Probably not. Across the globe, many people still don't have basic telephone or television service, much less access to the Internet. And although agencies that assist the homeless and disadvantaged are beginning to come online, most of those people are more concerned with day-to-day life in the real world than with what's going on in cyberspace.

Your neighbors may never get wired, but if some people have their way, your whole house might be. There are already prototype cars and appliances connected to their own individual Web pages by wireless transmitters. The theory is that these devices can be remotely operated, and in some cases repaired, via the Web. Is this dream of cyber-efficiency and convenience realistic? Check out [http://www.philauter.home/appliances/dishwasher.html] in about two years to find out. You'll probably see "HTTP/1.0 404 Object Not Found." Who knows? You might find out how long my dishes have before the dry cycle.

Bibliography

Note: The Web is an active environment and, unfortunately, page addresses often change. Although links to Web sites referenced in this chapter were accurate at the time of publication, pages may have since moved. To find a "lost" link, enter key words from the references into a search engine such as Alta Vista (http://www.altavista.digital.com/).

3Com. (1998). *3Com V.90 technology: Technical brief.* [Online]. Available: http://www.3com.com/56k/why56k/white.html.

Adobe Systems Incorporated. (1998). *Adobe PageMill 3.0: Easily build Web pages.* [Online]. Available: http://www.adobe.com/prodindex/pagemill/ feature1.html.

Berners-Lee, T. (1998). *Bio.* [Online]. Available: http://www.w3.org/pub/WWW/People/Berners-Lee.

Cailliau, R. (1995). *A little history.* [Online]. Available: http://www.w3.org/pub/ WWW/History.html.

CommerceNet/Nielsen. (December, 1997). *Electronic commerce on the rise according to CommerceNet/Nielsen media research survey.* [Online]. Available: http://www.commerce.net/news/press/121197.html.

Digital Equipment Corporation. (1997). *About Alta Vista.* [Online]. Available: http://www.altavista.digital.com/av/content/about.htm.

Eddings, J. (1996). *How the Internet works* (2nd Ed.). Emeryville, CA: Ziff-Davis.

Electronic Frontier Foundation. (1990). *New foundation established to encourage computer-based communications policies.* [Online]. Available: http://www.eff.org/pub/EFF/Historical/eff_founded.announce.

Gaffin, A. (December 11, 1996). *EFF's guide to the Internet*, v. 3.20. (Formerly *The big dummy's guide to the Internet.*) Electronic Frontier Foundation. [Online]. Available: http://www.eff.org/pub/Net_info/EFF_Net_Guide/netguide.eff; also from online archives at ftp.eff.org, gopher.eff.org, http://www.eff.org/, and elsewhere. (Published in hardcopy by MIT Press as *Everybody's Guide to the Internet.*)

Georgia Tech Graphics, Visualization & Usability (GVU) Center. (1997). *GVU's 8th WWW user survey.* [Online.]. Available: http://www.gvu.gatech.edu/ gvu/user_surveys/survey-1997-10/.

InterNIC. (1998). *About the InterNIC.* [Online]. Available: http://rs.internic.net/internic/index.html.

Meroz, Y. (December 7, 1995). *Commercialization of the Internet.* [Online]. Available: http:// ils.unc.edu/yael/commerce.html.

Netscape Communications Corporation. (1998). *Secure sockets layer.* [Online]. Available: http://www.netscape.com/assist/security/ssl/index.html.

Real Networks Inc. (April, 1995). *RealAudio product announcement: Progressive Networks launches the first commercial audio-on-demand system over the Internet.* [Online]. Available: http://www.real.com/corporate/pressroom/pr/prodannounce.html.

Real Networks Inc. (February, 1997). *Progressive Networks announces RealVideo, the first feature-complete, cross-platform video broadcast solution for the Web.* [Online]. Available: http://www.real.com/corporate/pressroom/ pr/realvideo.html.

Reardon, S. (1997). *Supreme Court declares Communications Decency Act unconstitutional: Analysis.* [Online]. Available: http://www.ema.org/ html/at_work/cdauncon.htm.

Sun Microsystems, Inc. *The Java language: An overview.* [Online]. Available: http://java.sun.com/docs/overviews/java/java-overview-1.html.

Tan, J. (May, 1997). *Next generation Internet (NGI) initiative.* [Online]. Available: http://www.jhu.edu/~hac_ns/ngi/.

Wal-mart Stores Inc. (1998). *Wal-Mart Online security guarantee.* [Online]. Available: http://www.wal-mart.com/docs/security.shtml.

Zakon, R. (1997). *Hobbes' Internet timeline v3.1.* [Online]. Available: http://info.isoc.org/guest/zakon/Internet/History/HIT.html.

11

Electronic Mail

Gayle McCarthy, M.A.*

As advances in computer-mediated communication make our lives easier, we may have a difficult time remembering the days when e-mail was a promise of the techno-future and online services meant waiting for the next available customer service representative to answer your call. Yet, network communication technologies are changing the nature of how people communicate and share ideas (Mostafa, et al., 1994). A pervasive example of such a technology is electronic mail, more commonly known as e-mail. E-mail, where users can send messages to each other via a local area network or the Internet, represents one of the most widely-used functions of networked computers (Delta & Matsuura, 1998).

E-mail is the single most visible application of distributed computing in use today (Radicati, 1992). Millions of people use e-mail on a daily basis whether for personal use or business interactions. Indeed, many argue that e-mail has become more popular than its counterpart, traditional "snail" mail.

Background

Sending an e-mail message is considered to be the quicker, cheaper, and more environmentally-friendly alternative to sending a standard letter via the postal service. However, the United States Postal Service (USPS) recognized the imposing threat of potential business loss years ago and jumped on the electronic mail bandwagon with Mailgram, a joint service with Western Union Telegraph Company (WUTCO). The Mailgram message was sent via phone or telex, a communicating word processor with full-editing features. WUTCO then transmitted the message to the post office equipped with special printing

---

* Graduate Student, College of Journalism and Mass Communications, University of South Carolina (Columbia, South Carolina); Media Buyer, South Carolina Press Association.

machines nearest the receiver. After the message was printed, it was placed in an envelope and sent to the receiver's address in the next outgoing mail batch. This system was popular prior to the widespread diffusion of the fax machine.

In 1982, USPS introduced a service called Electronic Computer-Originated Mail, or E-COM (Panko, 1985). Operated solely by the USPS, E-COM processed large batches of messages via electronic input, as opposed to the single-message Mailgram. Messages could be sent as individually addressed letters, common-text messages that sent the same message to several different addresses, or by text insert messages that allowed the sender to cut and paste pieces of a particular message.

While the post office has attempted to implement technological advancements, people prefer the swifter communication offered by e-mail messaging. Modern-day electronic mail systems are typically characterized by two major components:

- A set of facilities to create and display electronic messages in human readable form.

- A set of facilities to reliably transport information from one user to another (Radicati, 1992).

The hardware needed to facilitate this process includes a computer, a monitor, keyboard, modem (a device that allows a computer to receive data over telephone lines or cable), phonejack, and cables to connect all components together. E-mail is both synchronous and asynchronous communication. The term synchronous refers to e-mail messages that may be sent and received instantaneously, provided that both sender and receiver are logged on at the same time. Asynchronous, or "store-and-forward," communication refers to information that is transferred between sender and receiver. The message is stored and later retrieved by the receiver who is not required to be logged on when the message arrives.

E-mail offers the advantage of one-to-one as well as one-to-many communication exchanges. An example of a one-to-many communication exchange is a Listserv, an electronic mail distribution system that serves as a conduit of mail postings related to a specific topic. The result of a successful Listserv program is an Internet-wide system through which individual users may post messages or request feedback on particular topics. The messages are quickly made available to other subscribers who share an interest in that particular Listserv topic (Mostafa, et al., 1994).

Often used by professors at colleges and universities, Listservs allow messages to be redistributed to users (in this case, students) who have subscribed to the service. Students need not be living on-campus to receive the Listserv postings. Off-campus students and those attending classes at satellite campus locations may receive messages as well, provided that the computers facilitating the exchange are hooked up to the university mainframe, are equipped with modems, and the students have subscribed to the service.

One indication of the popularity of e-mail and other electronic communication technologies on-campus is the fact that many universities have reallocated budget dollars and are now spending more on CD-ROMs, online services, and other electronic information services than on printed journals and books (Folkerts, et al., 1998).

E-mail addresses divulge certain user-profile data. The following hypothetical address, JoeSchmoe@sc.edu, identifies the person (Joe Schmoe), specifies the person's affiliation (in this case, the University of South Carolina), and denotes the type of organization from which the e-mail is being sent (.edu, a university). E-mail veterans are usually less candid about divulging first or last names in their e-mail addresses; however, true user identities and profiles are not difficult to trace.

A critical factor in the technological convergence of e-mail systems worldwide is the use of the Internet. Direct Internet service providers (ISPs), companies that provide Internet access, have become so prevalent that competitive rates as low as $12.00 per month allow customers unlimited access to the Internet. Previously, America Online (AOL) subscribers were paying monthly fees of $9.95 for five hours of use or $19.95 for 20 hours, plus $2.95 for each additional hour (Lohr, 1997). As anyone can see, spending time in chat rooms sending e-mail back and forth could quickly become a costly event. AOL soon realized the need to reduce subscription fees for its customers. Once this occurred, AOL was swamped with calls, and users became frustrated because the lines were always busy. Throughout 1997 and early 1998, AOL launched efforts to increase capacity to better serve their customers.

Recent Developments

As of late, the largest issue surrounding e-mail and related Internet use concerns the issue of privacy. The Internet has been likened to an Orwellian world of privacy invasion and subliminal behavior control (Weber, 1996). As more and more companies and established businesses go online, more business transactions are being made. Even though encryption devices have been implemented to ensure network security, many potential online customers remain wary of completing a transaction and offering their credit card number as well as other personal demographic information. Organizations dissuade their employees from sending and/or receiving personal e-mails, such as the omnipresent forwarded joke-of-the-day or browsing through Web sites while on company time. Perhaps more important, any company reserves the right to view all material produced at work during business hours by its employee, including all e-mail.

One example of the lawsuits surrounding the privacy issue is an arbitration claim brought by a former First Boston bond analyst, Jay N. Patel, for defaming his name in association with an e-mail prank that disclosed confidential salary data of First Boston executives. Two anonymous e-mails were sent to hundreds of First Boston employees detailing confidential salary information of more than two dozen of the firm's highest-paid current and former executives (Former First Boston, 1996). Patel is seeking an undisclosed amount in damages from First Boston and Raymond Dorado, the company's in-house lawyer. As of mid-1998, this case is still proceeding and has not yet been documented in the Labor Arbitration Reports.

Another case involves Johan Helsingius, the operator of the world's largest anonymous e-mail service in Finland with about 500,000 users, most of them U.S.-based (Finnish e-mail, 1996). Helsingius' e-mail service, which automatically gave incoming e-mail an anonymous pseudonym and facilitated discussion of controversial political and personal issues over the Internet, has been shut down due to a court order that linked his service to child pornography. Although he has refuted these claims, Helsingius' service stands as a prime example of the debate over e-mail privacy, particularly in relation to children's advocacy forums.

A similar case was heard in 1995 by the U.S. Air Force Court of Criminal Appeals that involved a senior officer's misuse of his e-mail account with regard to privacy and child pornography. In *United States v. Maxwell*, Colonel James A. Maxwell, Jr., was convicted of service-discrediting misconduct by a general court-martial when he transported images of minors engaging in sexually explicit activities using his computer, e-mail, and America Online. Maxwell appealed his conviction on the grounds that his Fourth Amendment right to privacy had been violated when his e-mail accounts were searched. However, the military appeals court upheld Maxwell's conviction.

Interestingly, the issue of privacy may be somewhat of a non-issue. There is no mention of a right-to-privacy in the U.S. Constitution (Leibrock, 1997). Erik Larson (1992) discusses this topic at length:

> Our so-called right to privacy is a melange of constitutional interpretations and too specific legislation that fails utterly to take into account the passage of America into "cyberspace," the age of ephemera, where computers, fiber-optic superhighways, interactive cable television, smart-cards, and invisible telecommunications networks promote the constant, liquid transfer of personal information across all boundaries in a heartbeat (p. 15).

The problem is that technology is generating new legal boundaries (Samoriski, et al., 1996), and laws designed to protect the conventional provinces of broadcast, speech, and print are proving inadequate for the domain of cyberspace. Modern-day lawmakers have realized the significance of electronic privacy and responded by providing common law torts that recognize four arenas of invasion of privacy:

- Publication of truthful but embarrassing facts.

- False light or defamation.

- Appropriation.

- Intrusion into one's physical sphere of solitude or seclusion (Samoriski, et al., 1996).

The passage of the Electronic Communications Privacy Act of 1986 (ECPA) was an effort to address the newfound challenges brought about with technological advances such as electronic mail.

Under the rubric of the ECPA, more protection is granted to citizens in an effort to curb illegal hackers from gaining unauthorized access to specific e-mail accounts, particularly in the arena of business dealings. The principal feature of the law is the incorporation of language that makes it illegal to intercept the digitized (or data) portion of a communication (Samoriski, et al., 1996). The language stems from a 1985 report from the Executive Branch's Office of Technology sector that deals with new technology assessment and review. The report highlighted the disturbing fact that little or no protection was granted regarding electronic privacy, thus denying America's ability to store and safeguard business information.

Consumers are now taking matters into their own hands, and practitioners are following suit. For example, TRUSTe is a licensing system recently created so that online merchants disclose their practices and act as self-regulating watchdogs, ensuring that all online merchants abide by the same rules in business dealings. TRUSTe protects consumers from unethical marketers by first screening its licensees, and then requiring each merchant to sign a contract that calls for random spot audits. The consumer is thereby empowered with the choice of dealing with a given merchant from the outset. Thus, the power and choice of marketing practices are ultimately left in the hands of the consumer.

A more comprehensive technological approach was unveiled at Federal Trade Commission hearings by Tim Berners-Lee, the researcher credited with inventing the World Wide Web (Weber, 1997). Lee proposed the Platform for Privacy Preferences, or P3, that would allow a Web site to automatically transmit information regarding its privacy policies. Users can then program their browser software to permit entrance only to those Web sites that meet specific criteria as designated by the user.

Rivals Microsoft and Netscape Communications have joined forces in an effort to bolster consumer privacy rights. Software leviathan Microsoft agreed to support a privacy plan put forth by Netscape, which has been dubbed the Open Profiling Standard (OPS). OPS permits the PC user to disclose or withhold personal information from a given Web site. Profiles exist within the user's computer, which automatically exchanges data with a Web site in accordance to the user's preferences. OPS serves as a voluntary framework for companies and ensures businesses receive informed consent from customers about using their personal demographic information in future transactions. The participation of Microsoft in OPS dramatizes the urgency of the privacy debate for technology giants trying to stimulate the growth of electronic commerce (Clark, 1997).

Nevertheless, many online customers would prefer government regulation. A poll conducted for the *Privacy & American Business* newsletter found that 58% of computer users support passage of some kind of Internet privacy law (Weber, 1997).

Current Status

As of early 1998, approximately 15% of the U.S. population (about 30 million adults over the age of 16) used e-mail, up from about 2% in 1992 (Forrester Research, 1998). According to IntelliQuest Information Group, Inc. (1998), e-mail is the most popular online activity, with 75% of the survey population reported having used e-mail in the past month.

As of mid-1998, there were 123,689,000 e-mail boxes (accounts) in the United States (Electronic Mail & Messaging Systems, 1998). Eleven million of the market's 36 million business users use e-mail software (International Data Corporation, 1998), and 9.2 million households, equivalent to 27% of PC households or 10% of total households, now use Internet or online-based financial services (Demographics of, 1998). Future forecasts predict that nearly 50% of the U.S. population—135 million people—will communicate via e-mail by the year 2001 (Forrester Research, Inc., 1998). Worldwide use is predicted to grow at a faster rate, with the number of people with Internet e-mail access worldwide predicted to grow 800% to 450 million by 2001, up from 60 million in 1998 (E-mail growth, 1997).

Factors to Watch

Given the previous statistics, such rapid expansion will require Internet service providers such as AOL to increase lagging capacity and connection/transmission speeds. Much depends on what customers want and how people are using e-mail/Internet access in terms of predicting what path new technologies will take.

The next generation of e-mail may not be limited to text-only documents either. Video-broadcasting software for the Web-like Vosaic has morphed computer-mediated communication over the Internet into a makeshift TV set. Users are able to get six frames per second of video on a 28.8 Kb/s (kilobits per second) modem and full-motion video on a T1 connection. Improved voice-data-video capacity is also in the works. Steve Deering of Xerox Palo Alto Research Center and Van Jacobson helped invent Mbone, or multicast backbone, which is software that routes videoconferencing over the Internet. To date, videoconferencing hasn't reached an optimum adoption rate or "critical mass."

Critical mass, as defined by sociologists Oliver, Marwell, and Teixeira (1985), represents "a small segment of the population that chooses to make big contributions to the collective action while the majority do little or nothing" (p. 524). Early adopters of interactive media, such as the fax machine and videoconferencing, are paving the way for those who wish to adopt videoconferencing later when costs decrease and benefits of the medium are more readily presented.

In the same vein, e-mail has not yet reached critical mass, although statistical data suggest an accelerating adoption rate over the next several years. Many factors are involved in order to fulfill the prophecy of universal access, including required resources such as equipment and access devices and applied knowledge of the medium. Adopters must also make themselves available to reciprocate communication, termed communication discipline (Oliver, et al., 1985). In the same manner that a tennis game requires a willing player to hit the tennis ball back to his opponent, an e-mail user must check for messages on a routine basis and respond to received messages in a timely fashion in order to elicit maximum communication efficacy.

E-mail messaging has survived the battle for introduction into the communications arena, and it's here to stay. From private homes to business offices, e-mail has become the preferred method of sending quick, inexpensive, and seemingly ephemeral messages. While adoption is not yet universal, lowering costs of technology in general as well as increased efforts by politicians to make technological applications available to all citizens may someday make sending a handwritten letter with a stamp a thing that will be discussed only in history books.

References

Clark, D. (1997, June 12). Technology & health: Rivals Microsoft and Netscape team up to protect consumer privacy on the Web. *Wall Street Journal*, B14.

Demographics of the Internet. (1998, February 11). *Cyberatlas*. [Online]. Available: http://www.viss.com/newsdemographics.asp.

Delta, G. B., & Matsuura, J. H. (1998). *Law of the Internet*. New York: Aspen Law & Business.

E-mail growth/forecast statistics. (1997). *Computer industry almanac*. [Online]. Available: http://www.viss.com/newdemographics.asp.

Electronic Mail & Messaging Systems. (1998, February 9), 3.

Finnish e-mail service closed after charges of child pornography. (1996, September 3). *Wall Street Journal* (CD-ROM).

Folkerts, J., Lacy, S., & Davenport, L. (1998). *The media in your life: An introduction to mass communication*. Boston: Allyn and Bacon.

Former First Boston analyst files claim in e-mail incident. (1996, October 25). *Wall Street Journal*, 9B.

Forrester Research, Inc. (1997). *E-mail installed base*. [Online]. Available: http://www.viss.com/newsdemographics.asp.

Intelliquest Information Group, Inc. (1998). *E-mail usage*. [Online]. Available: http://www.viss.com/newsdemographics.asp.

International Data Corporation. (1997). *E-mail growth/forecast*. [Online]. Available: http://www.viss.com/newsdemographics.asp.

Larson, E. (1992). *The naked consumer: How our private lives become public commodites*. New York: Holt.

Leibrock, R. (1997, September 10). You are the product. *The Free Times*, 16-18.

Lohr, S. (1997, February 14). On-line service adds modems and posts quarterly loss. *New York Times*. [Online]. Available http://www.nytimes.com.

Mostafa, J., Newell, T., & Trenthem, R. (1994). *The easy Internet handbook*. Castle Rock, CO: Hi Willow Research & Publishing.

Oliver, P., Marwell, G., & Teixeira, R. (1985). A theory of critical mass I. Interdependence, group heterogeneity, and the production of collective action. *American Journal of Sociology, 91* (3), 522-56.

Panko, R. R. (1985). Electronic mail. In K. T. Quinn (Ed.), *Advances in office automation* (Vol. 1). London: Wiley Heyden Limited.

Radicati, S. (1992). *Electronic mail: An introduction to the X.400 message handling standards*. New York: McGraw-Hill.

Samoriski, J. H., Huffman, J. L., & Trauth, D. M. (1996). Electronic mail, privacy, and the Electronic Communications Privacy Act of 1986: Technology in search of law. *Journal of Broadcasting & Electronic Media, 40*, 60-76.

Weber, T. E. (1996, June 19). Net interest: Privacy concerns force public to confront thorny issues; One question: Can public data be too public? *Wall Street Journal*, B6.

12

Office Technologies

Mark J. Banks, Ph.D. & Robert E. Fidoten, Ph.D.[*]

Many of the technologies used in office settings and for office functions are described in other chapters. This chapter looks at some specific technologies, but also explores the larger picture of the use and impact of information and communication technologies in the workplace.

In addition to the typewriter, word processor, and more recently and inclusively, the personal computer, there are a number of other important office technologies that will be discussed in this chapter:

- *Desktop publishing* (DTP) allows a single worker to use a computer, printer, and appropriate software to incorporate research, art, photography, charts, graphs, writing, layout and design, and printing into documents such as newsletters, notices, and reports at a professional-quality level.

- The *facsimile*, or fax, machine transfers documents and images electronically over ordinary telephone lines to another similar machine.

- *Local area networks* (LANs), *wide area networks* (WANs), and *external* networks such as the Internet connect office devices, primarily computers, over appropriate network configurations. The *PBX* (private branch exchange) is another type of network used to connect telephones within an office.

- *Multifunction products* (MFPs) are automated devices that combine several functions into one unit, such as printing, scanning, fax, word processing, telephone, etc.

---

*Dr. Banks and Dr. Fidoten are Associate Professors, Department of Communication, Slippery Rock University (Slippery Rock, Pennsylvania).

- *Telecommuting* and *mobile offices* make use of several portable devices, most notably the notebook computer, the personal communications device, and, increasingly, the personal electronic "organizer."

- *Teleconferencing*, *e-mail*, *voice mail*, and *presentation programs* are among the technologies that augment the communication functions of the office.

Background

The old office technologies seem to hang on forever. Although handwritten documents gave way to the typewriter in the early 1900s, that technology and the telephone, dictaphone, and hand-delivered mail dominated the office environment through the first seven decades of the 20th century. The copy machine, which was added in the 1960s, represented one of the first major "modern" additions to the office. Although purported to be a major labor-saving device, the copy machine led almost immediately to an increase in the use of paper and file cabinets, and, in effect, an increase in the work load.

Writing Instruments

During the late 1970s and throughout the 1980s, a convergence of several technologies led to a mini-revolution in office technologies—innovations that would significantly change the nature of the office and its workers. These technologies included the personal computer. The change was so significant that Smith-Corona, vender of over 70% of typewriters in the United States, filed Chapter 11 bankruptcy in mid-1995, and, to survive, the company had to venture into other electronic products. Sales of typewriters fell to less than 800,000 in 1995. Although the typewriter is still part of most offices, it has been relegated to those few chores that are difficult to do on a computer, such as single mailing labels, filling out forms, and writing postcards (Elsberry, 1995; Deutsch, 1998).

The office computer began as a sluggish, space-taking configuration in the late 1970s. IBM's Mag Card and Office Systems and Wang word processors were among the first to appear. Soon, personal computers began to supplant the dedicated word processor, and today there are few electronic typewriters or word processors used in offices, although both typewriters and word processors continue to serve a small niche.

Mailing Functions

In the late 1970s and 1980s, the nature of mailing changed. Companies increased their need for faster delivery of mail, and overnight carriers such as FedEx emerged.

Facsimile (fax) technology has been around since the mid-1800s and was used for specialized functions such as news services during the first half of this century. During the 1950s and the 1960s, fax technology went through several improvements, leading to the proliferation of business and personal use after 1980. Ever-higher-speed facsimile technologies continue to evolve. In recent years, fax technologies are being incorporated into desktop and portable computers, moving from analog to digital systems, and finding increased use through the Internet (Harper, 1998; Wetzel, 1997).

Probably the most significant evolution of mailing technologies has been the expansion of electronic mail (see Chapter 11). This technology developed out of the convergence of personal computer technology and networking, both within organizations as LANs and among organizations through networks such as the Internet. Although long-distance carriers such as MCI offered early e-mail services, the Internet has played a significant role in the tremendous growth of e-mail technology in the 1990s.

Telephony

Few things were as unchangeable over the decades as plain old telephone service (POTS). Although developments such as direct dialing, easier access to international calling, and switched networks progressed significantly throughout the century, the end user saw little change in the way the telephone was used. In the United States, one telephone company, AT&T, owned it all, and it wasn't until the breakup of AT&T's monopoly in the early 1980s that telephone service providers and equipment manufacturers were able to introduce their own equipment and to vary the functions of telephone service. This led to a host of add-on technologies, again, centered on the desktop or portable computer. Among these technologies were fax machines, private branch exchange, voice mail, and automated call routing. On the wireless front, portable and cellular telephones proliferated, as well as pagers and, more recently, personal communication devices. Videoconferencing also grew out of telephony and satellite communication. (Videoconferencing is explored in detail in Chapter 24.)

Recent Developments

The technologies named above are not unique to the office environment, and several of them are described elsewhere in this book. The converging application of office technologies in the workplace have led to at least three major developments:

- The so-called "paperless office."

- The compression of office activities.

- The "virtual office."

The Paperless Office

The "paperless office" is a misnomer. Few offices, if any, will end up paperless. But many of the functions and activities that relied on the printed form in the past are becoming replaced by office technologies that allow them to be put into electronic form.

Part of the paperless office is the development of "Intranets," private networks that have the look and feel of Internet Web sites. Much of the software used for these Intranets is the same as that used for the Internet. A 1997 study by International Data Corporation shows that Intranets "represent a fundamentally new phase in information systems," with most medium to large companies having already built them (Study shows, 1997, p. 98). The researchers also predict, that by 2000, there will be 4.7 million Intranets (Planning workshop, 1998). Applications of these Intranets are as varied as the imagination.

- Regional or international sales forces can examine new products.

- Employees can get updated company news or check the lunch menus for the week.

- Personnel offices can make applicants' résumés available to members of a work team or department.

- Work groups can collaborate, often at great distances.

- Production can become more automated, using electronic inventories for just-in-time ordering.

- Management can share its decision-making process by making information about decision factors available online.

- Employees can communicate via e-mail, voice mail, and internal videoconferencing.

- Files of information can be shared widely and made immediately accessible.

- Archives can be stored and accessed by everyone.

- Training and development multimedia programs can be accessed from individual workstations.

- Expert systems can be tapped for assistance with work problems.

- Voice recognition systems may greatly reduce the need for typing, keying, or writing.

A comparable emerging application is the so-called "Extranet," which provides special links between companies and customers, suppliers, or clients. Instead of just using the Internet for such links, Extranets are more secure, while still allowing external browsers access to internal information systems (Moody, 1997).

Indeed, the very definition of "office" is changing rapidly. Because of the proliferation of not only the "wired" technologies such as Intranets but also "wireless" technologies such as cellular, personal communications systems, and portable computers, flexibility in the configuration of the office has led to several developments.

Compression of Activities

Early office computers were used almost exclusively by secretaries, and executives avoided them because of this clerical identity. As the technologies evolved through the 1980s and 1990s, however, computers in the office became used by more and more people at all levels of work, including executives. This has led to some job compression. For example, memos can be conceived, written, and printed or mailed in one basic operation by the originator. Telephone calls reach the desk of the recipient because automated voice call routing has eliminated most or all of the intervening human steps. The same is true with voice mail messages, which no longer need be written. In some specialties, such as desktop publishing, what used to take several steps in several places for writing, artwork, photography, typesetting, layout, and printing is now compressed into a single workstation where the job can be done by a single person in one place.

One of the significant recent office developments has been the appearance of multi-function products. These are automated devices that combine several functions into one unit. Such functions may include printing, scanning, faxing, word processing, telephone answering machines, and other computer functions such as data processing, networking, and CD-ROM. In 1995, there were more than 600,000 MFPs in use, 1.3 million in 1996, and over 2.2 million estimated for 1997. Researchers predict the gradual replacement of individual office machines by the convergence of several technologies into one MFP (PC use, 1997).

The Virtual Office

The "virtual office" has emerged as a feasible solution for contemporary and future work environments. Many types of traditional office work that required a fixed physical setting can be relocated to a wide variety of alternative sites. The employee's home, automobile, client/customer locations, or even temporary hourly/daily space can substitute for traditional centralized office space (Weston, 1997). An estimated 11.1 million workers were classified as telecommuters in 1997, according to the International Telecommuting Association (The checkoff, 1998).

Telecommuting technologies permit freedom of location, instantaneous interaction, and fast response and spontaneity. From a business perspective, the "virtual office" provides substantial economic benefits. Enterprises are partially relieved of relatively high-cost real estate investment or rental. "When your file cabinets are electronic, you don't need a building to put them in" (The virtual office, 1997, p. S6). The burden of providing office space is often shifted to the employee or independent contractor. Thus, the home takes on a new multi-faceted role. Capital investment shifts from fixed physical space to communication facilities, computing and associated equipment, and communication lines and related costs.

A major implication of this approach is the need for homes or telecommuting sites to be equipped with multiple high-speed communication lines, as well as wireless cellular technology. The speed and quality of residential communication lines shifts from primarily providing voice-oriented facilities to one that provides rapid data and image transport as well as videoconferencing capability.

Employers must often provide up-to-date computers with organizationally standardized and compatible software, fast modems, communications services, network access, and other related facilities. Further, it is essential that security be given additional emphasis, since there is markedly increased difficulty in maintaining control and limiting unauthorized access to proprietary information.

Since the office can also move into mobile virtual locations, employers may also provide laptop portable computers, modems, and fax facilities so that office workers can have almost infinite flexibility in reaching clients, colleagues, and "headquarters."

With the increasing attractiveness of telecommuting for employees and employers, special software is essential. Companies are beginning to provide specialized facilities for telecommuting. Service is being introduced in large metropolitan areas so that telecommuters may link from their home to a local switch, and then on to their company's server or office (Focal Communications, 1998).

The distribution and "filing" of vital organizational records has also changed drastically. Organizing and managing files and databases has become much more complex, and there are significant implications for meeting and maintaining files required for various business purposes and government requirements. Since organizational records, correspondence, legal documents, and other essential materials often exist solely in electronic media format, it is essential that increased emphasis be placed on their management and security. Traditional paper trails that may be required for legal, tax, and internal control purposes are transformed into "virtual" trails. Organizational systems now require additional automatic recording of transactions, conversations, data communications, and even voice conversations from remote locations. Conversations are often not only recorded to satisfy the ubiquitous "quality purpose," but to ensure that organizational records are complete, accurate, and timely. New security systems are required to prevent tampering, electronic manipulation, or deletion of vital communication-generated information. Use of Internet virtual office software products may pose some degree of security risk, thus requiring special concern when choosing this type of vehicle for office meetings (Taking virtual office, 1998).

A major office activity is the ever-present need to conduct meetings. Traditionally, meetings required physical presence. Telephone conference technology has been commonly utilized. Despite repeated introductions, however, video telephony has not been widely accepted or deployed. Videoconference facilities have been introduced in various manifestations over the past two decades, but these, too, have had limited acceptance. The missing elements in these technologies have been the lack of touch, body language, and the atmosphere of the setting.

The communication technology market has introduced new groupware products to solve the problems of meetings with participants in separate locations with separate schedules. Examples of this software include TeamWave Software Ltd.'s Workplace 2.0 software and Microsoft Corporation's NetMeeting. Although these are promising tools, reports comment that while "the 'virtual office' could revolutionize the way that workers collaborate, the virtual office market is virtually a vacant lot right now" (A new way, 1997, p. 77). Tools such as these are essentially operating system independent and can accept input from Windows, Mac OS, and UNIX environments. "Groupware enables the collaboration and group decision making that are high priorities as companies strive to get employees to work in teams. It also plays to the competitive demands of a global economy by breaking down barriers of time and space" (The virtual office, 1997, p. S6).

Despite the many economic and practical advantages posed by virtual offices, many new sets of problems emerge. The social fabric of an organization is significantly changed. Traditional authority structures must be completely reworked, since hierarchical organization is less appropriate. Even the modern so-called flatter organization may not be effective as new sets of work relationships materialize. Projects and activities can be built around the expertise required, wherever that expertise may be geographically located. Reporting structures may be based upon areas of specialized knowledge rather than traditional levels of authority. However, lack of employee interaction and managing distractions can pose new threats. Training and development programs may be needed to help employees understand their new independence and heightened individual responsibilities (Avoiding a "virtual disaster," 1998).

Current Status

As much of the previous discussion shows, office technologies are in a constant state of change. Because so many technologies are involved, they do not change at the same rate, and their changes are seldom coordinated. The changes occur at three levels: internal independent office technologies, wired services, and wireless services.

Independent office technologies—computer workstations, multifunction stations, and desktop publishing—continue to be the largest sector of office technology. Many offices and personal users may again need to upgrade as a result of:

- The introduction of Windows 98.

- The proliferation of CD-ROMs.

- The adoption of even more powerful processors that can accommodate not only word processing, but also larger files downloaded from the Internet and files that have audio and video in addition to data information.

Another significant development is the increased integration of telephone systems with computer systems using digital technologies.

The growth of Internet usage is probably the largest business application of wired services being used for research, communication, and, increasingly, as a means of promotion and advertising. Faster access through ISDN and other high-speed lines makes this medium even more convenient and versatile enough to also accommodate videoconferencing (Lavilla, 1997; White, 1997). On another front, "wireless networks" are increasing slowly, mostly because of the high cost. Although there are some companies, such as United Parcel Service, that use wireless data interconnection regularly, most have adopted a wait-and-see approach pending more facile technology and lower costs (Girard, 1998).

One of the more controversial developments is the increased use of the Internet for long distance telephony, which has an adverse effect on traditional long distance telephone companies. This use of the Internet is expected to grow to 10 million users by 2000 and carry 28% of all voice traffic by 2010 (Migdal & Taylor, 1997). Recently, the Federal Communications Commission (FCC) explored the possibility of regulating and taxing Internet telephony just as traditional telephone is regulated, although no official action had been taken by mid-1998.

Factors to Watch

There are several trends that seem to address the emergence and future of office technologies. These trends relate to flexibility and versatility, standardization and the human interface, adoption and market issues, and larger social issues (Rifkin, 1995).

Flexibility and Employee Adaptability

One of the most important outcomes of emerging office technologies is that they increase flexibility, both for workers and employers. This is apparent in most of the technologies described in this chapter. Workers are freed from being tethered to the desk, are given much greater access to information and communication through the technologies, and, through job compression, are able to do a greater variety of tasks than ever before possible. Because of increasing flexibility, jobs are becoming redefined.

One of the authors was recently in the waiting room of a county health department. A salesman was talking loudly on his cellular phone to a client while his family was waiting for travel immunizations. There was a woman with two children who were playing somewhat noisily, but not inappropriately. The salesman asked her to keep her children quiet. Without missing a beat, the woman fired back that this wasn't a phone booth.

On a recent vacation in Southern France, one of us saw a gentleman at an outdoor cafe using a cellular phone and a small notebook computer, while eating lunch and sipping wine. The May 16, 1994 cover of *New Yorker* magazine depicted a businessman on a midtown street with a desk work surface strapped across his waist. It had a cellphone, notebook computer, and other electronics. Although the latter example is a parody, the first two represent some of today's outcomes of emerging office technologies.

Samsung has introduced a Smart Phone that looks like a wireless phone, but opens up into a personal organizer with a screen and built-in modem. Mobile telephony via satellite is now available, though expensive. No doubt, availability will increase and costs will diminish as the several proposed global low-orbit satellite systems come into being. These examples lead us to consider both the advantages and the disadvantages of what's to come. One is given to pondering whether the technologies give freedom with one hand, and enslave with the other.

Several employee adaptability issues emerge because of this trend. They include questions of privacy for employees and information protection for employers, harassment and liability for those who use the technologies to communicate, copyright infringement, and employer monitoring of worker productivity and communications (Seifman & Trepanier, 1996).

In her *Wall Street Journal* column on work and family, Sue Shellenbarger (1998) wrote that the traditional "mental wall" between work and home disappears for most telecommuters, resulting in "integration...the hottest work-life buzzword since 'juggling' " (p. B1). Forecasters call this one of the top 10 trends of the next decade. The benefits and hazards of this trend are fairly obvious. For some workers, home life suffers by the constant presence of work. For others, it gives the worker greater control over work, thus resulting in more personal satisfaction.

Standardization and Human/ Machine Interaction

Communication technologies in the office are never stagnant. Because these technologies come from several fronts, including computers, telephony, and wireless technologies, there is rarely standardization. In the past, the telegraph, telephone, radio, television, and other "traditional" media became relatively standardized. Even when some systems are only partially standardized, as in the Apple and DOS computer platforms, some technology emerges to increase the compatibility of these systems, such as software to provide "transparency and fluidity" among the many tools available. This will likely be the case, as office technology attempts to bridge the gap between the wired and wireless worlds. It takes time to produce this effect, however, and it is complicated by the emergence of newer, incompatible technologies.

Probably the most prominent evidence of standardization is the emergence of a digital standard where all forms of information—video, audio, text, and data—can be converted into binary digits and integrated into digital systems. Virtually every office technology is moving toward digital (Miller, 1996). This growing digitization may be well received in the international arena where incompatible standards in television and other technologies have existed for a long time. Meanwhile, current users may have to resort to makeshift technologies or be hampered by having to make choices among incompatible technologies.

Like the present, the future holds some advantages and some disadvantages in technologies and applications. Among the advantages are that office technologies will continue to become smaller, lighter, faster, and more versatile—with greater capacity. As capacity increases, costs may go down. As workers unite to promote and demand better attention to their welfare, technology will become more ergonomic. As workers decrease their reliance on the tether to the office, technologies will enable more mobility, while providing greater and faster access to more information, and more flexibility in interacting with other workers electronically. All these technologies will converge into a more seamless approach, possibly by enabling the worker to complete several functions from a single, portable, easy-to-use machine.

Several problems remain, many related to the human/machine interface. Electronic "sweatshops" have emerged that use electronic means to track and impose sometimes oppressive worker productivity expectations. The process of employing information workers as independent contractors has the potential to make workers' environments resemble the factory sweatshops of earlier times. Workers can also be exploited when they are required to work at loosely regulated "home offices." Other factors include a rising incidence of injuries related to the use of some of these technologies.

As access to information explodes, the information worker is deluged with more and more data, much of which is either of questionable value or outright erroneous. One of the dark sides of the information explosion is information overload. There will be an increasing need within organizations for specialists who can manage information for employees.

Adoption and Market Issues

There are so many office technologies introduced that the prediction of their adoption is virtually impossible. Media companies sank considerable funds into the development of personal communications technologies only to find slow adoption. On the other hand, just a few years ago, no one would have predicted the runaway success of the Internet both for business and personal use.

As organizations consider the adoption of new technologies, particularly networking technologies, the investment is considerable, and they must consider factors such as:

- How long before obsolescence affects the technology?

- How will standardization come into play?

- How much will it cost to both implement and "debug?"

- How amenable will the technologies be to both users and clients?

Many a company has angered customers by its klutzy automated telephone call routing. Other adoption factors include ease-of-use, relative improvement over older technologies, and positive return on investment.

Social Issues

Sensitivity to user tolerance and patience levels is an essential part of office technology adoption and design. Often, adoption focuses on economic benefits without due regard to the social and psychological needs of the users or clients. For example, studies of telecommuters show that there are often considerable social consequences to the feelings of isolation and not belonging to the work team. New measures of "social payback" are required if organizations are to examine the total value of an investment. Emerging technologies must be evaluated based on the many intangibles that provide the "service difference" between competing organizations.

Bibliography

Avoiding a "virtual disaster." (1998, February). *HR Focus*, 11.

Baran, N. (1995). *Inside the information superhighway revolution.* Scottsdale, AZ: Coriolis Group Books.

Branscomb, A. W. (1994).*Who owns information? From privacy to public access.* New York: Basic Books.

The checkoff. (1998, March 24). *Wall Street Journal*, A1.

Crowley, D., & Heyer, P. (1995). *Communication in history* (2nd Ed.). White Plains, NY: Longman.

Deutsch, C. H. (1998, March 23). Using a key still works; Smith Corona's future rests in putting its name on other products. *New York Times*, C1.

Elsberry, R. (1995, December). Test of time. *Office Systems 95.*

Finley, M. (1995). *Techno-crazed; The businessman's guide to controlling technology—Before it controls you.* Princeton, NJ: Peterson's/Pacesetter Books.

Focal Communications offers a telecommuting solution. (1998, February 2). *PC Newswire*, 202.

Girard, K. (1998, February 23). Wireless revolution fizzles. *Computerworld, 32* (8), 6.

Goodman, D. (1994). *Living at light speed; Your survival guide to life on the information superhighway.* New York: Random House.

Harper, D. (1998, February). Telephony and faxing on the Net; New technology allows low cost communications over the Internet. *Industrial Distribution, 87* (2), 94

Heap, N., Thomas, R., Einon, G., Maron, R., & MacKay, H. (Eds.). (1995). *Information technology and society.* Thousand Oaks, CA: Sage.

Lavilla, S. (1997, July 28). VDONet rolls out VDOphone 3.0. *PC Week*, 112.

Migdal, J., & Taylor, M. (1997, January 27). Thief or benefactor? *Telephony,* 46.

Miller, S. E. (1996). *Civilizing cyberspace; Policy, power, and the information superhighway.* Reading, MA: Addison-Wesley.

Moody, G. (1997, May 29). Get a little more from your Extranet. *Computer Weekly*, 54.

A new way to collaborate. (1997, October 20). *PC Week*, 77.

PC use drives sales of multifunction products. (1997, December 11). *Purchasing, 123* (9), 66.

Planning workshop. (1998, January 5). *Industry Week, 247* (1), 13.

Rifkin, J. (1995). *The end of work; The decline of the global labor force and the dawn of the post-market era.* New York: Putnam.

Seifman, D., & Trepanier, C. (1996, Winter). Evolution of the paperless office: Legal issues arising out of technology in the workplace. *Employee Relations Law Journal, 21,* 5-36.

Shellenbarger, S. (1998, February 18). Forget juggling and forget walls; Now, it's integration. *Wall Street Journal*, B1.

Study shows Intranets become entrenched. (1997, November 10). *PC Week, 14* (47), 98.

Taking virtual office public. (1998, January 12). *PC Week*, 66.

The virtual office; Groupware creates consultants offices without walls. (1997, December 19). *San Francisco Business Times*, S6.

Weston, C. (1997, December 30). Taking offices to a hire level. *The Guardian*, 17.

Wetzel, R. (1997, November). The argument for Internet-based facsimile. *Telecommunications, 31* (11), 32.

White, L. (1997, February 13). Now showing on a screen near you. *Computer Weekly,* 58.

13

Document Printing Technologies

Edward M. Lenert, Ph.D. & Costin Jordache*

In the United States, printing is a $67 billion industry, but document printing utilizing personal computers makes up only a small part of the overall revenue of the printing business. The office segment of printing amounts to a mere 3% of all the pages printed annually in the United States (Bliss, 1997). However, the potential for growth in this area is almost unlimited and requires a closer look. At a time when countless pages of documents are electronically published on the World Wide Web, one might wonder, "What is the outlook for printing and the documents it creates?"

There are two broad views. On the one hand, some foresee that the technologies for printing documents on paper are at the threshold of a sharp decline. For example, Nicholas Negroponte (1995) in his widely read book, *Being Digital*, argues that document production and distribution will change dramatically as we shift from reliance on analog "atoms" to digital "bits." In this view, hypertext, with its dynamic links, becomes the ultimate form of the document's historical evolution (Barger, 1996). It is suggested that somehow the old form of information, such as the paper document, will dissolve before our eyes, and in its place will be quick and ubiquitous access to pure information via a network.

On the other hand, some perceive that the prediction of the document's demise, while holding some truth, is nevertheless an overstatement. Looking back, we can recall how, in the 1970s, many futurists predicted a "paperless" office, one in which typewriters and

---

* Dr. Lenert is Assistant Professor, Newhouse School of Public Communications, Syracuse University (Syracuse, New York). Mr. Jordache is a graduate student, Department of Radio-TV-Film, University of Texas at Austin (Austin, Texas).

file-cabinets would be displaced from their central role in world commerce. These artifacts of an earlier era would soon be available at bargain prices at rummage sales around the world. Paper documents would slowly fade from use as the adoption of digital documents, which are easier to store and retrieve, became more widespread.

These erroneous predictions were based in part on certain assumptions that, while true at the time, quickly changed. For example, in the 1970s, office printing technology was slow and noisy, and the range of output was limited to black characters in a limited selection of typefaces. What the futurists did not anticipate was the arrival of a technology for printing documents that was fast, quiet, relatively inexpensive, and directed at the individual user.

As it turned out, the anticipated transition to a paperless office was delayed—perhaps indefinitely—by the widespread diffusion of laser printer technology. When first introduced by Xerox in 1977, a laser printer cost $350,000. It was quiet, fast, and offered a variety of typeset appearances. As demand grew, technology improved and prices dropped. "Personal on-demand document production" in the business environment became increasingly lucrative as printer manufacturers realized that paper documents were still a requirement, even in the digital age.

The 1970s dream of a paperless world dimmed as, in the 1990s, more digitally produced paper documents were turned out by an increasing number of inexpensive printing technologies. As anyone who has used one knows, nothing seems to kill trees faster than a laser printer. So the futurists of the 1970s were only partially right. While typewriters are of significantly reduced importance in the office of the 1990s, file cabinets continue to sell very well, and they are still as full of documents as they were decades before.

The arrival of the paperless office has receded into the indefinite future. Today, most businesses rely on paper documents, even though the exchange of paper documents may be assisted by electronic technologies such as fax. People have not abandoned the traditional printed document simply because the technology and economics of document production has changed. As Brown and Duguid (1996) pointed out, documents are much more than mere containers for information. Indeed, documents are one of the oldest forms of communication, intricately interwoven into the fabric of any society. The traditions and practices associated with printing and paper documents are not easily brushed away, even in a digital era. Individuals and businesses still have a need for a "hard copy." Society's demand is still "Show it to me in print!"

This chapter examines some of the current developments in the document publishing industry with a focus on "the social life of documents." A short background on printers sets the foundation for discussing recent developments for printers and the documents they produce. The chapter concludes with several factors to watch as technology, economics, and politics combine to decide the fate of that most ancient social form—the document. The next section discusses the evolution of modern document production technology, which evolved in four stages.

Background

The force of new printing technologies brought about the predicted demise of the typewriter. The first of these technologies, called impact printers, dates back to the 1940s. These printers are nothing more than electrically driven typewriters. Essentially, the technology involves a "daisy wheel" moving serially across the paper. Following instructions from the computer, a small hammer strikes the shaped image character on the wheel at a selected print location. The print head then makes an impact through the printer ribbon onto the paper, producing a distinct "click." You can imagine the deafening sounds produced in an office full of these printers!

In the 1970s, dot matrix printing was introduced. While it generally produced lower-quality output, it was faster and cheaper than daisy wheel printing—but not any quieter. Using this technology, the printer receives instructions from the computer, which are converted into electrical signals that follow a path to the print head. The print head usually contains nine or 24 individual printer pins, also called wires. The signal from the computer causes an individual pin to "fire," strike the page through the ribbon coated with ink, and leave a dot. The dots, taken together, form the images of completed characters (White, 1997). Using a measurement called dots per square inch (dpi), dot matrix printers are rated by the number of pins and the quality of their output. Additionally, like other printers, they are rated by the number of pages per minute (PPM) they can print. Most computer stores no longer sell dot matrix printers, and their use is limited to specialized roles, such as printing multipart "carbon" copies and forms.

More recently, inkjet printers have reduced the demand for dot matrix-produced documents. Inkjet technology involves a print head which contains four ink cartridges and as many as 50 ink firing chambers, each nozzle smaller than a human hair. When an electrical signal from the computer flows through a resistor in the printer head, it heats the ink in the firing chamber to 900°F for a very short period of time—measured in millionths of a second. The heat expands the ink, fusing through the nozzle to form a droplet at the tip of the nozzle, and is forced on to the paper. The amount of ink that is ejected in this action is approximately one millionth of a drop of water from an eyedropper. As the resistor cools, a sucking action pulls fresh ink into the firing chamber (White, 1997). Modern inkjets are faster than dot matrix printers, and produce high-quality output with relatively little noise. A high-quality inkjet printer costs about $300.

The fourth category of personal document technology is the laser printer. As with dot matrix and inkjet printers, to a laser printer, everything printed is a dot. Here is how it works. The computer sends the data to the printer, which is a computer description of the image to be printed. The printer then converts the data, a binary representation of an image, into an image of dots to print on the page.

The principal standards for laser printers today are Hewlett-Packard's Printer Control Language (PCL) and Adobe's PostScript. "Page description language" is the technical name for the commands given by the computer that instructs the printer where to print the dots. In the case of PCL, the page is described in a series of ASCII statements.

For example, the page description language command, "<ESC>(s1S," sets the printer font to italics.

The dots are then copied onto a photosensitive drum by a laser beam. A rotating mirror directs the laser beam from side to side onto the drum. The drum itself is rotating and begins each rotation with a negative charge, one that it not attractive to the "ink," called toner, of a laser printer. In essence, toner is a very fine plastic powder which can be "melted" permanently onto paper. The drum's initial negative charge causes the toner to be repelled. However, each spot on the drum hit by the laser loses its charge. Without the negative charge, toner sticks to the drum wherever the laser hit it. At this point, the paper, which has been positively charged, moves through the printer. As it comes in contact with the drum, the toner on the drum is transferred to the paper. This takes place because the paper has a positive charge. A typical toner cartridge can usually print between 2,500 to 4,000 pages. In a final step, the toner is fused to the paper by heat at approximately 400°F. The paper then exits the printer and is stacked in an output bin (Smith, 1995).

A major breakthrough occurred in 1983 when Canon developed technology that permitted the widespread commercialization of personal laser printers. Canon's LBP-CX laser printer engine made it possible to market a printer for about $7,000. Using Canon's technology, Hewlett-Packard introduced its original HP LaserJet printer in 1984.

However, early HP laser printers suffered from significant limitations. The printer could print text at a relatively quick eight pages per minute, but the page output was in reverse order. Not a big deal if the document was five pages in length, but what if it was 50 pages? Still worse, graphics were coarse, solid areas were closer to dark gray rather than black, and the early font handling capabilities were not very flexible.

HP was not the only company producing personal laser printers. The introduction of the Apple Macintosh, with its graphical user interface, increased the demand for high-quality printer output. The 1984 Macintosh was the first personal computer bundled with a WYSIWYG ("What-you-see-is-what-you-get") word processor called MacWrite. Building upon a growing demand, the Apple LaserWriter quickly became a leading choice for graphics-intensive applications because of its superior performance, especially when combined with the Mac's superior graphical interface. Much of the Apple printer's superiority was due to its use of the much more flexible and powerful Adobe PostScript page description language (Smith, 1995).

Recent Developments

In the past two decades, document printing technology has evolved a great deal. Perhaps the most important recent development is affordable color printing. This technology, available with inkjet and laser printers, has increased the potential for creating diverse and specialized documents. At the business level, many clients no longer accept black-and-white presentations and require use of color especially in charts and graphs.

Consequently, businesses and consumers are using printers to create various forms of documents that compete in quality with specialized color printing houses (Hersch, 1997).

The adoption of personal, color document printers evolved gradually as the technology had to overcome strong resistance from some quarters. For example, in the mid-1990s, technology managers at major corporations strongly opposed the purchase of color printers. Color printing, it was argued, was a costly and unwarranted expense. After an initial outlay of $7,000 to $10,000 for a slow 300 dpi color printer, operation costs often ran to an exorbitant $1 per page (Milne, 1998).

Color page printers also performed poorly over a network—duty cycles and speeds were slow. Additionally, the monochrome print quality of color page printers was not as good as that of monochrome laser printers. Since 1997, however, vendors have improved color page printers' network capabilities. Now, network users can use the color printer as a monochrome substitute without being penalized in print quality or cost (Stafford, 1998).

As the technology has improved, so have the economics of color printing. A rapid drop in price has overcome most of the resistance to adoption of color printing. Presently, a high-quality color printer, like the HP LaserJet 4000, can be purchased for about $1,000.

Figure 13.1
Color Laser Printing Technologies

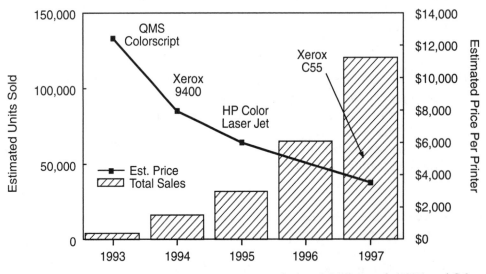

Source: Lyra Research (1997) and Other Sources

Color printing of documents can be done in several ways. Color laser printing is the most common technology used by the business sector to produce high-quality color documents. Although it is the most expensive, color laser printing offers good performance in document appearance as well as print speed. Most color laser printers offer resolution of 1,200 dpi, a wide palette of colors, and printing speeds of up to 17 pages per minute.

Liquid inkjet color printers are very popular and offer a low initial cost (some are priced below $300), but they are not well-suited to a multi-user office environment because of slow performance and low duty cycles. In addition, the quality of liquid inkjet plain-paper printouts, although good, does not compare with other printing technologies (Milne, 1998). However, inkjet's low price range has made the color printer an attractive purchase for home users.

Solid-ink color printers, which melt solid-ink sticks into a liquid that is transferred onto the page and dried rapidly, are a third method of color printing. Its good print speed, vivid colors, and low cost per copy has given the printer greater acceptance in office environments. On the other hand, the document's "waxy" texture as well as the printer's low resolution prohibit very fine characters, curves, or edges from being printed adequately (Milne, 1998).

The final color-printing category uses dye sublimation technology which can produce true continuous tone images and photographic quality. The printer converts solid ink to a gas, which is applied as a transfer roll and passed across a thermal print head. The color output is the best of all printing technologies, but comes at a price. The average machine costs approximately $14,000 today (Milne, 1998).

Another development in the document printing industry has been the introduction of multiple function devices. These workstations combine the functions of a color printer, copier, scanner, and fax machine into a single unit.

Machines such as the Xerox WorkCentre 450c allow document services to be controlled right from the desktop, and tasks such as color printing, copying, faxing, and output management are seamlessly integrated. Surprisingly, a multifunction unit with all of these features can be purchased for around $400. A multifunction unit allows the user to create and print documents, and then instantly "publish" them, either by fax or by posting on the World Wide Web.

Finally, current printing technology is even allowing for a little nostalgia. Some have criticized current laser printers for putting out digital documents that are "a little too perfect" because of their regularity of appearance. In response, Beowolf, Inc. has released a font that mimics the look of metal typography (Mendelson, 1995). The page description language that creates the font also creates small irregular imperfections that imitate those found in metal type set by hand. By using such a "sincerity font," laser printed documents can now bear the look of those typed on a 1956 Royal manual typewriter.

Current Status

While most new computer systems purchased today come equipped with a color inkjet printer, that may change. According to some estimates, "the color laser market is on the verge of exploding." One technology research firm, Dataquest, forecasts that color laser printer shipments will grow tenfold, from 32,600 units in 1995 to 304,000 in 2001. Other analysts put the figure at 600,000 by 2001 and call the number "conservative." The value of these shipments of color laser printers will grow about fivefold, from $178 million to $1.1 billion, during the same period (Kovar, 1997).

The rapid diffusion will be spurred, among other things, by much lower prices and by the growing popularity of the World Wide Web. Graphic intensive Web pages demand color printers that can quickly replicate vivid color and complex textures on paper. However, color printers have not successfully penetrated the mid- to large-scale organization, in part because of their high cost, typically $0.10 and higher per page, versus $0.01 on average for monochrome printers. Also, color printers still operate at speeds below that of their black-and-white counterparts (Hersch, 1997). At the same time, retail channels have revealed a boom in sales of low-priced color inkjet printers. Approximately 82% of businesses use color inkjet printers for overhead transparencies and business graphics. Some 16.3 million color inkjets are predicted to ship in 2001, up from more than 10 million in 1996 (Stafford, 1998). Attractive prices, cost-effectiveness, and improvements in speed are making these units household PC items as well as corporate standards.

Factors to Watch

The effects of digital printing and personal document technology are evidenced at all levels of the Umbrella Model discussed in Chapter 1. At the level of the individual user, there is increased demand for high-fidelity color in everyday printing. One factor driving this demand is the increasing diffusion of consumer digital cameras, which allow for "personal digital imaging."

From a hardware and software perspective, it is estimated that over one million of these filmless cameras will be sold in 1998, and there is a high level of interest in adopting this technology among the 20 million U.S. households with PCs and color printers. Consumer surveys show that instant access to "hard copy"—a printed photograph—is one of the most important features driving adoption of digital cameras. In addition, there is substantial interest in the software that can be used by consumers to manipulate and alter their photos in the manner of a "digital darkroom." This means an increased demand for photorealistic personal color printers and specially treated paper for printing. One can imagine that traditional chemical-based imaging companies such as Kodak fear that their core photoprocessing business may one day be superseded by inexpensive home printing of photographs from digital cameras (Cemacity, 1998).

At the organizational level, we can expect to see a continuing need to increase the "interoperability" of the analog world of paper and the digital world of electronic documents. Xerox, for example, has developed optical character recognition (OCR) software called TextBridge designed to make it easier for HTML (hypertext markup language) authoring packages to publish paper documents on the Web. It is unlikely that the greater use of the World Wide Web will mean that fewer documents will be printed. People will still want "personal hard copy" of important documents they see on the Web. In response to users' requests, Microsoft has incorporated a feature into Explorer 4.0 that allows the user to print only a portion of a document that may be viewed on the Internet (Bliss, 1997).

In addition, leading companies such as Adobe are working to adapt their printer languages, including the widely-used Acrobat and PostScript, to develop greater compatibility with the Internet and its related resources (Nordling, 1997). Also on the horizon are low-cost voice-recognition technologies with the ability to convert speech directly into text. For example, the most recent edition of IBM's word processing software, WordPro, seamlessly incorporates voice recognition of the user's speech (Fontana, 1998). A printed document, such as an office memorandum, is a likely candidate for such digitally converted speech. In the realm of the business enterprise, look for important developments in what is called "document imaging," which is building bridges between and among the once-separated worlds of computer graphics, color visualization, and color reproduction. In addition, document imaging is also abridging the formal distinction between prepress and press stages of document production (Schonhut, 1997).

Finally, at the social system level, look for a continued struggle to establish a new balance of power between the producers and users of documents, especially in the area of copyright law. The emerging digital technologies of printing and publication greatly facilitate the reproduction, incorporation, and alteration of documents. These changes herald a coming vigorous fight over the definition of "digital property rights" (Stefik, 1996). In other words, how will the legal system that governs publishing react to technological change?

Consider, for example, the long-established "right to browse" information without purchasing it, even in a commercial establishment such as a bookstore. This is because it was not considered possible to infringe an author's copyright by merely reading; only copying could infringe a copyright. In the new environment of the Internet and WWW, such settled questions are now open to reinterpretation. Why? As publishing interests argue it, a "copy" or electronic image of the document you are reading from the Internet is made in three places: on your computer screen, in your computer's memory, and in your browser's cache files on the hard disk. In essence, your electronic "browsing" of the information now may constitute an unavoidable infringement on the author's copyright.

In other words, emerging digital printing and document technologies are contributing to important changes in the traditionally analog world of print, and thereby challenging once-settled social understandings of what it means to publish. Indeed, the notion of what constitutes a document is becoming increasingly complicated and amorphous. Soon documents containing full-color non-textual media and even dynamic (e.g., personalized or hypertext) elements may no longer be considered a specialty form (Hearst, 1996), with various consequences for existing social, economic, and political arrangements.

Bibliography

Barger, J. (1996). A *hyper terrorist's timeline of hypertext history*. [Online]. Available: (http://www.mcs.net/~jorm/html/net/timeline.html).

Bielawski, L., & Boyle, J. (1997). *Electronic document management systems*. Upper Saddle River, NJ: Prentice-Hall.

Bliss, J. (1997, May 19). Massey sees Internet spurring printer sales. *Computer Reseller News*. [Online]. Available: hhtp://techweb.com/se/linkthru.cgi?CRN19970519S0015.

Brown, S. J., & Duguid, P. (1996, May 6). The social life of documents. *First Monday*. [Online]. Available: (http://www.firstmonday.dk/issues/issue1/index.html).

Bury, S. (1996, November). Color within reach: Desktop color printers. *Adobe Magazine*, 31- 35.

Cemacity. (1998, January 14). *Digital photography faces bright future*. [Online]. Available: http://cemacity.org./cemacity/gazette/files2/digphoto.htm.

Fontana, J. (1998, March 5). Lotus takes SmartSuite to net level. *Internet Week*. [Online]. Available: http://www.techweb.cmp/internetwk.

Hearst, M. A. (1996, May). Research in support of digital libraries at Xerox PARC: The changing social roles of documents. *D-Lib Magazine*.

Hersch, W. S. (1997, November 17). No limit: The need for speed—Speed, color growth areas in printer arena. *TechWeb*. [Online]. Available: hhtp://techweb.com/se/directlink.cgi?VAR19971117S0055.

Hurtado, R. (1998, March 8). A font of good fortune for a business printer. *New York Times*.

Kovar, J. F. (1997, December 15). Color lasers on the upswing for 1998. *Computer Reseller News*, 185.

Levy, D. M. (1994). Fixed or fluid? Document stability and new media. *Proceedings of the 1994 European Conference on Hypermedia Technology*. ACM Press.

Mendelson, E. (1995, September 12). Beowolf fonts mimic the look of metal typography. *PC Magazine*, 49.

Milne, J. (1998, March 1). Coloring inside your network lines. *TechWeb*. [Online]. Available: hhtp://techweb.com/se/directlink.cgi?VAR19980301S0025.

Negroponte, N. (1995). *Being digital*. New York: Knopf.

Neuman, S. (1996). *Computer buyer's guide*. Emeryville, CA: Ziff Davis.

Nordling, T. (1997, Autumn). It's a colorful, wired world. *Adobe Magazine*. [Online]. Available: http://adobe.com/publications/adobemag/autm97na.html.

Schonhut, J. (1997). *Document imaging: Computer meets press*. New York: Springer Verlag.

Smith, N. E. (1995). *Getting the most from your HP LaserJet*. Plano, TX: Wordware Publishing.

Stafford, J. (1998, March 2). Primary colors: Color page printers gain popularity in mainstream business. *TechWeb*. [Online]. Available: http://techweb.com/se/directlink.cgi?VAR19980302S0031.

Stefik, M. (1996). *Internet dreams: Archetypes, myths, and metaphors*. Cambridge, MA: MIT Press.

White, R. (1997). *How computers work*. Emeryville, CA: Ziff Davis.

14

Virtual & Augmented Reality

Frank Biocca, Ph.D., Ozan Cakmakci, Jeff Czischke, Jeff DeVries, Hsuan-Yuan Huang, Kristian Kind, Kristi Nowak, & Marcia Witt*

V irtual reality (VR) is not a single technology, but a cluster of rapidly-evolving technologies. Most commonly, virtual reality refers to a collection of technologies that use 3-D computer graphics, real-time simulation techniques, and a wide array of input and output devices to create illusions of being in a virtual environment (Biocca & Delaney, 1995; Durlach & Mavor, 1995). For some, VR is best defined by its goal—the experience of "being there" in an environment. This environment can be the cockpit of an F15 fighter plane, the surface of a gold atom, the inside of a future building, or inside the undulating walls of a heart artery.

The quality of a VR experience is gauged by its psychological effectiveness. The more successful implementations of virtual reality give users the compelling sensation that they are actually in the environment created by the technology. This is called a sense of presence (Steuer, 1994; Lombard & Ditton, 1997). In general, the more expensive immersive virtual reality systems tend to deliver more presence, but the sensation of presence is not determined solely by the technology (Biocca, forthcoming).

Why create virtual reality? Research and development of virtual reality technology has been spurred by the widely-shared belief that information technologies that make the best use of human sensorimotor systems can enhance the user's ability to learn, make decisions, or perceive distant environments. This phrase for this enhancement of human abilities is often called intelligence augmentation in the VR community (Biocca, 1996). Many VR designers accept the perceptual theory that the human senses have evolved to

---

* M.I.N.D. Lab, Michigan State University (East Lansing, Michigan).

allow humans to rapidly absorb environmental information (affordances), so that our ancestors could efficiently move, hunt, and survive in a 3-D physical world (Carr & England, 1995; Gibson, 1966; 1979). Therefore, some believe that computer-based information will be easier to perceive and understand if the interface presents data in a way that fully simulates the perceptual experience of a human moving and acting in the physical world.

This concept is sometimes called direct manipulation of information. One of the goals of virtual reality designers is to create the ultimate, natural interface, one that will augment human training, performance, and intelligence (Biocca, 1996). While some see great promise in VR as a training tool, others in the film, amusement park, and game industries see it as the "future of fun," a powerful medium to deliver new, interactive, eye-popping, and bone-jarring entertainment (Hawkins, 1995). Others see 3-D virtual environments as an easier way to travel, meet others, and interact with information on the Internet (Damer, 1998).

Simulating the Perceptual Properties of the Physical World

Like many media before it, VR creates illusions for the senses. These illusions are accomplished by a computer carefully orchestrating sensory information presented to the human body via output devices (displays) and monitoring input devices that sense body motion and user action (sensors) (see Figure 14.1). The most advanced virtual reality systems create rich, multisensory, 3-D environments—illusions for the eyes, ears, hands, and other senses. The most compelling VR systems are *immersive*, so termed because they immerse the senses of the user in computer-controlled stimuli. Figure 14.1 diagrams the range of possible input and output devices for a virtual reality system. Each output or display device creates illusions for one of the human senses. The computer tracks the motion and perspective of the user. It is then able to update the environment in real time according to where the user is looking or moving. For example, when a user walks or turns his or her head, the computer calculates what the user should be seeing and hearing at that exact spot, and quickly displays it to the user with little noticeable time lag. The result is the illusion of being present inside an exciting virtual world.

Let's consider what a person might experience when using the signature peripherals of an immersive VR system: a head-mounted display and a data glove. When users put on head-mounted displays, they feel they have entered a 3-D computer graphic landscape that stretches as far as their eyes can see—in all directions, all around them. Parts of this computer environment may behave like the real world. Users may walk through 3-D rooms and hallways and reach out to grasp, move, or even throw virtual objects such as a virtual baseball. In some sophisticated systems, the user might feel the texture and heft of the virtual baseball through output devices that simulate the sensations of touch and force (Biocca & Delaney, 1995; Durlach & Mavor, 1995). Today, consumer VR environments, especially those on the Internet, are still cartoonish and incomplete, but the technology is beginning to evolve rapidly. The most expensive systems used in flight simulation and advanced industrial training can create very compelling illusions.

Figure 14.1
The Range of Input and Output Devices for Virtual Reality

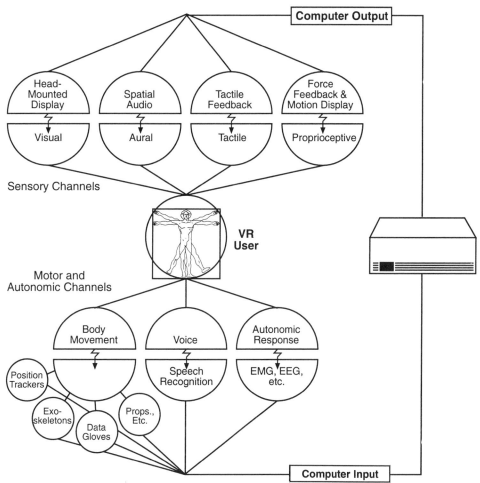

Virtual reality hardware can be thought of as an array of possible input and output devices. These devices transmit information to or from a VR user. Each is coupled to one of the user's senses and motor channels. Each output device serves as a sensory channel, and each input device is linked to a user's motor or auto-nomic channels. No single VR system incorporates all of these input and output devices, but most have been used in one or another system. In the ideal VR system, powerful sensory illusions can be simulated (*Source*: Biocca & Delaney, 1995).

Another goal of virtual reality is to completely immerse the user. An important component of immersion into a virtual world is that the user will forget that the experience is mediated. This is best achieved when the user is progressively embodied in the interface; that is, when more sensory and motor systems are directly engaged by computer-mediated stimuli (Biocca, 1997). Some designers feel this is difficult to do when a user must wear a bulky helmet, awkward gloves, or lots of wired sensors (Krueger, 1991). Technological advances such as light shutter glasses, optical sensors, and fully immersive virtual rooms are an attempt to reduce the amount of encumbering equipment a user must wear to experience high-quality perceptual illusions.

Types of VR: Portals to Virtual Worlds

There are various types of virtual reality hardware and software. Each system may support different degrees of sensorimotor immersion and may be specialized in its applications because of cost and complexity. (This classification extends a taxonomy suggested by Louis Brill. See more extended discussion in Biocca & Delaney, 1995.) Almost all share a common feature. Most allow the user to interact with a 3-D computer graphic environment in real time. The level of the 3-D illusion, sensorimotor interaction, and engagement differ predictably according to the quality and cost of the system. The variety of hardware and software systems can be thought of as different *portals* into the 3-D virtual environment. In networked virtual reality, this is a useful analogy. Two people may meet inside the same 3-D virtual environment, with one experiencing it via an inexpensive portal like a common desktop PC, while the other may be experiencing it through a high-quality, sensory-rich portal such as an immersive VR system. The types of VR are classified in Table 14.1.

There are a number of applications for virtual reality. Virtual environments or virtual worlds may be fantasy "worlds that are not and never can be" (Brooks, 1988, p. 1) or semi-realistic analogs of some physical environment. For example, in Epcot's *Aladdin*, a prototype fantasy ride created by the Walt Disney Imagineering team, users peered into suspended head-mounted displays, gripped carpet-covered handle bars, and suddenly felt as if they were flying on a magic carpet through an ancient, Arabic village (Pausch, et al., 1996). But in architectural, engineering, and other design applications, designers and their clients can see, touch, or walk around virtual models of buildings, planes, automobiles, and other products that do not yet exist in the physical world. In university and government laboratories, scientists experience analogs of physical environments they cannot experience directly with their senses: They walk around computer graphic molecules, peer into the center of galaxies, or fly through a model of a human blood vessel. We discuss applications of virtual reality in greater depth later.

Background

Jaron Lanier, a VR entrepreneur and artist, is most often given the credit for the term virtual reality (Lanier & Biocca, 1992). He first used it to describe a narrow set of immersive simulation technologies used by NASA, the U.S. Air Force, and a few research centers in the late 1980s. Although scientists and engineers working on immersive virtual environments and 3-D simulations were not always comfortable with the term, it gained widespread popular currency by the mid-1990s.

Over time, the meaning of the term virtual reality broadened. Various combinations of input and output devices created different classes of VR—sometimes referred to as different types of "worlds" or "environments." Marketers attempted to capitalize on the media attention to virtual reality. This led to a proliferation of systems that laid claim to the popular "virtual reality" label. As a result, by the late 1990s the term, "virtual reality," has come to describe a wide variety of technologies, some of which may be little more than inexpensive video games, while others are expensive research systems offering very compelling, multisensory experiences. Widespread use of the term has led to a search for terms that are more specific to particular applications. People use terms such as synthetic environment, artificial reality, experiential computing, simulation, augmented reality, or virtual environment. This multitude of terms may also reflect the idea that virtual reality is not necessarily a reflection of reality, as the term would apply, but a description of the user's experience.

Table 14.1
Types of VR Interfaces

Classification	Description	User Experience	Typical Applications	Example Companies and Systems
Window Portals	A simple PC and mouse are used to interact with a 3-D environment.	Limited field-of-view and limited sensorimotor interaction provide modest, but satisfactory, experience.	Internet VRML worlds, consumer computer games	Any PC, but especially those with 3D graphics boards.
Mirror Portals	User's image is captured by video equipment. The moving image of the user is superimposed on the computer environment which is projected onto a screen in front of the user.	Users see a video image of themselves on a large screen often with other people. Cutout images interact with the environment. Can be compelling, but sense of presence is modest.	Walk-thru museum or art experiences. Arcade gaming applications.	Mandalla. Myron Krueger's many intallations.
Panoramic Portals	Large wide-screen projection system fills user's visual field. Often includes 3-D stereographic glasses and tracked, hand-held input devices.	One or more users stand in front of large 3-D window on the virtual world. Motion effects and sense of depth can deliver higher levels of presence.	Corporate product visualizations. Scientific and medical visualization.	Paradigm's Immersadesk and PowerWall.
Virtual Rooms	Takes panoramic portals to their logical extreme. Users walk into a physical room that is a large display system. The motion and perspective of the user are continuously updated by tracking devices as the user interacts with the virtual environment.	Sense of space can be quite compelling because the visual system is immersed by the virtual environment which is viewed through 3-D glasses. Expensive system ideal for multiple users.	Scientific visualization. Corporate product and design visualization.	Paradigm's CAVE (University of Illinois), C2 (University of Iowa).
Vehicular Portals	Users enter a mock vehicle (i.e., cars, planes, submarines, tanks, flying carpets, etc.) where they are allowed to operate input devices that control their vehicle inside the virtual environment.	Sense of presence and motion can be quite compelling because the interface can faithfully reproduce a lot of detail of being inside the "real" vehicle.	Military and corporate training. High-end location-based entertainment rides.	Iwerks. Evans & Sutherland.
Immersive VR Portals	Users wear displays that fully immerse a number of the senses in computer generated stimuli (e.g., vision, hearing, touch). These systems often use the distinctive head-mounted, steriographic display, 3-D spatial audio, and sometimes include data gloves.	Arguably provides a significant jump in the level of presence because of the tight sensorimotor integration and the full immersion of at least the key senses of vision and hearing. Some purists argue that these environments represent the only true VR.	Scientific and medical visualization. Corporate product and design visualization. Military training. Some low-quality systems used for gaming.	Usually assembed from component suppliers: e.g., Virtual Research, N-Vision, Retinal Displays.
Augmented Reality Portals	The most advanced augmented reality systems use hardware similar to that found in immersive VR. But rather than fully immerse the user in a virtual world, augmented reality systems overlay 3-D virtual objects onto real world scenes.	Few systems yet achieve a truly convincing integration of a stable 3-D virtual object and the real world, but many show promise with less compelling implementations and schematic virtual displays.	Medical imaging. Battle display systems. Manufacturing and equipment maintenance.	Some low-end systems, but best are university and corporate prototypes: e.g., University of North Carolina at Chapel Hill, Columbia University, MIT, Boeing.

Source: Biocca & Associates

Virtual reality is an evolutionary rather than revolutionary technology. It owes its emergence to developments of many other technologies, especially other media. Commentators note that VR's roots plunge deep, stretching back through the history of communication media (Biocca, Kim, & Levy, 1995; Hamit, 1993; Rheingold, 1991). Centuries of human effort have attempted to simulate the world with perceptual illusions and representations—from coal on cave walls to the development of perspective painting, the camera obscura, dioramas, the stereopticon, and 3-D film.

Clearly, more recent developments have crossed a threshold and created novel experiences with the techniques we call VR. Key VR components have emerged from more than 40 years of government funded research, especially the development of enabling technologies for military and flight simulation and scientific visualization (Rolfe & Staples, 1986). In the 1960s, Ivan Sutherland's (1965; 1968) seminal contributions to the development of computer graphics and early 3-D, head-mounted displays marked a milestone in the emergence of modern VR systems (Rheingold, 1991).

The SuperCockpit program at Armstrong Aerospace Research Laboratory at Wright-Patterson Air Force Base (Ohio) was a significant site for government-sponsored VR research. Research conducted in the 1970s and 1980s under the direction of Thomas Furness extended flight simulator technology by improving head-mounted displays. In 1977, this work led to experiments with the VCASS helmet that immersed pilots into an abstract virtual environment. Pilots looked out on a VR world that displayed the terrain below, other planes, superimposed gauges, and other information. Various military-funded projects helped develop other key components of VR technology: for example, advanced simulation (Evans & Sutherland), distributed simulation (SIMNET), and telerobotics (Utah Arm, Sarcos).

At the same time the military was working to incorporate VR into training, research was underway at various universities to demonstrate civilian applications. Under the direction of Fred Brooks, projects at the University of North Carolina dramatically demonstrated the utility of VR techniques in molecular modeling, architectural visualization, and medical visualization. At the Universities of Wisconsin and Connecticut, Myron Krueger's research in the 1970s and 1980s on "artificial reality" demonstrated ways in which interactivity with a virtual world could be both entertaining and educational (Krueger, 1991).

A project at NASA in the late 1980s inadvertently helped diffuse VR technology. NASA scientists headed by Mike McGreevy and Scott Fisher created a VR system on a much tighter budget than those created by the military (Fisher, et al., 1986). Working with programmers and equipment suppliers from the computer game industry, they used "off-the-shelf" components to put together less expensive versions of the kinds of displays pioneered by the Air Force (Rheingold, 1991). Publicity surrounding this project—fanned by entrepreneurs such as Jaron Lanier of VPL and 3-D visualization companies such as Autodesk—helped fuel the media excitement over VR technology in the early 1990s. These demonstrations of VR technology caught the imagination of the public and raised hopes that immersive commercial VR systems could be built affordably for everyday applications.

A great deal of excitement and press attention greeted demonstrations of virtual reality in the early to mid-1990s. A number of startup companies rushed to capitalize on the potentially huge market for virtual reality experiences. In general, success was mixed. But the push for 3-D graphics on home computers, arcade games, and corporate visualization has been steady.

VRML—The Emergence of Popular Formats for Networked VR

The birth of the World Wide Web had dramatic impact on the diffusion of popular consumer forms of virtual reality. Virtual reality modeling language (VRML) was conceived in the spring of 1994 at the first annual World Wide Web Conference in Geneva, Switzerland (Bell, et al., 1995). A special interest group was formed to discuss the creation of a standard for 3-D design on the Web, and a mailing list became the forum for discussion (VRML Hypermail Archive). From these humble beginnings, the specifications for VRML 1.0 were set.

The Internet entered the consumer mainstream in 1995. For a number of years, VR enthusiasts had been eager to bring a universal form of networked, window VR to the Internet. Networked virtual reality was, until then, limited to expensive military simulations such as SIMNET that linked tank simulators from around the world for battle tactical training on a virtual battlefield. The result was a limited but successful form of VR called virtual reality markup language (VRML-pronounced "vermal"). "One of the key innovations brought about by VRML was the idea that virtual scenery could be incrementally assembled from lists of descriptive files gathered together from arbitrary locations on the Internet" (Rockwell, 1997, p. 29). VRML was designed to create a 3-D environment that allowed users to navigate through a scene. It was a language similar to HTML (hypertext markup language), but was used to model 3-D virtual worlds on the Internet. VRML-based virtual worlds, called scenes, are described with text codes that represent the objects and more complex aspects of an online world. Based on the ASCII format of Silicon Graphics' Open Inventor language (Pesce, 1995), Silicon Graphics was one of the most successful companies in 3-D design, and their format was well accepted by those suggesting standards.

The emergence of VRML was accompanied by speculation and enthusiastic predictions about its potential for distributed VR on the Web (Magdid, et al., 1995). The immediate reality was that VRML remained a relatively exclusive technology because of its demands for bandwidth and powerful computers. But many VRML viewers could easily be incorporated into common Web browsers. Because of limitations in bandwidth, computer processing, and the limitations of the VRML 1.0 spec, the first incarnation of WWW VR was limited, slow, and not very interactive. With the increases in processor speeds and memory in personal computers, VRML slowly gained more widespread use.

Virtual Reality as a Cultural Phenomenon

The *idea of virtual reality*—the image of a jarring realistic, dream-like environment—exerted a powerful influence on the imagination of the media and the general public throughout the decade (Biocca, Kim, & Levy, 1995). Even when few consumers had experienced quality VR, many saw reports of compelling computer graphic worlds on television. This beckoning, virtual image of what VR might become is as much a product as the real, existing VR hardware and software.

The emergence of VR as a symbol for a new computer-based culture beginning its exploration of cyberspace was seen in the plots of movies such as *Lawnmower Man 1* and *2*, *Johnny Mnemonic*, *Disclosure*, *Total Recall*, and numerous other films. Television created series dedicated to the idea of VR, among them *Wild Palms*, *VR5*, and *VR Rangers*. Strongest among the young, this fascination and its associated cultural manifestations fueled consumer demand for VR experiences and products. In some of these visions, VR was portrayed as a playground for fantasy realization, a place where all forms of illusions were possible (e.g., the *Star Trek* Holodeck).

On the other hand, VR, and especially the word "virtual," became a lightning rod for those uncomfortable with the socio-cultural changes unleashed by networked computing. Darker, fearful depictions of virtual reality appeared to reflect the public's unease with the pace of technological change in the 1990s. For example, cartoonists, such as *Doonesbury*'s Gary Trudeau, used the image of VR by drawing dystopic views of the technology as a consumerist nightmare or a place where false experiences substituted for real life. In the popular culture, the word "virtual" was sometimes used negatively to suggest illusion and trickery. At the same time, VR engineers and computer scientists working at NASA and university research centers complained that the media's "VR hype" was confusing the public and threatened the short-term development of virtual reality by raising unrealistic expectations.

Recent Developments

VR Hardware

The recent development of VR hardware has been torn by two opposite, but not irreconcilable, trends. The first trend has been toward *progressive embodiment* (Biocca, 1997) of the user inside the virtual environment. By progressive embodiment, we mean that the user's senses and body movements are increasingly connected to the virtual environment and represented in that environment via sensorimotor feedback and by some 3-D representation of the user's body.

The other trend has been fueled by the desire to have high-end VR experiences with *minimal encumberment*, such as having to "suit up" with uncomfortable "goggles and gloves" for quality sensory experiences. At this time, the more the user is embodied, the more likely he or she will be encumbered by hardware that is attached to the senses or motor systems. For some, this is no more a problem than wearing equipment to play sports. But it presents restraints for casual or widespread use of the technology.

Display Technologies

Display technologies did not change radically during the mid-1990s. For example, VR's signature technology, the head-mounted display (HMD), slowly and steadily evolved (Melzer & Moffit, 1996). Using phase-sequential CRTs, tiled LCDs, and other techniques, HMDs from companies such as Virtual Research and Kaiser Optics got lighter, cheaper, and provided better resolution. The cost of spatialized audio also dropped as the technology found its way into consumer games (Begault, 1994).

Practical haptic ("touch") feedback devices showed the biggest jump. A force feedback device called the Phantom found its way into more applications. It provided compelling sensations of touching objects with a single finger or through a tool like a scalpel. Force feedback joysticks and steering wheels become more common in consumer games as well. More ambitious research on force and touch feedback continued to push the envelope, but few were incorporated into applications (Burdea, 1996).

One of the more dramatic developments in display technologies was the rise of augmented reality and wearable computer systems, discussed in greater depth below.

Sensor Technologies

Developments in sensor technologies reflected the general evolutionary trend toward progressive embodiment—getting more of the body into the virtual world. For example, the mid- to late-1990s saw the success of motion capture systems led by companies such as Ascension, Adaptive Optics, Polhemus, and Motion Analysis Corporation. The typical motion-capture systems track the motion of the head, torso, and limbs of a human or animal using dozens of optical, ultrasound, or electromagnetic markers. These markers are all placed at key motion points of the body. These systems provided accurate data to animate the bodies and faces of virtual humans and animals. Body trackers added motion realism to 3-D virtual humans and animals used in high-end film and video animation. For example, stunning results can be seen in some of the hyper-realistic computer animations found in the movie *Titanic*. Such systems were used less frequently for real-time animation of users interacting inside VR systems.

One of hardest areas of the body to capture for 3-D animation is the face. Face motion recognition and capture is critical to animation of the face of a user's avatar in future VR systems. (An avatar is an animated character representing a person in a virtual environment.) Commercial facial capture systems, such as Motion Analysis Corporation's Geppetto system, are now used for performance and video animation. Performance animation is like 3-D puppetry. The animated virtual human interacts through a screen with a live or televised audience.

A key problem has been how to allow users of immersive VR systems to walk freely around a large virtual world. In the early 1990s, the idea of omnidirectional treadmills was introduced (Biocca, 1992), but was received with some skepticism. In the mid-1990s, omnidirectional treadmills started to appear in military applications of virtual reality to allow soldiers to "walk" through virtual environments that included vehicular tank simulations.

Another challenge of virtual reality and 3-D graphics has been technology that can create 3-D "pictures" of objects and full scenes. Companies introduced expensive full-body scanners that created reasonably accurate models of the shape of the body and everyday objects. Several image-based techniques are also under development, offering the possibility of quickly generated 3-D computer graphic models of the outside world.

Computer Processing of Virtual Environments

Realistic 3-D virtual environments are most often composed of meshes of polygons. They are virtual 3-D wireframes of squares and triangles pasted with realistic pictures. In medical imaging, other techniques using little "cubes" called voxels are often used. Realistic 3-D graphics and high levels of interactivity in virtual environments require significant processing power to manipulate the geometry in real time. New graphics-intensive hardware continues to evolve to meet the increasing demand for 3-D computer graphics in more and more applications.

At the very high-end, companies such as Silicon Graphics and Evans & Sutherland pushed real-time 3-D graphics processing closer to immersive photorealism. Introduced at a cost of $1 million, SGI's Onyx2 Reality Monster could be acquired with sixteen 195-MHz R10000 processors, 8 GB of memory, 320 MB of frame buffer, up to eight Infinite Reality graphics pipelines, and 200 GB of storage. Such fast, high-bandwidth machines could generate upward of 80 million polygons per second. Users could employ such systems to traverse more visually realistic 3-D worlds in real time.

By the late 1990s, workstation-quality 3-D graphics were starting to find their way into the home. An early example was the humble game platform, Nintendo64, which boasted powerful graphics processors, optimized for game play, from Silicon Graphics Corporation. By the mid- to late-1990s, powerful 3-D chipsets moved from the graphics workstation to the average home computer.

Merging Virtual Reality with Physical Reality

Most 3-D virtual environments exist only inside the computer. It has been argued that, at some point in the evolution of virtual reality, 3-D virtual environments may merge with the physical environment (Biocca & Nowak, 1997). There are a number of applications where it is valuable to have 3-D virtual objects appear to exist side-by-side with the physical environment. For example, a technician might find it useful to have a 3-D schematic diagram of a machine superimposed on the actual machine while it is being repaired or operated (Feiner, MacIntyre, & Seligmann, 1993). This subclass of virtual reality where 3-D computer illusions appear side-by-side with physical objects is called augmented reality (Azuma, 1997).

A key technological challenge of augmented reality is to find the best ways to integrate virtual objects with the physical environment. In the application areas of medical imaging, transparent, head-mounted displays are used along with powerful graphics computers and accurate tracking systems to superimpose the image of virtual objects onto the physical world (Biocca & Rolland, 1998). Doctors are able to see 3-D radiographic data of a patient superimposed directly on the patient's body, providing a kind of "x-ray vision." In industrial applications, workers looking through augmented reality displays can see exactly where wires and parts need to be inserted in new pieces of equipment. Boeing used augmented reality-based systems to assist technicians in wiring and connector assembly (Nash, 1997).

The ideal augmented reality system would be extremely small and portable so that the user could access relevant 3-D data at any time. Also, the ideal virtual reality interface is one that is merged seamlessly with the body of the user. Trends in personal computing technology show that computers are getting faster, demanding lower power, and shrinking in size—hence, becoming suitable for mobile computing applications. These trends and advances in broadband wireless networking infrastructure point to a new paradigm in computing: wearable computing. "A wearable computer, powered on at all times and worn like clothing, has the potential to be a true extension of the mind and the body" (Mann, 1997, p. 56). Augmented reality technology and virtual reality techniques are being merged with the growing areas of wearable computing.

Currently, wearable computer prototypes make use of virtual reality enabling display technologies such as heads-up displays that overlay computer-generated output (such as text or graphics) onto the physical world. They suggest an improved visual interface that enhances the user's reality (Starner, et al., 1997). Micro-Optical Corporation demonstrated an eyeglass-based display system prototype that looks and functions as a regular pair of eyeglasses (Spitzer, et al., 1997). Micro-Optical's aesthetically designed eyeglasses let the user maintain eye-contact with others while using the display, making this display well suited for a computer-supported work environment.

VR Software

High-end virtual environments are built using special application development interfaces made of libraries of functions typically written in C (a common computer programming language). With the uncertainty and instability of the VR consumer and entertainment market in the mid-1990s, leading companies sought stability by focusing on markets such as product design simulation and visualization, corporate training, or medical imaging. Division, Deneb, and Prosolvia AV have focused on the use of software tools for product visualization and plant design. For example, software from these companies has found widespread acceptance in automotive engineering. Multigen, a favorite in the military simulation industry, found markets in the automotive and gaming industries. Multigen was also the developer of one of the more innovative VR software products of the mid-1990s, an application development tool that allowed users in an immersive VR environment to build a virtual town by locking objects together as if they were Lego blocks.

Companies working the low end of the VR development market have had to change strategy with the arrival of VRML. Early leaders such as Superscape and VREAM (now Platinum Technologies) have focused more of the their R&D on VRML-related tools.

VRML 2.0—An Emerging Standard

VRML inched closer to becoming a de facto standard for WWW 3-D. The VRML Consortium was created in 1996, and charter members include many of the major companies interested in 3-D graphics on the WWW: Microsoft, SGI, Sun, Sony, NTT, and others. VRML offered the promise of content that would play on multiple platforms and a much wider audience for VR developers.

The introduction of VRML 2.0 in August 1996 continued the trend toward standardized platforms on the Internet. VRML is becoming more competitive as a standard for interactive 3-D renditions of objects on the Web. VRML 2.0 has taken Web-based 3-D to the next level, as the characters and objects on Websites and in chat rooms can now move and interact.

There is a continued effort to add more continuity to VRML worlds. Living Worlds is a joint proposal by Paragraph International, Black Sun Interactive, and Sony Corporation. The goal of Living Worlds is to bring all Web-based environments to one standard. Their proposal would allow users and 3-D objects to more easily move between environments. Users would be able to employ their favorite avatar in all of the worlds and not need to modify avatars for each new environment. Also, it would standardize motion (for example, the way to wave in one world would be the same in another), so users would not have to spend time learning new procedures for interaction.

Another effort, the Open Communities proposal, is working in conjunction with VRML. This effort is led by Mitsubishi Electronic Research Laboratories and is attempting to provide an application that would work independent of whatever language is employed to implement the visualization of the community (Rockwell, 1997). This would be a step to make the international infrastructure of the Internet truly accessible by cyber-citizens of all nations.

Networking Virtual Environments

The science fiction novels *Neuromancer* and *Snowcrash* helped popularize the image of cyberspace as a worldwide 3-D virtual environment populated by thousands of strange individuals beaming from world-to-world. The vision has merged with the practical aims of Computer Supported Cooperative Work (CSCW) and Computer Supported Collaborative Learning (CSCL) to promote the goal of large collaborative virtual environments (CVEs). In these environments, work teams physically distributed anywhere around the globe can use telecommunications systems to meet inside a virtual environment to work on practical problems such as automobile design, architectural reviews, contract negotiation, and medical consultation. This is currently one of the most challenging and active areas of research in VR. CVE has the promise to provide novel ways of learning, teleworking, and interpersonal communicating in the near future (Benford, et al., 1995; Bowers, et al., 1996; Leigh, et al., 1997; Slater, 1998).

Social virtual environments, like collaborative virtual environments, are a form of geographically-distributed, multi-user virtual environments. The technology of social virtual environments may differ little from the more expensive cousin, collaborative virtual environments, although they tend to be implemented in the lower-cost VR environments. The two differ more in the function of the virtual space in the same way that a bar or dance hall differs from a corporate conference room. The first distributed interactive virtual worlds were text-based, called MUDs (multiple user dungeons) or MOOs (MUD, object oriented), or other object-oriented games that evolved from the board game *Dungeons and Dragons* (Waters & Barrus, 1997). These have evolved into 3-D computer graphic environments populated by virtual beings.

A great deal of interest has begun to focus on how users and artificial intelligence is embodied in virtual environments (Benford, et al., 1995; Biocca, 1997). When a user is embodied and controls a computer graphic character, this virtual human representation is called his or her avatar. The Hindu term, *avatar*, originally meant the incarnation or the embodiment of a deity or a spirit in an earthly form (Vilhjálmsson, 1996). But sometimes the virtual human representation is controlled not by a human, but by the computer. All beings in 3-D virtual environments are known as agents, although other specialized terms are sometimes used including bots, boids, etc. The term agent refers to more than just 3-D artificially intelligent entities (Petrie, 1996), but this discussion is focused on that subclass of agents that are embodied in animated virtual beings. Agents can act as guides to the environment or task (e.g., Sense8's Guardian Angel), advertisers (The SpokesBot, Blaxxun Interactive), or play various other roles such as simply "populating" a world, providing an "opponent," or entertaining in some way.

There is tremendous interest in the design of agents and avatars, as well as the social/psychological implications of these representations of us. In many 2-D and 3-D social environments such as Worlds Chat, World Away, and Online Traveller, users can choose some of the features of how their avatar will look. The very choice itself signals something about the user. In other worlds such as the Palace, avatars are almost graded into a class system where "newbies" are marked by a particular look such as smiley face. Visitors to these 3-D environments can observe all kinds of social behavior, including gossip, political activity, group singing (Online Traveller), "sexual" encounters, gang attacks, and commerce.

Applications

Virtual reality technology is becoming a standard working tool in a number of application areas. Although early VR developers thought entertainment would lead the development of immersive virtual reality, this has not occurred. With some significant exceptions, entertainment and consumer applications are carried via lower-cost VR interfaces such as window portals. More immersive VR systems remain the domain mostly of professional applications. Table 14.2 shows the key areas of significant penetration of high-end virtual environments in various business and professional areas.

Applications of VR in design and training are among the most promising (Roberts, 1995). Among professional applications, the U.S. military has developed some of the most advanced uses of virtual reality, going back to the early days of flight simulation. VR simulations for training continue to be the major area of investment and the leading area for advanced applications of virtual reality technology. Because of the success of virtual reality technology in flight and battle simulation, the Department of Defense has expanded its commitment to the use of virtual reality (see the work of the Simulation, Training and Instrumentation Command [STRICOM]). Training applications of virtual reality have diffused from the military to the corporation.

Table 14.2
Selected Professional Applications of High-End
Virtual Reality

Application Area	Application	Examples	Sample Clients, Suppliers, & Research Centers
Manufacturing	Product Visualization	Engine designs, vehicles, packaging, architectural walkthroughs.	EDS, GM, Ford, Caterpillar, Prosolvia Clarus
	Plant Simulation & Design	Vehicle and human workflow, assembly line design.	Deneb Robotics, Division
	Product Marketing	Product simulations in tradeshows, corporate site, or Internet.	Evans & Sutherland, Ford, EDS
Adult Training	Military Combat Training	Networked tank simulation, fighter plane simulations, dismounted infantry.	DIVE, Boston Dynamics, Naval Postgraduate School, STRICOM
	Flight Simulation	Exact replicas of passenger & fighter plane performance.	Boeing, US Air Force, Evans & Sutherland
	Manufacturing Training	Assembly line work, system maintenance	Adams Consulting, Motorola, Boeing
Medicine	Radiographic Imaging	3D fly-throughs of arteries and other body spaces.	HT Medical, Stanford, University of Pennsylvania
	Medical Training	Simulations of injections, intubation, surgery, and other procedures.	HT Medical, Greenleaf Medical, SRI
	Veterinary Medicine	Animal simulations, medical procedures.	Michigan State University, University of Pennsylvania
	Treatment of Disabilities	Prosthetics.	Greenleaf Medical, Biocontrol Systems
	Surgery	Surgical planning; laproscopic surgery.	HT Medical, DARPA
	Psychotherapy	Phobia treatments, self-representations & role playing, autism, patient support.	Georgia Tech, North Carolina State University, Starbright Foundation
	Telepresence Surgery	Remote telepresence surgery.	SRI, DARPA
Scientific Visualization	Physical, Small-Scale Data Visualization	Fly, walk around, and/or touch atoms, molecules, genes, etc.	University of North Carolina, University of Illinois, many others.
	Large-Scale Data Visualization	3-D fly/walk throughs of large natural environments, earth weather systems, solar system and universe.	University of Illinois, Sandia Labs, UC San Diego

Source: Biocca & Associates

The most successful application areas for high-end virtual reality tend to have these common patterns:

- The problem is important enough economically or socially to warrant expensive solutions (e.g., military training, medicine).

- The professionals in the area already have some form of commitment or work-station computer graphics (e.g., radiological data in medicine, CAD/CAM data in manufacturing, engineering, and architecture).

- New interfaces like virtual reality can significantly improve the performance of key professionals (e.g., radiologists, doctors, engineering review).

Rocky Market for Consumer VR

A great deal of excitement and press attention greeted demonstrations of virtual reality in the early- to mid-1990s. A number of startup companies rushed to capitalize on the potentially huge consumer market for virtual reality experiences (e.g., VPL, Forte Technologies, CyberMaxx, and Virtual I/O). In general, there was a pattern of mixed success. The push for 3-D graphics on home computers, arcade games, and corporate visualization has been steady. However, consumer adoption of advanced input and output devices, especially those essential to immersive virtual reality, has been slower in the home market than some entrepreneurs expected (e.g., head-mounted displays and data gloves).

Companies that overextended themselves pursuing the consumer market with immersive VR technologies were hurt when the predicted growth did not occur. Identified early on as a threat to consumer VR (Biocca, 1993), the problem of simulation sickness associated with the immature technology raised liability fears and fueled sensational articles (Gross, Yang & Flynn, 1995). The London-based Virtuality, a leader in the location-based entertainment market, folded in 1996. Virtual I/O, Forte Technologies, and Victor Max, all manufacturers of head-mounted displays targeted for consumers, were restructured. Mature companies including Sega and Nintendo also had difficulties with consumer immersive VR products. Industry observers felt that some technology, such as Nintendo's Virtual Boy, was poorly implemented. Stripped down to a simple red monochrome interface to meet consumer price points, this VR toy may have been too "watered down" to be successful. On the other hand, award-winning hardware such as Virtual I/O's head-mounted display was introduced without adequate content support. Consumers had an adequate product, but little to do with it.

The VR startups that survived into the late 1990s tended to be lean, avoided overextending themselves chasing the consumer market, and concentrated on narrower markets such as military, training, medicine, and manufacturing and design visualization. Small companies such as hardware manufacturer Virtual Research and software companies such as Multigen and Sense8 attempted to remain profitable by focusing on smaller, but growing industrial and professional markets for VR products.

Current Status

In general, the development of virtual reality technology is becoming clearer as some market strategies fail and others succeed. This section reviews some of these trends.

The trend toward the use of 3-D graphics, a key element to many forms of virtual reality, is very strong. The steady drumbeat at SIGGRAPH96-98, a major industry conference on computer graphics, was 3-D. Jon Peddie Associates (1998) predicts a boom in 3-D for the desktop. According to Peddie, a market analyst, "We haven't seen this much activity in the graphics industry for a number of years. A lot of it is happening under the surface, with chip makers aggressively marketing themselves to OEMs, but the results are there for PC users to see. We're seeing better and better 3-D graphics every month, and the growth in the market is astonishing."

Developers of virtual reality on the World Wide Web suffer from a problem common to WWW applications: "How do we make money with this stuff?" For example, social virtual environments were greeted with great enthusiasm, but have been disappointing commercially. Social virtual environments are being created by a variety of different corporations, including Microsoft, Fujitsu, IBM, Mitsubishi, Softbank, Intel, Sony, Nippon Telegraph and Telephone, Black Sun Interactive, Ubique, OZ Interactive, Worlds Inc., Chaco Communication, Onlive Technologies, ParaGraph International, The Palace, and many others. Few projects were profitable as of mid-1998.

One way to be profitable is through sponsorship. Pioneering environments such as OnLive Traveller introduced their 3-D social world with full audio capability and big name sponsors such as MTV and others. But few consumers have the technology and know-how to visit the world. Low bandwidth connections often caused technical problems. SuperScape and other companies with free worlds supported by sponsors have also not found this route to be very profitable.

Some creators of social virtual environments are looking at membership fees and other "services" to provide revenues. Although it is free to enter and chat in most of these worlds, membership allows the user to enter "special rooms" not accessible to non-members. Users might even be willing to pay for other forms of customization. For example, only members can create and use their own avatar (non-members have to use those created by designers) or have a name (non-members will be identified as "guest").

Exclusivity is another appeal. Members do not associate much with non-members, frequently called "newbies," and prefer to inhabit the "members only" rooms. Any time a "guest" attempts to go to a forbidden room, they are reminded that they cannot enter, and are offered the chance to purchase membership. Damer (1997) explains as he describes one experience in Worlds Chat,

> I noticed a shimmering doorway and asked Blue Bear, "what is
> that?" He and I both tried to enter, but were bounced back and
> given the message, "access to this area requires an upgrade to

Worlds Chat 1.0 Gold," and given the Web site of Worlds Inc. to order it. "Drat!" I said, but thought that it was very good marketing (p. 39).

At the VRML '98 conference, a number of companies were considering using VRML in advertising. Social virtual environments already have billboards with links to information about the companies that support them. The most blatant form of this type of commercialism is in AlphaWorld, where billboards advertise everything from books to links to credit card applications. Also, agents inside the virtual world may push products. For example, in AlphaWorld, users might be approached by the immigration explorer who solicits them to become a member. After a few minutes online, a user might receive a message like: "Immigration Officer: Hey 'Blue Bear' did you know that Active Worlds citizenship costs only U.S. $19.95 a year and gives you many additional features? For more information go to the Help menu and select Registration." The arrival of obtrusive advertising to this medium has been greeted with the same unease as the rise of advertising in the WWW and e-mail.

Factors to Watch

Because virtual reality is a highly protean cluster of technologies, it is best to separate the factors to watch into medium-term and long-term trends. Medium-term trends deal with this generation of technologies, while long-term trends deal with prospects for the evolution of this and future generations of VR technologies.

Medium-Term Trends

The attention to VRML and collaborative virtual environments suggests that most analysts think the future success of VR is increasingly linked to network applications. If we accept this premise, then we can see a future pattern to the diffusion of VR. Consumers and some business users are likely to first experience and adopt VR using networked window portals. As networked window portals become more common and as bandwidth increases, users are likely to demand more interactivity with the 3-D world they see on the screen. This can only be achieved by moving to higher-quality portals. With the diffusion of HDTV and the reintroduction of more consumer-friendly input and output devices (such as force reflective joysticks, 3-D mice, etc.), the sensorimotor quality of the portals is likely to increase.

Work on a networking standard similar to VRML is likely to continue and intensify in importance. Size and download speed will remain issues in the near future because of continued bandwidth limitations. Beside the use of databases to control download time, research groups are also developing methods for streaming VRML and compressing the language into binary code. Both of these methods in conjunction could cut a file's size significantly.

As 3-D on the Web becomes increasingly mainstream, it may become necessary to further integrate VRML with HTML. At the moment, a browser interprets VRML worlds much like an HTML page—by reading the ASCII text and placing the described objects in a browser window. This results in a distinct separation between HTML and VRML content. In 1998, working groups were organized to investigate how 3-D content can be integrated into the new cascading stylesheet form of HTML. Some possible applications of this would be the use of VRML scenes in Webpages just like a normal bitmap. This would allow for layering and closer combination of media elements.

In October 1996, the Internet-2 initiative was announced. Internet-2 is an alliance of research universities, telecommunications companies, and government organizations who seek to provide a general increase in bandwidth up to one gigabit per second, provide various levels of service (e.g., slow lanes and fast lanes), and to push the envelope for high-bandwidth applications. The desire to support high-bandwidth virtual reality applications using high-quality portals, so called tele-immersion applications (e.g., telemedicine, virtual laboratories, collaborative work teams, etc.), is an important factor in Internet-2 development. The vision for Internet-2 is the growth of regional gigabit networks. Initially, only universities will be connected, but eventually it will be expanded to other types of organizations.

The standards process has become critical to creating universal formats for virtual reality. The key force behind the ultimate utility of VRML rests on allowing users to travel easily from world to world, regardless of who authored it or the location of the server. The success of initiatives that standardize world formats such as the Living Worlds Specification and the Open Communities proposal will be factors to watch. On the other hand, other 3-D standards such as Java3D threaten to create more competing 3-D formats and slow the movement toward easy authoring and interoperability. In 1998, Microsoft and Silicon Graphics Corporation announced they were working on Fahrenheit, a project that might lead to other competing standards for networked 3-D graphics. The emergence of widely-accepted standards will signal the cementing of a coherent WWW community for 3-D graphics.

The integration of VRML with database technologies is also something to watch. Oracle is currently working on integrating database technology with VRML (Davison, 1998). There are many reasons why this could be a very important development, some tying in directly with the need to have information embedded in 3-D objects such as the case with Living Worlds. Databases would allow VRML to save information about the scene or world as it changes so, as users visit a world, objects and animations would remain consistent from session to session. Some other obvious advantages for integrating database technology would be security (being able to attach security preferences to specific areas or objects), VRML catalogs and other marketing applications, and controlling the size of the world or scene by dividing it into smaller chunks. As a user navigates, the VRML file could communicate with the database and only download objects that relate to the users' location in the world.

In the United States, the 1997 Defense Technology Area Plan listed simulation and modeling as one of the five most important areas in information technology critical to defense needs. The number of commercial firms involved in virtual training and simulation has greatly increased in recent years, with over 150 companies in the state of Florida alone (Boris, 1997). Analysts expect to see major developments in cutting-edge military systems, as VR interfaces become a means of fully integrating military intelligence, battle operations, and training systems. Annual government funding for simulation, a factor in training systems, is predicted to increase from $169 million in 1989 to over $280 million by 2003 (Zyda & Sheehan, 1997).

Virtual reality is evolving as an expressive medium. We are currently at a stage where today's VR storytellers are working on the theory and methods that will define the new medium (Meyer, 1995). Some VR story development may borrow from traditional narrative forms such as the novel and film. For example, in spring 1998, a team headed by Dr. Bernie Roehl of the University of Waterloo staged the first live VRML Shakespeare performance on the WWW. But the inherent capacity for interactivity on the part of the user will really change the way authors and designers think about narrative structure. In her book, *Hamlet on the Holodeck*, Janet Murray (1997) foresees a "cyberdrama" art form that embodies the artistic qualities of traditional narratives with the participatory and immersive nature of virtual reality. "[Cyberdrama] will not be an interactive this or that, however much it may draw upon tradition, but a reinvention of storytelling itself for the new digital medium" (p. 207).

It is a belief that virtual worlds will become more interesting as more people can easily create them. Advances in design tools have made it easier for users to create 3-D objects and VRML worlds. Some social environments such as AlphaWorld allow users to build their own environments. Now that high-end workstations such as those made by Silicon Graphics are not as necessary to create 3-D objects and animations, powerful tools are migrating to common platforms. As tools become easier to use and the machines to run the more powerful VR software become less expensive, technological diffusion and world building increases. This is especially advantageous for implementation in schools.

Because VR can be a powerful technology, its potential effects have been widely speculated in popular and academic writing (Turkle, 1997), but only a few have been systematically studied. One of the most studied is a variety of perceptual-motor disorders called simulation sickness (Biocca, 1993; Kolasinski, 1995). These disorders can cause users of VR systems to experience symptoms that resemble motion sickness. Other forms of intersensory conflict caused by the imperfections of VR systems can alter body perception as evidenced by visual-motor discoordination (Biocca & Rolland, 1998). Press reports of these problems (Gross, Yang, & Flynn 1995) and fears of liability risks may have influenced the limited growth in the consumer segment of the VR industry.

Although virtual reality is used to treat mental illness (Hodges, et al., 1995), some are concerned that prolonged immersion in fantasy worlds might lead to mental health problems, such as confusion about reality and identity. For example, television has been studied for such effects since its invention. TV's real-life presentations may influence physiological and unconscious cognitive mechanisms so that judgments of the real world may be influenced by memories of television events as well as real events (Shapiro & Lang, 1991). One might conclude then that VR would provide even "greater sensory resemblance to natural reality. This could lead to more 'reality monitoring' errors, especially if stress at the time of the experience biased judgments of the experience toward a categorization of 'real' before storing it in memory" (Shapiro & Mcdonald, 1995, p. 339). Research indicates that some subtle memories produced from virtual experiences are difficult to discriminate from real and imaged experiences (Hullfish, 1995). These and other findings suggest that VR is likely to cause the reiteration of all the social and psychological concerns that have accompanied the use of other media such as television and video games.

Long-Term Trends

Something like virtual reality has been the long-term goal of people developing communication media for hundreds of years (Biocca, Levy, & Kim, 1995). Media have steadily evolved toward greater sensory realism to the point of full embodiment, an interface where the user's sensorimotor system is surrounded by a media environment (Biocca, 1997). The long-term trend is toward a medium that delivers very high levels of presence, something akin to the goal of virtual reality. However, when fully implemented, it might have a different name (e.g., Holodeck, reality field, Simstim, etc.).

Some see education as active, situated in concrete settings and problems, and geared toward the development of mental models of the world. Some point to a new paradigm for distance education which incorporates virtual reality as part of a Webcentric course curriculum. An advocate of this vision, Dr. Veronica Pantelidis (personal communication, March 2, 1998), sees education as extending beyond the scope of traditional academic institutions and tells her students that "we are educating individuals not for the 21st century but for the 22nd century. Many young people alive today will live to be over 100 years old. We need to give them knowledge and education in life skills that will sustain them through the whole century. Simulations such as VR are a wonderful way to do it."

Immersive virtual reality, augmented reality, and wearable computing reveals an increasingly blurred boundary between the human body and technological extensions. Virtual reality, augmented reality, wearable computing, and other advanced technologies increasingly integrate the body with the computer interface. At the same time, users of virtual environments are increasingly represented by a 3-D virtual body—that is, they are embodied in virtual environments. One might say that users of advanced computer systems are some form of cyborg (Gray, et al., 1995). Researchers and designers have commented on psychological opportunities and dilemmas caused by the increasing integration of technology with the human body (Gray, et al., 1995; Biocca & Rolland, 1998).

On the other hand, VR may only make clear that computer users may already be cyborgs. Human bodies are already augmented by wearable technologies such as wristwatches, cellphones, laptops, and PDAs to assist in daily living. A wearable augmented reality computer has the potential to replace all these devices and provide additional functionality such as improving our sensory capabilities by extending visionary capabilities through audio augmentation (Foner, 1997). Enthusiasts of wearable computing feel it may offer better coupling of the machine and human through the use of flexible input and output devices. As wearable computing and augmented reality become more socially acceptable and as computer devices are fashionably integrated into our clothing, we may see increased cultural interest in the idea of the cyborg, leading many to contemplate the subtle tradeoffs of what has been called the cyborg's dilemma (Biocca, 1997).

Acknowledgments

We would like to thank for the following individuals for their help in preparing this and a longer review article available at the M.I.N.D. Lab.

John Barrus
Mark Billinghurst, Hitlab, University of Washington
Thomas Blackadar,Personal Electronic Devices, Inc.
Bruce Damer, Contact Consortium
Jason Leigh, University of Illinois at Chicago
Igor Pandzic, University of Geneva
Bob Rockwell, Blaxxun Interactive
Bernie Roehl, University of Waterloo
Cliff Shafter, Virginia Tech
Steve Schwartz, Xybernaut Corporation
Jolanda Tromp, University of Nottingham

Bibliography

Argus VR International corporate Website. (1998, February 28). [Online]. Available: http://www.argusvr.com/ focus/apps/training.htm.

Azuma, R. T. (1997). A survey of augmented reality. *Presence: Teleoperators and Virtual Environments, 6* (5), 355-385.

Begault, D. R. (1994). *3-D Sound for virtual reality and multimedia.* New York: AP Professional.

Bell, G., Parisi, A., & Pesce, M. (1995). *VRML 1.0 specification.* [Online]. Available: http://vrml.wired.com/ vrml.tech/vrml10-3.html#Mission Statement.

Benford, S., Bowers, J., Fahlon, L., Greenhalgh, C., & Snowdon, D. (1995). User embodiment in collaborative virtual environments. In *CHI '95 conference proceedings on human factors in computing systems*, 242-249.

Biocca, F. (1992). Virtual reality technology: A tutorial. *Journal of Communication, 42* (4), 23-73.

Biocca, F. (1993). Will simulation sickness slow down the diffusion of virtual environment technology. *Presence, 1* (3), 334-343.

Biocca, F. (1996). Intelligence augmentation and the vision inside virtual reality. In J. Mey & B. Gorayska (Eds.). *Cognitive technology.* New York: Elsevier/North-Holland.

Biocca, F. (1997). The cyborg's dilemma: Embodiment in virtual environments. *Journal of Computer Mediated Communication.*

Biocca, F. (forthcoming). *Presence of mind in virtual environments.*

Biocca, F., & Delaney, B. (1995). The components of virtual reality technology. In F. Biocca & M. Levy (Eds.). *Communication in the age of virtual reality.* Hillsdale, NJ: Lawrence Erlbaum Associates.

Biocca, F., & Levy, M. (1995a) (Eds.). *Communication in the age of virtual reality.* Hillsdale, NJ: Lawrence Erlbaum Associates.

Biocca, F., & Levy, M. (1995b). Communication applications of virtual reality. In F. Biocca & M. Levy (Eds.). *Communication in the age of virtual reality.* Hillsdale, NJ: Lawrence Erlbaum Associates.

Biocca, F., & Nowak, K. (1997). The MIND Lab's five long-term trends in information technology. *Michigan Forward on New Technology.* Michigan Chamber of Commerce Newsletter.

Biocca, F., & Rolland, J. (1998). Virtual eyes can rearrange your body: Adaptation to visual displacement in see-through, head-mounted displays. *Presence, 7* (3), 262-277.

Biocca, F., Kim, T., & Levy, M. (1995). The vision of virtual reality. In F. Biocca & M. Levy (Eds.). *Communication in the age of virtual reality.* Hillsdale, NJ: Lawrence Erlbaum Associates.

Biocca, F., Nowak, K., & Lauria, R. (1997). Virtual reality. In C. H. Sterling (Ed.). *Focal Encyclopedia of Electronic Media.* Boston: Focal Press.

Boris, B. (1997, November 7). Governor praises central Florida technology. *Training Consortium & Simulation Technology Consortium News Release.* [Online]. Available: http://www.techware.com/tstc/news/19971107.htm.

Bowers, J., Pycock, J., & O'Brien, J. (1996). Talk and embodiment in collaborative virtual environments. In *CHI '96. Conference proceedings on human factors in computing systems,* 58-65.

Brooks, F. (1988). *Grasping reality through illusion: Interactive graphics serving science* (TR88-007). Chapel Hill, NC: Department of Computer Science, University of North Carolina.

Burdea, G. (1996). *Force and touch feedback for virtual reality.* New York: Wiley.

Carr, K., & England, R. (Eds.). (1995). *Simulated and virtual realities: Elements of perception.* London, UK: Taylor and Francis.

Damer, B. (1998). *Avatars! Exploring and building virtual worlds on the Internet.* Berkeley, CA: Peachpit Press.

Davison, S. (1998). Continuing a commitment to a growing technology. An interview with Tony Parisi of Intervista. *VRML Developers Journal, 1* (1) 38-40.

Durlach, N., & Mavors, A. (1995). *Virtual reality: Scientific and technological challenges.* Washington: National Academy Press.

Feiner, S., MacIntyre, B., & Seligmann, D. (1993). Knowledge-based augmented reality. *Communications, 36* (7), 53-62.

Fisher, S., McGreevy, M., Humphries, J., & Robinette, W. (1986). *Virtual environment display system. Proceedings 1986 workshop on interactive 3-D graphics.* Chapel Hill, NC: Department of Computer Science, University of North Carolina.

Foner, L. (1997, October). *Artificial synesthesia via sonification: A wearable augmented sensory system.* Poster session presented at ISWC, Cambridge, MA.

Gibson, J. J. (1966). *The senses considered as perceptual systems.* Boston: Houghton-Mifflin.

Gibson, J. J. (1979). *The ecological approach to visual perception.* Boston: Houghton Mifflin.

Gray, C. H., Figueroa-Sarriera, H., & Mentor, S. (Eds.). (1995). *The cyborg handbook.* New York: Routledge.

Gross, N., Yang, D. J., & Flynn, J. (1995, July 10). Seasick in cyberspace. *Business Week Science & Technology: Virtual Reality, 3432,* 110.

Hamit, F. (1993). *Virtual reality and the exploration of cyberspace.* Carmel, IN: SAMS Publishing.

Hawkins, D. (1995). Virtual reality and passive simulators: The future of fun? In F. Biocca & M. Levy. (Eds.). *Communication in the age of virtual reality.* Hillsdale, NJ: Lawrence Erlbaum Associates.

Hedberg, S. (1996, December). Agents for sale: First wave of intelligent agents go commercial. *IEEE Expert,* 16-23.

Hodges, L., et al. (1995). Virtual environments for treating the fear of heights. *IEEE Computer, 28* (7), 25-34.

Hullfish, K. (1995). *Virtual reality monitoring: How real is virtual reality?* [Online]. Available: http://www.hitl.washington.edu/publications/hullfish/.

Kolasinski, E. M. (May, 1995). *Simulator sickness in virtual environments.* Alexandria, VA: U.S. Army Research Institute for the Behavioral and Social Sciences, Department of the Army.

Krueger, M. (1991). *Artificial reality.* New York: Addison-Wesley.

Lanier, J., & Biocca, F. (1992). An insider's view of the future of virtual reality. *Journal of Communication, 42* (4), 150-172.

Leigh, J., Johnson, A., & DeFanti, T. (1997). *CAVERN: A distributed architecture for supporting scalable persistence and interoperability in collaborative virtual environments.* [Online]. Available: http://www.evl.uic.edu/spiff/covr/cavernpapers/.

Lombard, M., & Ditton, T. (1997). At the heart of it all: The concept of presence. *Journal of Computer-Mediated Communication.*

Magdid, J., Matthews, D., & Jones, P. (1995). *The Web server book.* Chapel Hill, NC: Ventana Press.

Mann, S. (1997). Smart clothing: The wearable computer and WearCam. *Personal Technologies, 1* (1).

Melzer, J. E., & Moffit, W. (1996). *Head-mounted displays: Designing for the user (Optical and electro-optical engineering series).* New York: McGraw-Hill.

Meyer, K. (1995). Design of synthetic narratives and actors. In F. Biocca & M. Levy. (Eds.). *Communication in the age of virtual reality.* Hillsdale, NJ: Lawrence Erlbaum Associates.

Molitoris, J., & Taylor, T. (1997, December 3-6). *Advanced simulation, battle managers, and visualization.* Paper presented at the 1995 Winter Simulation Conference, Arlington, VA.

Murray, J. (1997). *Hamlet on the holodeck: The future of narrative in cyberspace.* New York: Free Press.

Nash, J. (1997, October). Wiring the jet set. *Wired, 5,* 129-135.

Parks, M., & Roberts, L. (1997, February). *Making MOOsic: The Development of personal relationships online and a comparison to their off-line counterparts.* Presented at the Western Speech Communication Association. Monterey, CA.

Pausch, R., Snoddy, J., Taylor, R., Watson, S., & Haseltine, E. (1996, August). Disney's *Aladdin*: First steps toward storytelling in virtual reality. *Computer Graphics.*

Jon Peddie Associates. (1998). *Latest study from JPA shows 3-D market is fastest-growing segment in semiconductor industry.* [Online]. Available: http://www.jpa.com/about/pr 3d97.html.

Pesce, M. (1995). *VRML: Browsing and building cyberspace.* New York: New Riders.

Petrie, C. J. (1996, December). Agent-based engineering, the Web, and intelligence. *IEEE Expert,* 24-29.

Reveaux, T. (March, 1995). *Aladdin* let VR out of the lamp. *VRWorld, 3* (2), 40-42.

Rheingold, H. (1991). *Virtual reality.* New York: Summit Books.

Rhodes, B. J. (1997). The wearable remembrance agent: A system for augmented memory. In *Proceedings of the First International Symposium on Wearable Computers.* Cambridge, MA.

Riva, G., Melis, L., & Bolzoni, M. (1997). Treating body-image disturbances. *Communications of the ACM, 40* (8), 69-72.

Roberts, B. (1995) Reality at last; Virtual reality technology. *PC Week, 12* (45), E9.

Rockwell, R. (1997, March). An infrastructure for social software. *IEEE Spectrum,* 26-31.

Rolfe, J., & Staples, K. (1986). *Flight simulation.* Cambridge, MA: Cambridge University Press.

Shapiro, M. A., & Lang, A. (1991). Making television reality: Unconscious processes in the construction of social reality. *Communication Research, 18* (5), 685-705.

Shapiro, M. A., & McDonald, D. G. (1995). I'm not a real doctor, but I play one in virtual reality: Implications of virtual reality for judgments about reality. In F. Biocca & M. Levy. (Eds.). *Communication in the age of virtual reality.* Hillsdale, NJ: Lawrence Erlbaum Associates.

Slater, M. (1998). *Announcement of a workshop on presence in shared virtual environments.* [Online]. Available: http://vb.labs.bt.com/SharedSpaces/Presence/.

Spitzer, M. B., Rensing, N. M., McClelland, R., & Aquilino, P. (1997). Eyeglass-based systems for wearable computing. In *Proceedings of the First International Symposium on Wearable Computers.* Cambridge, MA.

Starner, T. (1996). Human powered wearable computing. *IBM Systems Journal, 35,* 3-4.

Starner, T., Mann, S., Rhodes, B., Levine, J., Healey, J., Kirsh, D., Picard, R., & Pentland, A. (1997). Augmented reality through wearable computing. *Presence, 6* (4).

Steuer, J. (1994). Defining virtual reality: Dimensions determining telepresence. In F. Biocca & M. Levy (Eds.). *Communication in the age of virtual reality.* Hillsdale, NJ: Lawrence Erlbaum Associates.

Sutherland, I. (1965). The ultimate display. *Proceedings of the International Federation of Information Processing Congress, 2,* 506-508.

Sutherland, I. (1968). A head-mounted three dimensional display. *FJCC, 33,* 757-764.

Turkle, S. (1997). *Life on the screen: Identity in the age of the Internet.* New York: Touchstone.

Vilhjálmsson, H. H. (1996, May). *Avatar interaction.* [Online]. Available: http://hannes.www.media.mit.edu/people/hannes/project/index.html.

Waters, C., & Barrus, J. (1997, March). The rise of shared virtual environments. *IEEE Spectrum,* 20-25.

Zyda M., & Sheehan J. (1997). *Modeling and simulation: Linking entertainment & defense. ET Research Group Internet Home Page.* [Online]. Available: http://www.npsnet.cs.nps.navy.mil/npsnet.

15

The Digital Revolution in Home Video

Bruce Klopfenstein, Ph.D.*

This chapter reviews perhaps the most significant new and existing home video technologies: the videocassette recorder (VCR), camcorder, laserdisc, and DVD (formerly digital video or versatile disc). As is the case with broadcasting, home video technologies are in a transitional stage from the past's analog technologies to the very near future's digital technologies. Home video also has an established history of market success (VCR), market failure (RCA CED videodisc and Sony Betamax), and limited adoption (laserdisc) (Klopfenstein, 1989a). The camcorder is only now beginning to become another household fixture, especially in households with children—41% of 1997 households with children owned a camcorder versus 19% of households without children (Statistical Research, 1997).

Background

Videotape technology was originally developed for the broadcast industry in the 1950s (Klopfenstein 1985, 1989a). By 1970, Sony had developed the first compact videotape recorder, the U-Matic 3/4-inch VCR, which used an easily-inserted, book-sized cassette. Too expensive for consumers, it found a home as a video-training device in schools and businesses. Sony refined the technology and introduced its famous half-inch Sony Betamax in 1975 for $2,295, with the price dropping to $1,300 in 1976.

* Associate Professor of Telecommunications, Bowling Green State University (Bowling Green, Ohio).

A rival Japanese manufacturer, Japan Victor Corporation (JVC), developed its own incompatible half-inch VCR, the VHS (video home system). JVC designed the VHS to work at a slower speed than Beta, made its cassette larger to hold more tape, and made the tape thinner. Although VHS produces a lower-quality picture than Beta, VHS could record for longer periods. RCA obtained a license to market the VHS and introduced its VCR in August 1977 at a price $300 less than the Betamax. Within two years, VHS controlled 57% of the U.S. market. VCR sales rose from 400,000 in 1978 to four million in 1983 and nearly 12 million in 1985, with VHS accounting for 90% of the market. The future of Beta was sealed when several electronics manufacturers that had produced Beta models switched to VHS in 1986, leaving Sony virtually on its own.

In the early 1980s, the VCR was an expensive item that seemed destined for elite households (Klopfenstein, 1989b). A less expensive alternative that could exceed the video quality (but could not record) was the videodisc player (VDP). One VDP was developed by RCA, which needed a new product to follow the success of color TV in the market (Graham, 1986). Another was the "Laservision" format developed by MCA (in partnership with Philips) as an outlet for its movie inventory. The Laservision videodisc could hold up to 54,000 separate video images, or up to one hour of moving pictures, on one side. Philips's Magnavox rushed its $695 Magnavision to market in Atlanta in late 1978. Two hundred disc titles, mostly old movies, were available. Technical problems cropped up with both the hardware and the discs (as many as 90% were defective), and the player price was soon raised to $775.

RCA introduced its $500 Selectavision VDP in March 1981. Blitzed by the more expensive but versatile VCR, RCA announced in 1984 that it was abandoning its VDP at a loss of $580 million. About 550,000 players had been sold. Far fewer laser VDPs had been sold to consumers. Laser advocates announced in late 1984 that they too would retreat from the home market. A third VDP format, Matsushita's VHD, found some success in Japan but none in the United States.

While about 40% of U.S. households had a VCR by the end of 1986, most experts had predicted only five years earlier that the VDP would be more popular than the VCR. What many experts did not appreciate was the usefulness of a player that could also record programs from television. While many lamented the lack of a standard, the competition between the Beta and VHS formats actually led to lower prices and additional features. The VDP with limited software was no match for the VCR, which, ironically, also had more prerecorded software than both VDP formats (Klopfenstein, 1985).

Recent Developments

VCR Use

Current VCR user research is surprisingly difficult to locate in the public domain. One recent study showed that, as of 1997, about 89% of TV homes also owned a VCR, and penetration is projected to rise to 93% by 2000. VCR ownership is particularly high among the most affluent and best-educated households. And with the reduction of VCR prices, even the lowest-income homes are beginning to acquire their first machines (Everything about, 1998).

Average VCR use totals about six hours per week. This is divided between 1.5 hours of recording programming and 3.5 to 4 hours for playing both home-recorded and rented or purchased tapes. One-third of program recordings are never played back. Over half of taping (55% to 60%) is done while the TV set is off, 25% with the set tuned to the same channel being taped, and 15% to a different channel. Most recording occurs in prime-time (30%), the weekday daytime hours (30% to 35%), and fringe hours (13%). About 60% of recordings are of shows aired by the big three networks or their affiliates. Serial dramas are the most commonly-taped genre, accounting for half of all shows taped (Everything about, 1998).

While the VCR had the potential to be disruptive to the U.S. system of commercial broadcasting because users could easily eliminate commercials on playback, most VCR use remains playback of prerecorded tapes (e.g., movies). On any given night, only between 1% and 5% of the Nielsen television program ratings pie is made up of those taping shows. The 1.3 million viewers who taped the NBC hit program *ER* in 1995 made up about 3% of the 40 million who viewed it (Bash, 1995). This is quite instructive because it tells us something about the media consumer's interest in controlling his or her media environment and serves as a wake-up call to those who continue to predict great things for new media services into the home.

Video Rentals

Early studies showed that the primary use of VCRs was to record programs off the air for later viewing, a practice known as time shifting, but the VCR's biggest impact was upon the U.S. film industry. In 1985, 100 million people went to the theater to see a movie; about 103 million movie cassettes were rented each month. Video rental revenues in 1995 reached about $8 billion with cassette sales climbing over $7 billion (Rental stores, 1995). Home video revenues appear to have reached a high plateau (King, 1996), and the enormous number of tapes now available would seem to assure the VCR some life despite the encroachment of DVD.

The home video market witnessed a relatively bad year in 1997. A widely-documented downturn in the rental business sent video retail stocks plummeting (see Table 15.1), and played a role in retailers' aggressive lobbying for longer windows of video exclusivity before titles are released to pay-per-view (retailers want 60 days versus the actual average of 38 days). Despite differing views among researchers, a wide cross-section of Hollywood studio executives and retailers believed that consumer video rental activity declined significantly in 1997. The Video Software Dealers Association's VidTrac program showed rental revenues running 7% behind 1996, while Warner Home Video expected results to show a 3% to 5% drop for the year. Other measurements contradicted this, claiming the market actually grew as much as 5% (A year-long battle, 1998).

Table 15.1

Home Video Leading-Chain Stock Price
Declines

Chain	Share Price*	
	Dec. 1996	Dec. 1997
Hollywood Entertainment	$19.63	$10.25
Movies	$6.00	$1.13
Movie Gallery	$15.00	$3.13
Viacom (Blockbuster)	$37.13	$38.75
Video Update	$4.19	$2.38
West Coast Entertainment	$10.38	$1.50

* Monthly high.

Source: *Video Business* (1998)

The most often cited explanation for the decline was a shortage of box-office block-busters. Prominent Wall Street investment analyst Tom Wolzien believes as many as five million rentals a month "evaporate" from the market because more households are watching movies on cable PPV and digital satellite systems. Hollywood studios and video retailers reacted by working together to offer consumers more in-store copies of new releases (A year-long battle, 1998).

Two other studies confirm significant sales declines for 1997 as well as the first quarter of 1998. Alexander & Associates' Video Flash, a weekly phone survey of 1,000 consumers, estimates that for the first 12 weeks of 1998, consumers spent $2.2 billion buying videos, down 3.1% from the amount they spent in the comparable period of 1997. For 1997, Video Flash puts consumer spending at $9.3 billion, down about 10% from $10.38 billion in 1996. VideoScan, which tracks point-of-sale data from 16,000 retail stores around the country that account for about 70% of all video sales, says purchases for the first quarter of 1998 were down 5% from 1997. Annual sales for 1997 fell by a similar percentage from 1996 figures (Arnold, 1998). According to VideoScan, the top seller in 1997 was *Bambi*, with *Ransom* taking the top spot in 1997 rentals (A year-long battle, 1998).

Table 15.2

1997 Top Home Videocassette Rental and
Sale Titles

Top Video *Rentals*		Top Video *Sales*	
Title	Studio	Title	Studio
1. Ransom	Buena Vista	1. Bambi	Buena Vista
2. Scream	Dimension	2. Space Jam	Warner
3. Phenomenon	Buena Vista	3. 101 Dalmatians	Buena Vista
4. Jerry Maguire	Columbia TriStar	4. Hunchback of Notre Dame	Buena Vista
5. Liar Liar	Universal	5. Sleeping Beauty	Buena Vista
6. First Wives Club	Paramount	6. Jerry Maguire	Columbia TriStar
7. Michael	Warner	7. Men in Black	Columbia TriStar
8. Jungle 2 Jungle	Buena Vista	8. The Lost World	Universal
9. Absolute Power	Warner	9. Beauty and the Beast: Enchanted Christmas	Buena Vista
10. A Time to Kill	Warner	10. Star Wars Trilogy: Special Edition	20th Century Fox

Source: *Video Business* (1998)

Consumers purchased nearly 700 million videocassettes in 1995, an 18% increase from 1994 according to one market research study (King, 1996). Research data from 1993 cited by *American Demographics* showed that families with children under age 6 spent the most on videotapes. These households spent more than $55 in 1993 on blank tapes to capture baby's first smiles and prerecorded tapes to entertain the child. As infancy wanes, purchases taper off. Families with children aged 6 to 17 spent $43 on videotapes in 1993, and those with only teens and no younger children spent $41 (Mogelonsky, 1995). Walt Disney Home Video's *The Lion King* shattered records with 20 million copies sold in its first week of video release in 1995 (Snow, 1996).

According to a number of research sources, video stores supply about 54% of movie studios' revenues by paying more than $60 per video. The store sees a profit after renting the tape 25 times, which usually happens as demand remains high up to six weeks after a hit is released. About 8% of store revenue comes from fees charged for late tapes. In 1995, Media Group Research estimated total video revenue would grow about 8% per year over the next few years, with rentals rising about 3% per year and cassette sales leaping 15% per year (Rental stores, 1995). That prediction looked shortsighted in 1997.

Although viewing movies at home has become a way of life, the movie theater still has some advantages over home tape viewing:

(1)   Going to the movies is a social occasion.

(2)   Movies appear in theaters before they appear on cassette.

(3)   Theaters offer the large, wide screen.

As noted at the conclusion of this chapter, home theater technology threatens to erode the non-social reasons for going to the movie theater.

Digital VCRs

A digital VCR standards group representing about 50 companies announced it had agreed on technical specifications for recording transmission signals from the U.S. high-definition TV (HDTV) system. JVC jumped ahead of digital VCR competitors by rolling out its own digital VHS (D-VHS) format in 1994. D-VHS players will initially work with Thomson Consumer Electronic's digital satellite system (DSS) set-top boxes, which are deployed as part of Hughes' DirecTV digital broadcasting satellite (DBS) system. VCR manufacturers believe the VCR is on track to survive its 25th anniversary and thrive beyond the millennium, if only due to the lack of consensus on a rewritable DVD format as a replacement for analog tape. The prospects for advanced digital VCR formats such as D-VHS or W-VHS (a high-definition format) are negligible in 1998, but they may be introduced along with a new M-DVD format (VCRs proliferate, 1998). In January 1998, JVC demonstrated a prototype D-VHS deck operating in the high-definition mode needed for terrestrial DTV (digital television), but offered no timetable for announcing a standard. At the same exhibition, Hitachi touted the merits of its current D-VHS model, co-developed with Thomson, for recording high-definition DSS signals (Novel VCRs, 1998).

The JVC and digital VCR consortium formats are incompatible in many respects, including tape size; JVC is seeking to leverage the existing worldwide VHS standard it created for the analog world. Other manufacturers are expected to emerge as suppliers of D-VHS players. JVC's digital VCRs use "bitstream recording" involving error correction and digital signal processing. The D-VHS players rely on accompanying hardware such as set-top boxes containing MPEG decompression and integrated circuits for playback. As a result, JVC has said its new digital VCRs may be priced only about $350 more than its analog machines (Krause, 1995a).

While we wait for pure digital VCRs to arrive on the scene, a recent breakthrough will allow VCR technology some new life. New VCRs with 19-micron heads offer nearly the same picture quality when recording at slow speeds as at fast speeds. In addition to precision tracking, the high-performance materials used in these heads will help triple tape economy, while ensuring maximum picture quality. (For more on VCR recording technology, see Philips Consumer Electronics Company, 1996.)

Smart VCRs are expected to be on the market in 1998. These VCRs will automate many aspects of installation including setting the clock and locating the available channels, and they will even accept voice commands (an interface likely to begin showing up almost as an afterthought on many home appliances in the next few years). Not only are VCRs available that skip commercials in recording and/or playback, Thomson markets one that also fast forwards past the prerecorded promotional material at the beginning of most feature films on cassette (Cole, 1998). Sanyo's has a feature called Speed Watch that allows users to watch a tape at two times the normal tape speed while the audio remains at normal levels. It seems likely that only the huge installed base of videotapes will assure some future for the VHS VCR, but its fate is tied to the success of recordable disk technology (VCR near end, 1998).

Camcorders

The new digital camcorders rival professional broadcast equipment. For as little as $1,000, a digital camcorder using the standard digital video format is now available, and it will allow still-frame digital photography as well. At the Winter CES (Consumer Electronics Show), for example, Panasonic debuted two VHS-C camcorders with digital photography capabilities: the PV-L858 ($999.95) and PV-L958 ($1,099.95). Both models can transfer still images to a computer through an RS-232C connection. Both also feature flip-out 3.2-inch LCD monitors and "super-stretch" time-lapse recording for up to six hours on a 40-minute compact VHS tape (Director, 1998).

Entry-level camcorders are loaded with new features. These home movie machines have shrunk to the point that the smallest of them is hardly bigger than a deck of cards. Picture and audio quality on today's camcorders vary quite widely, although even an inexpensive unit will produce more-than-acceptable video and sound under normal shooting circumstances.

Camcorders, of course, come in several incompatible formats:

- Full-size, inexpensive VHS, although these are becoming obsolete.

- VHS-C—Compact tapes that, with an adapter, play back on a VHS VCR.

- 8mm.

- The new digital video format.

There are also high-resolution versions of the VHS and 8mm formats, called Super-VHS (or S-VHS), S-VHS-C, and Hi8mm. Digital gives the best picture quality, and the price is beginning to be competitive with other camcorders. Perhaps as a sign of things to come, digital "tapeless" camcorders are now available for professional users.

Each format has its advantages and disadvantages. Full-size VHS camcorders are the most stable without technological intervention (small camcorders can have "image stabilization" circuitry built in), and the tapes are completely compatible with any VHS VCR. They're also much bigger and bulkier than the others. VHS-C tapes only record up to half an hour on the fast speed and have an adapter to play back the tapes in your VHS VCR.

About the size of an audiocassette, 8mm tapes can record for two hours at standard speeds. When you play back your tape, you have to use the camcorder, plugging it into your TV, VCR, or receiver, adding wear-and-tear on the camcorder. VHS and 8mm (high and "regular" resolution) have comparable quality to each other. Sound quality can vary widely, but if you get a hi-fi camcorder, your audio will be very good. All camcorders have a microphone built into the unit, but a remote microphone can help eliminate extraneous noises in recording.

The elite trade publication *Television Digest* reviewed the state of camcorders in late 1997 (Camcorder brand-share, 1997). Digital video camcorders had no significant impact in the overall U.S. camcorder market in 1997. This compares with a 55% share for digital video camcorders in Japan, which is often a harbinger of what is to come in the area of consumer electronics in the United States. The VHS-C format still undercut all 8mm formats, and less expensive VHS camcorders retain about 15% share of the market. The price differential between digital video and high-end Hi8 camcorders is negligible in Japan, where *Television Digest* says status-conscious consumers and early adopting buyers play a more crucial sales role than in the United States. Copyright protection is also less of an issue in Japan than it is in the United States, and Japanese model digital video camcorders have a pure digital-dubbing link to PCs and some VCRs.

Personal Computing and Home Video

The personal computer (PC) may be about to become an important element in home video. At earlier points in its history, the PC had no more to do with mediated communication than a typewriter or photocopier; now, newer multimedia PCs are at the forefront of a possible computer/television merger. Continued advances in processing speed and personal mass storage bode well for computer applications in media including video. About 45% of U.S. households had computers at the start of 1998 (PCs in over 45%, 1998; Lanctot, 1998), and various manufacturers offer combination TV/PCs intended for the home's current television viewing room. In terms of new technology, the trend of the Internet and World Wide Web is clearly toward multimedia applications (audio, video, and animation). The day is rapidly approaching when we may be able to access hundreds, even thousands of video titles through video servers made available from Hollywood studios, telephone companies, and cable TV companies (Klopfenstein, 1997).

Analog Laserdisc

The venerable laserdisc is a technology that, despite many impressive technical achievements, has never really caught on. It found its place in educational media centers and was used in intensive training applications in both government and industry. Had it found success in the consumer market, less-expensive and fuller-featured machines would have evolved more quickly. Given the introduction of the DVD, the days of the analog laserdisc appear to be numbered, and the industry has announced it will no longer track aggregate laserdisc player sales. The compact disc showed that a well-entrenched technology (LP records) could be replaced in a surprisingly short time.

DVD

In 1995, consumer electronics manufacturers including old VCR rivals Sony and Matsushita bombarded the media with descriptions of their next-generation recording medium, DVD (formerly "digital videodisc" and "digital versatile disc"). They touted DVD and its players as digital replacements for VCRs and VHS tapes, laserdiscs, video game cartridges, and compact discs (CDs), both audio CDs and computer CD-ROMs. As they did with videocassette tapes, DVD vendors proposed different standards: it was Sony's Multimedia Compact Disc (MMCD) versus Matsushita's DVD called Super Density DVD (SD-DVD) (D'Amico, 1995). A format war was avoided when industry players agreed to support a format that combined a Toshiba design with a Sony/Philips (the original CD partners) encoding scheme (Braham, 1996).

DVD is a variation on the now-ubiquitous compact disc. Introduced in March 1997, DVD can be used to store movies, music, video games, and multimedia computer applications. This new disc is identical in shape and size to the CD, but DVD has a great deal more storage capacity. The 1998 vintage DVDs can hold 133 minutes or 4.7 gigabytes of video per side, and, unlike CDs, both sides of DVDs can be used to store data. Furthermore, double-layered DVDs capable of holding 241 minutes or 8.5 GB of video per side are expected to be available within a year, with increased storage at least theoretically possible after that. These higher-capacity discs are more likely to be used in computers, however. For example, if stereo music is the stored information, a single DVD can hold the contents of more than a dozen CDs. The movie studios consider the DVD a major opportunity because it will allow them to sell many of their existing films all over again, just as music companies have done with the audio CD.

A key to the early success of the DVD will be the perceived difference in the sharpness of the user's television picture. The CD was clearly an improvement over easily-scratched LP records. The DVD creates images with smaller and more varied pixels (720 pixels per horizontal line versus the standard 240). This allows clearer shapes and forms that are more detailed than today's best VCRs can produce. If viewers can easily see the difference, this will bode well for DVD. Frank Vizard (1997) came to that conclusion in his review for *Popular Science*. Given the steady increase in television picture resolution, these differences will become more visible. An easily-overlooked problem, however, could be the durability of DVDs and how well they can stand up to the rigors of video rental. On the other hand, DVDs may allow longer archiving than is possible with videotape.

The industry trade publication *Computer Shopper* defined the various DVD formats as of spring 1998:

DVD-Video. The DVD Forum's DVD-Video standard—the basis for today's DVD movie discs—is being challenged by the controversial new Divx format, supported by retailer Circuit City and several leading movie studios. Promoted as a new way to rent movies and as a possible future mechanism for music and software distribution, Divx DVDs will cost about $5 and allow unlimited playback for 48 hours. Viewers would then discard the disc or use a modem-equipped Divx player to buy more time. Today's DVD-ROM drives and set-top players won't play Divx media.

DVD-ROM. Second-generation DVD-ROM drives have little trouble reading the various flavors of CDs, as well as DVD-Video and interactive DVD-ROM titles created exclusively for playback on a PC. But there's no guarantee that they'll be capable of reading any future DVD format that emerges.

DVD-Recordable (DVD-R). The DVD Forum's first DVD-R specification defines a write-once format storing 3.95 GB of data per side, but two rival proposals define 4.7 GB capacities. A universal standard isn't likely soon. Today's DVD-R drives cost well over $10,000, making them of interest only to DVD content creators.

DVD-RAM. Rewritable DVD is caught in a battle between the DVD Forum's announced 2.6 GB-per-side DVD-RAM specification and at least three competing proposed formats, including a 3 GB-per-side version, DVD+RW, backed by Sony and Philips. As of mid-1998, no resolution is in sight, and the Forum is working to upgrade to its existing specification. Current DVD-ROM drives probably won't be able to read at least a few of the proposed formats. A number of DVD-RAM units are shipping, but shifting standards might give these early models short life spans.

DVD-Audio. The DVD Forum recently announced a draft DVD-Audio specification defining discs that can hold up to 30 hours of six-channel sound. A final draft is expected in late 1998, but it could be derailed by a competing Sony/Philips proposal called Direct Stream Digital. Today's DVD-ROM equipment might lack the copy-protection circuitry needed to play DVD-Audio discs (State of the DVD union, 1998).

Table 15.3 summarizes DVD's key advantages and disadvantages. It is based on Johnson (1995), but remains valid as of this writing.

DVD offers other advantages, including new flexibility in home video use. During playback of movies, DVDs have the capability of providing a choice of viewing options. First, it displays the standard television picture in a 4:3 (width to height) ratio. In the 4:3 format, widescreen movies are "panned-and-scanned," where only part of the picture is selected in the same way most movies are displayed from broadcast and tape sources. By pressing a button on the player's remote control, users can switch to letterbox for viewing on a big screen TV or to provide high-resolution pictures on advanced widescreen (16:9) sets. DVD movie discs have the capability of presenting soundtracks in eight different languages and up to 32 distinct subtitles. Another feature announced by Toshiba is a built-in parental control system that allows selection of the ratings version to be viewed: PG, PG-13, R, or NC-17. The player then automatically shows a version of the movie edited to that ratings level by the producers of the film.

The potential of DVD includes a broad range of multimedia and computer applications. Because DVD consists of a suite of disc types, each with increasingly higher storage capacities, the format holds tremendous growth potential for data-intensive home and business applications.

Somerfield (1996) sees a number of obstacles for DVD. The short-term success or failure of the DVD rests on the availability of a large number of movie discs, until the time comes when recordable discs are available. The movie studios that control the production and sale of movies on video have their own agendas. First, movies released on video in this country may not yet have made it to theaters on other continents. The studios are concerned that a DVD released here could be sold abroad and hurt foreign theatrical sales. To prevent this, the studios want the discs encoded to prevent playback in certain regions of the world. This "regional coding" system has not yet been designed.

Table 15.3
DVD—Advantages and Disadvantages

Advantages	- It provides better video and audio quality. - It is backward compatible with existing CDs. - Movie discs will be more convenient, more durable, and potentially less expensive than cassette tapes. - Viewers will be able to cut directly to particular scenes with no need to rewind. - When used as a computer CD-ROM, the disc will have far more storage capacity. (The same is true for DVD audio.) - Formerly a liability, the cost differential between a plain DVD player and higher-end VCR is already negligible.
Disadvantages	- There is limited software available, and it will take years to approach the number of titles available on VHS. - The disc players do not record. This statement, taken with the previous one, make software availability a limiting factor in the adoption of DVD players as was the case with the laserdisc around 1980. - Experience has shown that CDs and CD-ROMs are only marginally durable enough to withstand the abuses of public library and commercial rentals. - Current CD-ROM drives will not play DVD discs. - DVD systems will eventually be challenged by delivery of movies on demand over cable or phone lines.

Source: Johnson (1995)

The studios also want copyright legislation similar to the Audio Home Recording Act that delayed the launch of the digital audiotape (DAT) format several years ago. That legislation took 18 months (from proposal to passage) to get through Congress. No such DVD legislation had been introduced as of mid-1998. To compound the issue, the MPEG-2 encoding that must be implemented to compress the movies to DVD is quite complex and can only be done by three facilities: Warner, Sony, and MCA.

Recent Developments

Two factors will pull DVD in different directions. First, if Disney and other studios continue to limit their releases on DVD, its value as a home video technology will be greatly diminished. On the other hand, as computer manufacturers begin to substitute DVD-ROM drives for CD-ROMs, the cost of DVD player manufacturing will go down. Incredible as it may sound, DVD-ROM player prices were expected to drop to $80 in 1998 (Kovar, 1998). It is not unreasonable to believe that, in the next five years, VCRs will remain the home video technology of choice, while a significant number of homes add a DVD player to their home entertainment systems.

DVD movie players and titles have been available in the United States since March 1997, while PCs equipped with DVD-ROM drives began shipping during the third quarter of 1997. The DVD-ROM market will enhance economies of scale in manufacturing and should lead to lower DVD player prices. The initial signals for DVD players in 1997 were mixed. While DVD player sales in the first year compared very well to those for CD players, the meaning of that comparison is limited by the lack of affordable CD players in that technology's first year.

Despite significant publicity surrounding the launch of DVD hardware and software, a survey from the Yankee Group research firm found consumer awareness of the new format (the first stage of adoption) surprisingly low. Only 28% of the more than 1,900 U.S. households surveyed by the Yankee Group were familiar with DVD. Among consumers who had heard about DVD, only 13% said they were very or somewhat likely to purchase a DVD player within the next 12 months (Yankee Group study, 1997).

According to *GameWeek* (McGowan, 1998), between 100,000 and 200,000 U.S. households had purchased DVD systems by the end of 1997. There were 350,000 DVD players shipped to retailers in 1997, according to the Consumer Electronics Manufacturers Association. In terms of software, in 1997, there were 1.5 million to 2 million DVDs sold in the United States at stores tracked by VideoScan, which provides home-video sales data for *Video Business* magazine and logs DVD sales at most major retail chains. Warner Home Video claimed that Warner alone shipped more than three million DVDs (worth $50.6 million wholesale) to retailers in 1997, including 92,000 copies of *Batman & Robin*. Most DVD-Video titles will play in DVD-ROM drives, more than 500 DVD-Video releases were available at the end of 1997, and the DVD Video Group predicts that total will rise to 1,500 by the close of 1998. The DVD system price point to reach mass consumer market acceptance is said to be $299 (Bismuth, 1998).

Cable and DBS are becoming far more serious competitors to home video than ever before. Much to the chagrin of the Video Software Dealers Association, Hollywood has been allowing shorter windows between the time a movie comes out on video and the time it's available on pay television. Hollywood stands to benefit if the CD audio library-rebuilding phenomenon is repeated in DVD. Just as CD adopters bought new copies of music they already owned in LP record format, so might VHS tape owners choose to buy better quality DVD copies of movies they already have on tape.

Factors to Watch

Although sales of DVD players in 1998 will pale compared with those for VCRs, we are likely to see the beginning of a product substitution of DVDs for VCRs. They cannot match the random access capability of the DVD, and more home video movie buffs will not miss having to rewind a tape. If the CD market is any indication, software manufacturers will keep DVD prices higher than those for prerecorded videocassettes. This may well be based on psychological rather than economic considerations. On the other hand, don't be fooled by ridiculous comparisons between 1997 to 1998 sales of DVD players and those of VCRs in the mid-1970s or CD players in the early 1980s (DVD posts, 1998). In real terms, the DVD player costs significantly less than 10% of what VCRs and CD players cost in their first year of introduction, so the perceived risk in adoption is dramatically less than what it was for its two predecessors. After videophiles purchase their DVDs, the next set of adopters will look at how many titles are available. With only a few hundred select titles available at any one store in 1998 and many blockbuster movies still only available on VHS, growth of the DVD player will remain restrained.

The Divx technology (see its proponent's home page at http://www.Divx.com/) may only serve to confuse the marketplace. In spring 1998, Circuit City said the Divx introduction will be delayed. Selected Circuit City markets will begin sales during mid-year, with a national roll-out predicted before the critical Christmas shopping season. The company expects to lose money on the format for two years (the consumer electronics retailing business operates on very thin profit margins, so this sounds very risky). The more promotion Circuit City gives to Divx, the higher consumer awareness of DVD itself will become.

This author is very skeptical about the prospects for Divx. Recent polls do not offer encouragement for Circuit City (Consumer poll, 1997). History has shown that announcements of participation by other market players (Hollywood studios, in this case) do not portend market success (Klopfenstein, 1985). Divx's prospects are further limited because Circuit City will have to depend on other retail competitors to carry its product.

DVD manufacturers will continue to add features, perfect the technology, and lower the player price. Recordable DVD machines are in development, although exact predictions of market introduction are all but impossible. A good rule of thumb is to ignore bold predictions about recordable DVD introduction dates and instead watch for the date when they are actually available. Until one standard recording technology is established, DVD recorders will not be an important factor. Indeed, if incompatible standards are sought, the life of the VHS VCR will be extended. The day is coming, however, when we will be transferring our home movies from analog tape to digital disc.

Four camps were promoting different rewritable DVD standards in late 1997:

- DVD-RAM, which is championed by Matsushita's Panasonic and approved by the industry's DVD Forum.

- DVD+RW, which is endorsed by Sony and Philips.

- The DVD-R/W format proposed by Pioneer.

- The DVD Multimedia Video File Format (MMVFF) proposed by NEC.

A major concern of three of the four competing camps is maintaining compatibility with some existing specific CD or DVD formats, at least in the first generation of rewritable DVD systems. The DVD-RAM format by Panasonic will play CD-PD and most other CD format discs. Sony and Philips say they see DVD+RW as a "natural extension" of the CD-RW format, and in Pioneer's DVD-RW system, compatibility with DVD-R is considered critical to its niche of potential customers. The first DVD+RW drives were scheduled for sample shipments in mid-1998, with retail shipments planned for the third quarter. The specifications call for a 3 GB per side disc that does not require a caddy. It will play most CD formats, including DVD-Movie, DVD-ROM, DVD-R, CD-ROM, CD-R, CD-RW, and CD Audio.

The DVD-RAM format that is supported by Panasonic, Hitachi, and Toshiba, and has the backing of eight of the 10 original DVD Forum members, features single- and dual-sided discs with 2.6 GB per side, both of which require a caddy. The format's strength is said to be its random access characteristics and backward compatibility with discs using most of today's CD and DVD formats. Panasonic, Toshiba, and Hitachi planned to deliver drives in 1998 at a price of about $799. Blank DVD-RAM media is sold in single-sided ($24.95) and dual-sided ($39.95) configurations. Both will require a caddy, although caddies for single-sided discs are removable, allowing the bare disc to be played in future versions of DVD-ROM and DVD-Video players. DVD-RAM promoters see their format first as a data storage device. Panasonic has also made it clear that the specifications will serve as a platform for audio and video recording systems coming in the next five years when:

- Greater storage capacities can be realized from new technologies, such as blue laser.

- MPEG-2 encoded chips are more economically manufactured (Tarr, 1997).

Pioneer released DVD-R drives in October 1997 for $17,000. This price could drop within a few years to less than $5,000. The initial price for blank DVD-Rs is $50. DVD-RAM drives will be introduced for less than $1,000, with blank discs at about $30 for single-sided and $45 for double-sided discs. Disc prices for both DVD-R and DVD-RAM will drop quickly, but DVD-R discs will probably be cheaper in the long run. Toshiba, Pioneer, and Hitachi expect DVD-RAM to be available in early 1998, which means it will probably appear in the middle of 1998 (DVD FAQ, 1998).

Finally, another standards battle to watch is that which is going on with DVD-Audio. This format would seem to elicit less interest than DVD-Video because music copyright holders have little to gain with DVD. Copy protection is a key concern, and few audio applications require the hours of recording capacity that DVD audio will allow. Indeed, the recording industry often sells multiple CD sets rather than fill one CD to capacity, presumably due to marketing rather than technical considerations.

The Technological Future of Mediated Communication

Speculating about the future of new media technologies is an interesting task. William R. Holm was president of the Society of Motion Picture and Television Engineers (SMPTE) at a time when speculation was rampant that "video cartridges" were on the verge of revolutionizing television. Videodiscs, videotape, and devices that played film cartridges on the TV set were all reportedly nearly ready to be unleashed on the consumer. Holm stepped back from the situation and took a long look. He made some simple observations in an address to a 1971 SMPTE conference in Montreal that are still appropriate today.

Holm defined technology as "science applied to the problems of society" (1971, p. 7). Drawing from this definition, Holm concluded, in part, that technology is not useful merely because it exists: It must satisfy needs so that people will buy which, in turn, may give birth to future technologies. It is reasonable to suggest that many new technologies will fail: The lessons from history have shown this. Look to history to try and understand where new media may be going. Only when new media provide potential adopters with a service that fills a need at a reasonable cost will they have a chance to be successful. When all is said and done, it is the technological interface between the user and the software that must become transparent.

Bibliography

A year-long battle to re-energize. (1998, January 5). *Video Business, 18* (1), 1, 7.

Bash, A. (1995, November 28). VCRs can't rescue faltering shows. *USA Today.*

Bismuth, A. (1998, March 30). PCs, players seek one DVD solution. *EETimes.* [Online]. Available: http://www.techweb.com/search/search.html.

Braham, R. (1996, January). Consumer electronics. *IEEE Spectrum, 33* (1), 46-50.

Camcorder brand-share dominance continues. (1997, December 29). *Television Digest with Consumer Electronics, 16* (33).

Cohen, J. (1991, December). Making sense of new electronics products. *Consumer's Research, 74,* 34.

Cole, G. (1998, February 5). VCRs get smart (about time). *Electronic Telegraph (London Daily Telegraph).* [Online]. Available: http://www.telegraph.co.uk:80/et?ac=000647321007942&rtmo=flfs-vDVs&atmo=flfsvDVs&pg=/et/98/2/5/ecvid05.html.

Consumer poll gives thumbs down to Divx. (1997, September 26). *TWICE (This Week in Consumer Electronics).* [Online]. Available: http://www.twice.com/domains/cahners/twice/archives/webpage_1084.htm.

D'Amico, M. (1995, June 5). Digital videodiscs. *Digital Media, 5,* 13.

Director, K. M. (1998, March). Zoom in, zoom out. *Videomaker.* [Online]. Available: http://www.videomaker.com/edit/mag/mar98/D4azoom.html.

DVD frequently asked questions. (1998, April 7). [Online]. Available: http://www.videodiscovery.com/vdyweb/dvd/dvdfaq.html.

DVD posts impressive gains during first year on market, far outpacing early sales of VCRs and CD players. (1998, April 3). *Business Wire.* [Online]. Available: http://www.digitaltheater.com/news/apr3.html.

Everything about television is more. (1998, March 6). *Research Alert, 16* (5), 1.

Graham, M. (1986) *RCA and the videodisc player: The business of research.* New York: Cambridge University Press.

Holm, W. R. (1971). *Socio-economic aspects of videoplayer systems: A perspective. Video cartridge, cassette, and disc player systems.* New York: Society of Motion Picture and Television Engineers.

Johnson, G. (1995, September 15). Agreement reached on a new format for video; Technology: Toshiba and Time Warner end battle with Sony and Philips on disc for use in devices to replace VCRs. *Los Angeles Times*, D1.

King, S. (1996, April 19). Home video rentals drop but revenues are still up. *Los Angeles Times*, 26.

Klopfenstein, B. C. (1985). *Forecasting the market for home video players: A retrospective analysis.* Unpublished doctoral dissertation. Columbus, OH: Ohio State University.

Klopfenstein, B. C. (1987). New technology and the future of the media. In A. Wells (Ed.)., *Mass media and society.* Lexington, MA: Lexington Books.

Klopfenstein, B. C. (1989a). The diffusion of the VCR in the United States. In M. Levy (Ed.). *The VCR age.* Newbury Park, CA: Sage Publications.

Klopfenstein, B. C. (1989b). Forecasting consumer adoption of information technology and services—Lessons from home video forecasting. *Journal of the American Society for Information Science, 40* (1), 17-26.

Klopfenstein, B. C. (1997). The future of new media technologies. In A. Wells & E. Haakanen (Eds.). *Mass media and society.* Greenwich, CT: Ablex.

Klopfenstein, B. C., & Sedman, D. (1990). Technical standards and the marketplace: The case of AM stereo. *Journal of Broadcasting & Electronic Media, 34* (2), 171-194.

Kovar, J. F. (1998, March 23). ATI gives VARs the option of DVD-ROM software. *Computer Reseller News.* [Online]. Available: http://www.techweb.com/se/directlink.cgi?CRN19980323S0139.

Krause, R. (1995a, April 17). Digital VCR specs diverging. *Electronic News, 41*, 20.

Lanctot, R. C. (1998, March 23). Household PC penetration disputed at 45%. *Computer Retail Week.* [Online]. Available: http://www.techweb.com/se/directlink.cgi?CRW19980323S0026

McGowan, C. (1998). *350,000 DVD-video players shipped in 1997.* [Online]. Available: http://www.gameweek.com/news/1_21/news1.htm.

Mogelonsky, M. (1995, December). Video verite. *American Demographics, 17*, 10.

Novel VCRs and camcorders make CES debut. (1998, January 19). *Consumer Electronics, 16* (36).

PCs in over 45% of U.S. homes in 1997. (1998, March 10). *Reuters news service.* [Online]. Available: http://my.excite.com/news/r/980310/06/business-pc.

Philips Consumer Electronics Company. (1996). *The videocassette recorder (VCR).* [Online]. Available: http://www.magnavox.com/electreference/videohandbook/vcrs.html.

Pietrucha, B. (1996, April 1). Video industries seek copyright, digital video legislation. *Newsbytes News Network.*

Rental stores keep packing in the crowds. (1995, October 23). *USA Today.*

Schoenherr, S. (1996). *Recording technology history: A chronology with pictures and links.* [Online]. Available: http://ac.acusd.edu/History/recording/notes.html.

Snow, S. (1996, March 7). Morning report; TV & video. *Los Angeles Times*, F2.

Somerfield, H. (1996, March 25). *Trouble in DVD paradise: Problems may delay digital videodisc launch.* [Online]. Available: http://e-town.myriadagency.com//html/news_html/articles/9613hsc.html.

Spiwak, M. (1995, December). Gizmo's holiday gift guide. *Popular Electronics, 12*, 6.

State of the DVD union. (1998, May). *Computer Shopper.* [Online]. Available: http://www.zdnet.com:80/cshopper/content/9805/297878.html.

Statistical Research, Inc. (1997). *1997 TV ownership survey.* [Online]. Available: http://www.sriresearch.com/pr970616.htm.

Tarr, G. (1997, December 8). *Recordable DVD erupts: What if they gave a format war and nobody came?* [Online]. Available: http://www.e-town.com/news/articles/dvd120897gtt.html.

VCR near end of long and rewinding road. (1998, January 10). *Atlanta Journal-Constitution.* [Online]. Available: http://www.accessatlanta.com/business/news/1998/01/10/ces2.html.

VCRs proliferate, as do their brands. (1998, January 5). *Video Week, 16* (33).

Vizard, F. (1997, August). DVD delivers. *Popular Science, 251* (2), 8-72.

Yankee Group study finds low awareness for DVD. (1997, December 15). [Online]. Available: http://www.yankeegroup.com/press_releases/lowAwareDVD.html.

16

Digital Audio

John Roussell, Ph.D.*

Audio is one of the fastest-growing areas of communication technology, with applications ranging far beyond home entertainment. As computer technology continues to find its way into consumer entertainment products, more demands are being placed on software and hardware that create, distribute, and store better sound.

In the near future, portable digital radio will be commonplace. The ability to send digital audio along with other information will transform listening into more of a multimedia experience. Listeners can receive a playlist display, updated sports scores, or weather information—all while enjoying their favorite station. Developments in high-definition television (HDTV) and digital versatile disc (DVD) will provide better quality pictures and audio, and video game makers have come a long way from the *Mario Brothers'* sound theme to include full three-dimensional surround-sound, thus augmenting the total gaming experience. In addition, Internet Web pages are incorporating more advanced sound to provide a richer multimedia experience that includes Internet radio stations, music on demand, samplings of the latest CDs, and even entire concerts. As the digital revolution continues to reshape the way we communicate, advances in audio storage, transmission, and playback capabilities will reshape what and how we hear.

Background

Evolution of Audio Media

When studying digital audio and its applications, it is important to understand how digital audio recording and playback is fundamentally different from analog technology.

* Assistant Professor, Department of Communication Design, California State University, Chico (Chico, California).

Before digital technology was invented, audio was recorded by capturing an analog signal and reproducing the sound wave generated by the original source. For example, the sound from an electronic signal produced from someone singing into a microphone would be electromagnetically stored on magnetic tape. Reproduction of the signal involves the recreation of the sound wave produced by the singer's voice. Thus, the term analog comes from the word analogous, which means resembling or comparable to the original.

Although the shapes of the sound waves are analogous, the reproduction is always slightly different from the original. Each copy of a copy is called a "generation." The more generations a copy is removed from the original, the more the signal deteriorates, and consequently the quality of the sound decreases (see Figure 16.1).

Figure 16.1
Analog Versus Digital Recording

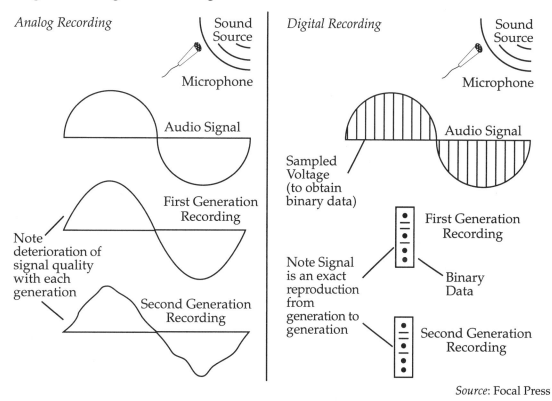

Source: Focal Press

In an analog-to-digital conversion, the same analog electronic signal from the person singing into the microphone is measured at certain intervals. A digital code is created and stored in sequence to become a digital representation (file) of that electronic signal. The number of times the signal is measured per second is referred to as the "sampling frequency." Upon playback, the digital file is fed into a digital-to-analog converter that reproduces the original analog signal.

In digital audio formats, multiple copies can be made without the deterioration of signal found in analog formats. The quality of sound at the final audio stage is dependent upon how many samples were taken during the original digital conversion. The more samples taken, the more accurate the representation of the electronic signal (i.e., the lower the distortion), but consequently, the larger the storage space needed for the digital file. In addition, the bit rate which is used to determine the accuracy of the sample also affects quality and file size. The audio on a typical compact disc is sampled at 44.1 kHz (or 44,100 times per second), using 16 bits of data for each channel. A stereo recording (two channels of audio) equals about 10 megabytes of data per minute.

In the age of digital communication and bandwidth limitations, a compromise between quality and portability has been at the heart of digital audio. This compromise has influenced decisions on both how digital audio will be used for traditional home audio entertainment and how standards for digital audio will be established for multimedia and Internet applications.

Digital Audio for Home Entertainment Systems

Compact Discs

Sony and Philips jointly introduced the compact disc (CD) in 1982. CDs were marketed as indestructible digital audio alternatives to the LP. Professionally, the CD player has replaced the turntable at radio stations (Gross & Reese, 1993). Although the claim of indestructibility has been refuted by millions of CD owners, the sound, ease-of-use, and random access capabilities have made CDs the overwhelming choice in home entertainment for prerecorded music.

The one disadvantage from the consumer's point of view is the inability to record onto the CD. Recent developments in the WORM (write once, read many) format have led to the release of CD recorders (CD-R). Other developments in digital audio recording have limited applications for the compact disc to what it does best—provide consumers with an affordable convenient source for playing prerecorded music (Reese & Gross, 1998). Although major home audio manufacturers have not given up on developing better CD-R systems, it is unclear what format consumers will ultimately embrace for their home digital audio recording needs.

Digital Audiotapes

The digital audiotape (DAT) format was first available in Japan in the late 1980s, but gained notice when Sony introduced it into the U.S. market in 1990. DAT's ability to record digital sound on a tape half the size of a standard audiocassette directly threatened compact disc sales by offering home digital recording. Music industry concerns led to the Audio Home Recording Act of 1992, which essentially placed a royalty on the sales of DAT tapes and players. The money generated from the royalties are divided between the artists (40%) and the record companies (60%) (Holland, 1995).

In addition, the 1992 Audio Home Recording Act required a record-limiting technology, called the "Serial Copy Management System" (SCMS), be built into all digital recorders. This prevents DAT machines from making copies of copies, essentially discouraging widespread bootlegging of digital audiotape recordings. DAT's primary uses now are:

- Backing up computer network data.

- Professional recording applications such as film and video production.

- High-end recording devices for concerts, such as Phish and The Grateful Dead, where capturing the event on tape is not only allowed but encouraged (Spielberg, 1997).

Current high-end home DAT recorders retail between $1,000 and $1,400 (Internet Consumer Electronic Guide, 1998).

Digital Compact Cassettes

In 1992, Phillips introduced the digital compact cassette (DCC) directly to the consumer market as a replacement for analog audiocassette recorders. The DCC never caught on, in part because of mixed and often contradictory reviews from both industry critics and consumers concerning sound quality. DCC was developed to be "backward compatible" with standard audiocassettes, so DCC players could play analog cassettes as well, even though analog cassette players could not play DCCs. A combination of poor marketing, lack of portability, and high prices limited the sales of both software and hardware, making the DCC an unprofitable venture. In 1995, prices for DCC units were slashed to half the price of a DAT machine in a final attempt to increase market share. In 1996, Philips finally gave up on manufacturing the recorder, but has promised to support current owners (Spielberg, 1997).

Minidiscs

One of the major reasons for the demise of the DCC was competition from Sony's digital portable minidisc (MD), which is a cross between a compact disc and a computer diskette. Sony introduced MD in November 1992. The technology shares certain storage and navigational features with the conventional compact disc player. Like a CD, the MD can store 74 minutes of music, and the user can randomly access any information on the MD within one second. The two main areas that set the MD apart from the CD are its ability to record, and its portability and ruggedness—it can be bounced around, and playback won't be affected. Both features make the MD a viable competitor to the popular CD. The 2.5-inch disc uses a digital compression technique known as Adaptive Transform Acoustic Coding (ATRAC), which allows the MD to hold the same amount of music as a CD even though it has only one-fifth of the storage space. There is a slight difference in sound quality between the MD and the CD because the ATRAC compression determines which sounds are inaudible and does not record them. This method can have a slightly noticeable effect on the perceived depth of the audio, but only with more revealing stereo playback systems (Spielberg, 1997).

Unlike the DCC, it appears that consumers are embracing the MD as a superior alternative to the audiocassette. MD may ultimately be the portable digital choice, not necessarily replacing the CD, but being a welcome addition to the consumer's home electronics audio system.

Recent Developments

Although audio home entertainment systems have continually improved, the recent coupling of digital audio with improvements in computers, Internet technologies, and video reproduction has created new competition for the next wave of audio entertainment. Computer-based audio has continually lagged behind home entertainment audio, as users have accepted commonly-available but inferior technologies for storing and generating sound for computer applications. Now, with the trend toward richer multimedia and entertainment applications for the computer, recent developments have been aimed at closing the gap between computer audio and the home entertainment system.

3-D Audio Technology

As innovations in video quality continue to develop, multimedia producers and developers have become more aware of the relationship between high-quality video and high-quality audio. Inadequate sound can ruin the experience for the multimedia participant, and, with each innovation, the consumer is less likely to accept a previous, inferior audio format. Thus, with the continued improvements in high-resolution digital video featured in the latest video games, DVD, and HDTV, the challenge is there for audio developers to keep pace and make comparable improvements in audio technology (QSound Labs, 1998).

In February 1998, the three leading manufacturers of high-quality audio technology—Seponix, Seiko/Nippon Precision, and QSound Labs—formed a partnership to develop the new QSound QS7777 chip. This chip will enable a listener to receive a full range of state-of-the-art 3-D sounds and correctly identify the position of the sound images as intended by the producer's original mix. As listeners watch the latest movie on DVD or play the latest video game, they will be able to experience the sounds as if in the center of the action. Visual and auditory orientation will compliment each other, rendering an almost life-like experience. The chip has been included in audio systems in Asia, Australia, and the Mid-East as of March 1998, and should be released in the United States and Europe in late 1998.

Digital Audio and the Internet

Since 1992, the number of Internet users has doubled each year. This has been accompanied by an increase in the number of Web pages and the development of more elaborate multimedia applications for those pages. These applications range from more graphic-intense, interactive components to real-time streaming audio- and video-enhanced Web pages. It also stands to reason that, with more people using the Web, there is greater demand for bandwidth, especially when sending and receiving large multi-media files. As a result, there has been a push for better compression and delivery of digital audio over the Net, as Internet radio stations and audio on demand strive to become an acceptable alternative to radio and CDs.

The ability to stream real-time audio has helped make sending and receiving audio one of the most popular uses of the Internet. Internet radio services distribute regular station broadcasts, as well as Internet stations to your PC. Ball games, concerts, and special events that may only be broadcast to a specific geographical audience can now be distributed around the world using the Internet. Other applications for sending digital audio over the Internet include national news, advertising, and public service announcements. For example, AP newsfeeds can be now be sent directly into the newsroom via the Internet (Medialink, 1996). National advertising spots and public service announcements can be stored in databases, where they can be accessed by subscribing clients, made ready for immediate airing, and easily stored for future use (Audioworld, 1997; Medialink, 1996).

Factors to Watch

The major factor to watch concerning digital audio is compression. As developments in digital audio applications for the Internet, full-digital radio, DVD technology, and HDTV progress, the sampling issues will remain critical. The higher the sampling rate, the better the quality—and the larger the file. The issue of delivering digital sound quality equal to the original listening experience can only be addressed with better compression. Improved audio compression capabilities also impact the latest video game applications, as players are demanding games that feature richer graphics and fuller sound. Finally, DVD and HDTV further underscore the importance of state-of-the-art audio to enhance the latest digital video developments.

Bibliography

Audioworld. (1997). [Online]. Available: http://www.audioworld.com/news_index.html.

Digital radio. (1998, February). [Online]. Available: http://www.magi.com/~moted/dr/.

Frequently asked questions about MPEG Audio Layer-3. (1998, March). [Online]. Available: http://www.iis.fhg.de/amm/.

Gross, L., & Reese, D. (1993). *Radio production worktext* (2nd Ed.). Boston: Focal Press.

Holland, B. (1995). Digital royalties for artists, labels static in 94. *Billboard, 107* (19), 10.

How to encode MP3s. (1998, March). [Online]. Available: http://www.iis.fhg.de/amm/techinf/coding/layer3/index.html.

Internet consumer electronic guide. (1998, March). [Online]. Available: http://www.ee.nus.sg/~zhu/consumer/consumer.html.

Introduction to digital audio. (1998, February). [Online]. Available: http://www.microsoft.com/hwdev/devdes/digitaudio.htm.

Krantz. (1996). Digital music, right off the Net. *Time, 148* (25), 76.

Luther, A. (1997). *Principles of digital audio and video.* Boston: Artech House.

Medialink. (1996, March). [Online]. Available: http://www.medialinkworldwide.com.

Merli, J. (1997). Local digital radio gets closer to reality. *Broadcasting & Cable, 127* (44), 46-48.

MIDI format. (1998). [Online]. Available: http://www.midifarm.com/info/fm.htm.

QSound Labs. (1998, March). [Online]. Available: http://www.qsound.ca/home0.htm.

Reese, D., & Gross, L. (1998). *Radio production worktext* (3rd Ed.). Boston: Focal Press

Rogers, D. (1998, March). *All about DCC.* [Online]. Available: http://www.lightlink.com/rogers/DCC-L/.

Rotenier, N. (1996). Radio days? *Forbes, 157* (9), 114.

Solari, S. (1997). *Digital and video audio compression.* New York: McGraw-Hill.

Spielberg, I. (1997). *Home theatre & stereo colossus.* Los Angeles: Vision Quest Works.

Stuart, J. (1998). Digital audio for the future. *Audio, 82* (4), 31-37.

Verna, P. (1996). Sony taps new digital stream. *Billboard, 108* (11), 59.

World Wide Web virtual library audio. (1998). [Online]. Available: http://www.comlab.ox.ac.uk/archive/audio.html.

ZAP format info. (1998). [Online]. Available: http://www.lysator.liu.se/~zap/tutorial/formats.html.

IV

TELEPHONY & SATELLITE TECHNOLOGIES

L ocal and long distance telephone revenues in the United States exceed those of all advertising media combined. Clearly, point-to-point transmission of voice, data, and video represents the single largest sector of the communications industry. The sheer size of this market has two effects: Companies in other areas of the media want a piece of the market, and telephone companies want to grow by entering other media.

Until 1996, federal regulations kept the telephone, cable television, and broadcast television industries separate from each other. The Telecommunications Act of 1996, however, changed the playing field, allowing levels of cross-ownership across these media that were unthinkable a few years ago in an effort to encourage competition and technical innovation.

These changes in the regulation and organizational structure of communication media have been forced by rapid advances in digital technology that are erasing the distinctions in the transmission process for video, audio, text, and data. Because all of these types of signals are transmitted using the same binary code, any transmission medium can be used for almost any kind of signal (provided the needed bandwidth is available). Furthermore, the advance of digital compression technologies reduces the bandwidth needed to transmit a variety of signals, further blurring the lines dividing communications media.

The immediate result of these technological innovations is more competition. The most dramatic site of new competition will be telephony, where consumers will soon have a choice of service providers, and service will consist of more than the transmission of voices. The chapters in this section explore different technologies used for point-to-point communication, thus far dominated by the telephone industry.

The first chapter in this section discusses the basics of today's telephone network in the United States, and the following chapter explores the most important technological innovations in the switched network used for the transmission of telephone signals. In addition to explaining how these technologies work and how much information they can transmit, this chapter discusses a variety of organizational, economic, and regulatory factors that will influence when and how each becomes part of the telephone network.

Satellites are a key component of almost every communication system. Chapter 19 explains the range of applications of satellite technology, including the history of the technology and the range of equipment needed (on the ground and in space) for satellite communication. Chapter 20 then explores one of the most important applications of early satellite technology—distance learning—which has since evolved to encompass virtually every communication medium.

The manner in which cable companies are attempting to enter the telephone service market in competition with the traditional telephone companies is explored in the discussion of cable telephony and data services in Chapter 21. The rapidly-evolving (and lucrative) cellular telephone industry is then reviewed in Chapter 22, along with explanations of the differences between traditional cellular telephony and newer incarnations such as personal communication services. A new chapter for this edition, Chapter 23, then explores the range of personal communication devices that are serving important niches in our daily lives.

The final chapter in this section offers glimpses of telephone technologies that add video to telephone service. The teleconferencing chapter discusses videoconferencing and videophone systems that are primarily designed to facilitate face-to-face communication over distances, and covers the rapid evolution of the group-based videoconferencing systems and the continued failure of one-to-one videophones.

In studying these chapters, you should pay attention to the compatibility of each technology with current telephone technologies. Technologies such as cellular telephone are fully compatible with the existing telephone network, so that a user can adopt the technology without worrying about how many other people are using the same technology. Other technologies, including the videophone and ISDN (Integrated Services Digital Network) are not as compatible. Consumers considering purchase of a videophone or ISDN service have to consider how many of the people with whom they communicate regularly have the same technology available. (Consider: If someone gave you a videophone, whom would you call?)

Markus (1987) refers to this problem as an issue of "critical mass." She indicates that adoption of interactive media, such as the telephone, fax, and videophone, is dependent upon the extent of adoption by others. As a result, interactive communication technologies that are not fully compatible with existing technologies are much more difficult to diffuse than other technologies. Markus indicates that early adoption is very slow, but once the number of adopters reaches a "critical mass" point, usage takes off, leading quickly to use by nearly every potential adopter. If a critical mass is not achieved, adoption of the technology will start to decline, and the technology will eventually die out.

One of the most important concepts to consider in reading this chapter (and the other chapters that include satellite technology) is the concept of "reinvention." This is the process by which users of a product or service develop a new application that was not originally intended by the creator of the product or service. Satellite technology is being reinvented almost daily as enterprising individuals devise new uses for these relay stations in the sky.

The final consideration in reading these chapters is the organizational infrastructure. Because of the potential risks and rewards, even the largest companies entering the market for new telephone services are hedging their bets with strategic partnerships and experimentation with multiple, competing technologies. In this manner, the investment needed (and thus the risk) is spread over a number of technologies and partners, with the knowledge that just one successful effort could pay back all the time and money invested.

Bibliography

Markus, M. L. (1987). Toward a "critical mass" theory of interactive media: Universal access, interdependence, and diffusion. *Communication Research, 14* (5), 491-511.

17

Local and Long Distance Telephony

David Atkin, Ph.D.*

Following the passage of the Telecommunications Act in 1996, many felt unbridled optimism concerning the prospects for competition in telephony (Telecom Act, 1996). However, 1998 may be remembered as the year we came to grips with the daunting industry consolidation wrought by that law. Such were the early returns on America's first major revision in telecommunications regulation in over 60 years, easily the most comprehensive experiment in media deregulation undertaken to date. Yet, as the act's second birthday approached, commentators expressed concern that genuine competition between wire providers remains a "long distance" off (Cauley, 1997).

That such media convergence is so difficult to achieve underscores its rapid departure from the era of plain old telephone service (POTS), already one of the most ubiquitous and lucrative communications technologies in the world. In the United States, the $250 billion in gross revenues from local and long distance companies exceeds the combined revenues of all advertising media, even surpassing the gross national product (GNP) of most nations. The profit potential of these two markets is the primary force behind the revolution in telephony, as other media companies seek to enter the lucrative telephone market, and telephone companies seek to enter other media. Using history as a guide, this chapter will outline the influence of changing regulations on the conduct of telephone companies, including implications for cross-media competition.

*Associate Professor and Assistant Chair, Department of Communication, Cleveland State University (Cleveland, Ohio).

Background

The Bell system dominated telephony in the century after Alexander Graham Bell won his patent for the telephone in 1876. Under the early leadership of Theodore Vail, Bell, Inc. staved off Western Union's bid to become a phone company. After leaving and then rejoining the company, Vail guided Bell through a series of acquisitions and mergers that became major avenues of growth for the company, which came to be known as American Telephone and Telegraph (AT&T).

After the original phone patent lapsed in 1893, over 6,000 independent phone companies entered the fray to provide phone service and sell equipment (Weinhaus & Oettinger, 1988). However, competition was not always a virtue. In Hawaii during the 1890s, competition between Mutual and Bell led to confusing allegiances of customers who could not be connected to the clients of the competing phone company. Similar concerns over gaps in standardization and interconnection prompted government oversight of telephony in 1910 (Mann-Elkins Act, 1910).

Meanwhile, acquisition of independents intensified after 1910, forming the building blocks for the regional Bell operating companies (RBOCs). Concerned over Bell's acquisition of an independent (Northwestern Long Distance Co.), the Justice Department threatened its first antitrust suit against AT&T in 1913 (*U.S. v. Western Electric*, Defendant's Statement, 1980). As a preemptive strike against antitrust remedies, AT&T Vice President Nathan Kingsbury sent a letter to the Attorney General. Known as the "Kingsbury Commitment," this letter outlined AT&T's promise to dispose of its stock in Western Union, cease acquisition of independents, and provide interconnection for independents to the Bell network (*Kingsbury Commitment*, 1913). Content with these concessions, the Justice Department ended its antitrust plans. Before long, AT&T was allowed to resume its acquisition of independent companies, pending Interstate Commerce Commission approval.

By the 1920s, Congress was actually in favor of a single monopoly phone system (Willis-Graham Act, 1921). In the meantime, Bell worked to accommodate the remaining independents, allowing interconnection with 4.5 million independent telephones in 1922 (Weinhaus & Oettinger, 1988). This industry rapprochement enabled Bell to focus its energies on new ventures, such as "toll broadcasting" on radio stations, such as WEAF (Brooks, 1976; Briggs, 1977).

Yet, fearing telco domination of the nascent broadcast industry, Congress formalized a ban on telco-broadcast cross-ownership in the Radio Act of 1927 (and the succeeding Communication Act of 1934). The 1934 Act also granted AT&T immunity from antitrust actions, in return for a promise to provide universal phone service; 31% of U.S. homes had a phone at that time (Dizard, 1989).

After investigating complaints concerning AT&T's market dominance in the 1930s, the Federal Communications Commission (FCC) endorsed the industry's structure, characterizing it as a "natural monopoly" (FCC, 1939). By the late 1940s, however, the Justice Department began to feel uneasy about the sheer magnitude of AT&T's empire and initiated antitrust proceedings against the phone giant in 1948, which culminated in a 1956

Consent Decree. Under that decree, the government agreed to drop its lawsuit in return for an AT&T pledge to stay out of the nascent computing industry.

The next year saw the first serious challenge to AT&T's notorious exclusive dealing practices, euphemistically known as the "foreign attachment" restriction. Under the guise of protecting its network from problems of incompatibility and unreliability, AT&T alienated several non-monopoly companies seeking to attach consumer-owned equipment to the network. Hush-A-Phone, Inc. petitioned the FCC to vacate this policy, which had been applied against the company's mouthpiece "hushing" device (*Hush-A-Phone v. AT&T*, 1955). On appeal, the court held that AT&T's ban of a device that emulates the natural cupping motion of one's hand "is neither justifiable nor reasonable" (*Hush-A-Phone Corp. v. United States*, 1956, p. 266).

This pattern of deregulation accelerated in 1968, when the FCC ruled against AT&T's ban of an acoustic coupler used to connect radio-telephones to the telephone network. When allowing connection of this "Carterfone," the FCC and courts signaled that equipment not made by AT&T's manufacturing subsidiary, Western Electric, could be used in its network (Carterfone, 1968). During the following year, MCI was given permission to operate a long distance line, despite AT&T's objections (Microwave Communications, Inc., 1969).

Having sustained these legal setbacks, then, various segments of the Bell network began to feel the pressure of competition. This exogenous shock was compounded by inflation during the 1960s, which further reduced AT&T's profitability. The company responded by postponing maintenance and reducing labor costs during the 1970s in order to assure profitability. In the meantime, demand for telephone service skyrocketed, prompting long waits for equipment and charges of sloppy service (Dizard, 1989).

Dissatisfied with this industry conduct, in 1974, the Justice Department initiated proceedings, reminiscent of its 1948 action, to dismember AT&T. In particular, the complaint against AT&T alleged that they:

(1)   Denied interconnection of non-Bell equipment to the AT&T network.

(2)   Denied interconnection of specialized common carriers with the Bell network.

(2)   Foreclosed the equipment market with a bias toward Western Electric.

(4)   Engaged in predatory pricing, particularly in the intercity service area (*U.S. v. AT&T, 1982*; Gallagher, 1992).

This time, however, with the aid of several interested non-monopoly firms, the Justice Department was in a much stronger position.

As demand for telecommunications services grew, the Bell system's regressive monopoly structure proved an impediment to growth and innovation. After enjoying government protection during the first part of this century, the Bell monopoly gradually fell into disfavor with regulators for inefficient and anti-competitive market conduct.

In defense of AT&T, it is fair to say they achieved the burdensome goal of providing universal service with the highest reliability in the world. The company employed a million workers in 1982, claiming that it lost about $7 a month on its average telephone customer (Dizard, 1994). While that plea may be debatable, such costs necessitated the practice of cross-subsidization via pricier long distance services. Known as the behemoth that worked, AT&T was the largest company in the world, subsuming 2% of the U.S. GNP in and of itself. By 1982, the company carried over a billion calls per day (*U.S. v. Western Electric*, 1982).

Whether or not AT&T was well compensated for its trouble, by 1982, even they viewed their regulated monopoly as an impediment to progress. AT&T's willingness to consent to divestiture was, arguably, as much a function of self-interest as exogenous government pressure. Dizard (1989) notes that they had the financial and political clout to litigate the 1982 decree for another 30 years, had they been so inclined. Yet, AT&T recognized, instead, that it was in their own interest to discontinue the "voice-only" monopoly in which they had become encased. So they negotiated a divestiture settlement, or consent decree, known as the Modified Final Judgement (MFJ) in 1982 (*U.S. v. AT&T*, 1982). Effective in 1984, AT&T's local telephone service was spun off into seven new companies, known as the regional Bell operating companies (RBOCs), sometimes referred to as the "Baby Bells." The divestiture allowed Bell a convenient vehicle to exchange pedestrian local telephony for entry into the lucrative computer market. Perhaps more important, it enabled them to cut most of their labor overhead without fear of union unrest or "bad press."

The question remains, though, whether AT&T's conduct was an anomaly of that particular monopoly, or characteristic of capital-intensive phone utilities in general. In gaining a better understanding of that question, it is useful to examine the post-divestiture conduct of the telephone industry, and the implications of their entry into allied fields.

A discussion of telephony is not complete without an understanding of the telephone network. The telephone network is often referred to as a "star" network because each individual telephone is connected to a central office. The heart of the network (see Figure 17.1) is the central office (CO), which contains the switching equipment to allow any telephone to be connected to any other telephone. Without such switches, it would be necessary to interconnect every possible pair of telephones with a dedicated wire, leaving major cities buried under miles of copper!

Central offices are, in turn, connected to each other by two networks. The local phone companies interconnect all central offices within a service area (known as a LATA), and long-distance companies provide a network of interconnections for central offices located in different LATAs.

Current Status

The telephone companies are now in a formidable position to dominate video and information services, as they control more than half of U.S. telecommunications assets

(Dizard, 1994). The industry handled 620 billion phone calls in 1994. In addition, the telephone companies have a long history of providing mass entertainment and information services a.k.a. dial-it lines. The value of AT&T and its former segments has tripled since divestiture, as the telecommunications sector will soon subsume $1 trillion worldwide, and a sixth of the U.S. economy (King, 1994). Taken together, telephony and information services account for roughly $1 of every $10 spent in the United States, or over $2,000 per household annually.

Figure 17.1
Traditional Telephone Local Loop Network Star
Architecture

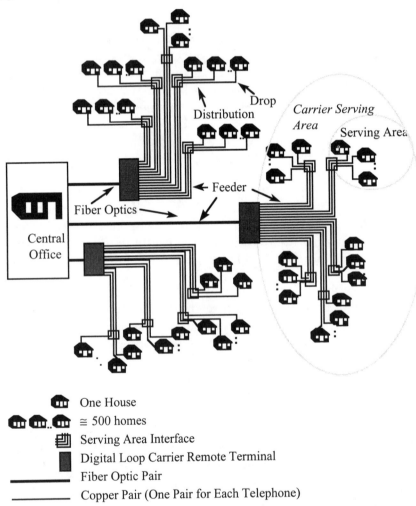

Source: Technology Futures, Inc.

While the Baby Bells are anxious to compete in long distance, incumbent long distance carriers express concern that local companies still command 98% of local revenues over their regions. The RBOCs retain control over 98% of the 155 million phone lines in the United States, possessing "bottleneck control of the local loop and they'll be able to squash competition" (Berniker, 1994, p. 38). Given that MCI pays $.46 of every dollar it earns to the RBOCs in access charges, they have an interest in entering the $25 billion access charge business.

At present, AT&T continues to dominate long distance, commanding roughly two-thirds of the market, and the Big Three (rounded out by MCI and GTE) hold more than 80% of the market. There are over 400 smaller long-distance players (e.g., Metromedia).

At the time of divestiture, the Baby Bells represented roughly 75% of AT&T's assets. Between 1940 and 1980, the number of independent companies decreased by 77%, owing chiefly to consolidation among themselves, although some were acquired by AT&T (Weinhaus & Oettinger, 1988). By 1982, there were 1,459 independent telephone companies, producing 15.9% of the industry's revenues. Although merger activity has slowed since that time, the United States now has over 1,200 telephone providers. The Baby Bells remain dominant, however, with yearly revenues placing each among the Fortune 500 (see Table 17.1).

Four of the Baby Bells have themselves been involved in mergers. Bell Atlantic and NYNEX have merged, and SBC (formerly Southwestern Bell) purchased Pacific Telesis. The service areas of the remaining five RBOCs are illustrated in Figure 17.2.

Although the Telecommunications Act of 1996 returns telephone regulation to the federal government, the Baby Bells were previously governed by the judiciary through Judge Harold Greene's review of the MFJ. The MFJ imposed several line of business restrictions on the RBOCs, including:

(1)   The manufacture of telecommunications products.

(2)   Provision of cable or other information services.

(3)   Provision of long distance services (*U.S. v. A.T.& T.*, 1982).

Selected waivers to the MFJ were granted in the years leading up to the act, including decisions allowing regional phone companies to transmit information services owned by others (*U.S. v. Western Electric Co.*, 1987) as well as their own such services (*U.S. v. Western Electric Co.*, 1991).

In 1988, the FCC reified telco-cable cross-ownership restrictions out of fear that telcos would engage in predatory pricing characteristic of capital intensive industries, and use their natural monopoly over utility poles and conduit space to hinder competition with independent cable operators. Under these FCC rules, former RBOCs were banned from providing cable service outside of their local access transport area (FCC, 1988).

Table 17.1

Top 10 U.S. Telephone Company Revenues

Company	1997 Revenues
AT&T	$53.3 billion
Bell Atlantic	$30.2 bilion
SBC Communications	$24.9 billion
GTE	$23.3 billion
BellSouth	$20.6 billion
MCI Communications	$19.7 billion
Ameritech	$16 billion
U S WEST	$15.4 billion
Sprint	$14.9 billion
WorldCom	$7.4 billion

Source: http://www.pathfinder.com/fortune/fortune500/ind157.html

Recent Developments

After a series of FCC and court rulings relaxed restrictions on telco entry into cable (Brown, 1993; *Cheasapeake & Potomac Telephone Co. v. U.S.*, 1992), the RBOCs requested that the MFJ be vacated in 1994 (Atkin, 1996). As 1995 came to a close, the Supreme Court was hearing arguments to lift the ban on telco-provided video on appeal in *Chesapeake*. The case was preempted when Congress and the President removed the ban as part of the Telecommunications Act. Since the law also removes the ban on telco purchases of cable companies in communities of over 35,000, we're likely to see new alliances between these industries. Although U S WEST was one of the most aggressive entrants into cable, they recently moved to split their phone and cable businesses into two companies. It seems, then, that the strategy of simultaneously pursuing an in-region and out-of-region service in the two industries hasn't succeeded. Telcos can also provide video programming in their own service areas, but the new law prohibits joint telco-cable ventures in their home markets.

It is clear, however, that the Baby Bells are much more interested in pursuing markets in telephony than in cable. Before being allowed to enter long distance markets, RBOCs must prove that they've opened their local phone networks to new rivals, following a 14-point checklist contained in the 1996 Act. The FCC set rules for implementing RBOC entry into long distance in August 1996, promulgating 742 pages of guidelines. The rules were immediately challenged by BellSouth, which was later joined by other Bells, GTE, and even the State of South Carolina, among others. These various challenges were consolidated in the Eighth Circuit Court of Appeals in St. Louis, which granted a stay of the FCC's order. In late December 1997, U.S. District Court Judge Joe Kendall issued a ruling that undermined those FCC rules, claiming that they trample on state's rights (Mehta, 1998). The ruling could allow RBOCs a quicker entrance into long distance, although it is under review by the Supreme Court at the time of this writing. The St. Louis appeals court had earlier imposed a stay on the FCC rules, which the Supreme Court refused to review during 1997 (*FCC v. Iowa Utilities Board*; *AT&T v. Iowa Utilities Board*; *MCI v. Iowa Utilities Board*; Felsenthal, 1998).

Figure 17.2
RBOC Service Areas

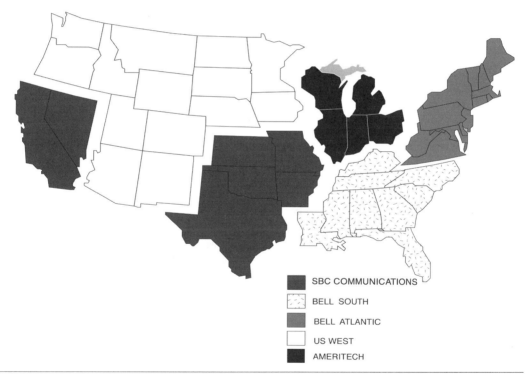

Source: J. Teeter

While this pending competition should lead to lower long-distance rates, local rates may not decline until AT&T, MCI, et al. get favorable terms for reselling Bell connections as their own services. New York City offers the first laboratory for local loop competition, as NYNEX competes against Tele-Communications, Inc. (TCI) and Cox's MFS Communications. Those cable-based competitors now control over 50% of Manhattan's "special access" market—which involves the routing of long-distance calls to-and-from local lines—and will soon face further competition from MCI (Landler, 1995). The Telecommunications Act of 1996 also maintains universal service mandates, but allows states and the FCC to decide how it should be funded.

With the new law resonating like the starting gun at a race, several companies issued threats about invading each other's markets. Naik (1996), for instance, reported the following actions a day after the act was signed:

- Three long-distance carriers announced that they had filed a complaint against Ameritech, for allegedly making it more difficult for customers to choose a new long distance company.

- AT&T issued a declaration of war on local phone companies everywhere, pledging to enter the market in all 50 states and win over at least one-third of the $100 billion sector in the next few years. It plans to offer service via alternative access providers.

All of this activity prompted Judge Harold Greene to wonder whether the new law could deter phone giants from essentially reconstructing the Bell monopoly (Keller, 1996). Counter to that concern, AT&T initiated a split into three different companies in 1995:

(1)   A communication services firm, which retains the name AT&T.

(2)   A network services firm named Lucent Technologies.

(3)   A computer unit, named Global Information Solutions.

Yet, if telco entry into other industries is any indicator, this move toward computing and video won't prove as revolutionary as some might fear. Years after winning the right to provide information services, the Baby Bells had little to show for their efforts, as most were in the process of "regrouping" (Singer, 1993). Ameritech recently offered ISDN (Integrated Services Digital Network) services in Ohio, including high-speed data and video applications, along with enhanced facsimile service—complete with mailboxes for storing and sorting fax messages. Pacific Telesis has offered voice mail subscribers customized reports, delivered on a daily basis (Pagano, 1992).

NYNEX and Dow Jones entered a joint venture to deliver text information over phone lines in New York (Potter, 1992). Dow Jones is also working with BellSouth to serve cellular phone customers in Los Angeles. Beyond that, the Baby Bells have engaged in a few other modest projects involving information delivery. But the scope of these activities pales in comparison to those involving video delivery (see Chapter 18), although cross media competition has yet to reach the levels envisaged under the act.

In addition to numerous video on demand and pay-per-view ventures, phone companies such as AT&T and MCI have been engaging in direct broadcast satellite (DBS) experiments (the latter in conjunction with Rupert Murdoch). MCI also recently entered an alliance with Microsoft for the delivery of online products and services.

Ancillary Services

While POTS still dominates telco activities, there has been a prolific expansion in other services, including call-waiting, call-forwarding, call-blocking, computer data links, and audiotext (or "dial-it") services. Little is known about the content and usage of these services, since such information is subject to non-disclosure restrictions under FCC rules. It is clear, however, that these chip-intensive—or "intelligent"—phone applications now comprise a multibillion-dollar industry.

Perhaps the most explosive growth has been in an area involving mass-audience applications of the telephone—800 and 900 services. Virtually nonexistent prior to divestiture, these services generated 10 billion calls in 1991, and over 1.3 million 800 numbers earned an estimated $7 billion in revenues for the industry in 1993 (Atkin, 1995). It is striking to note that this ancillary revenue source for telephony was nearly equal to that of the entire radio industry that year.

One study of audiotext (LaRose & Atkin, 1992) reported that, of 44 dial-it services available in one local exchange, one-fifth were sexually oriented; the remainder involved other types of entertainment (e.g., soap opera updates), while a few served information and self-help needs. Sexually-oriented services are still predominant (Atkin, 1993). Although industry leader Carlin Communications recently filed for bankruptcy, they attributed their financial difficulties to an expensive, but otherwise successful, challenge to phone indecency restrictions (*Sable Communications of California, Inc. v. FCC*, 1993).

Aside from dial-a-porn, audiotext presents hundreds of information options, including daily television listings, national and international news, celebrity information, recordings of dead (and undead) celebrities, sports scores, horoscopes, and updates on popular television soap operas (Ameritech Publishing, 1995; Atkin, 1995). Audiotext today is roughly a $1 billion per year industry (Neuendorf, Atkin, & Jeffres, 1998). Even so, the Psychic Friends Hotline, which targeted low-income minority households with extensive promotions on cable, went out of business in 1998; they apparently didn't see the end coming.

Similar entertainment applications are found with 900-number call-in polls, such as the Video Jukebox Network, which allows home viewers to select music videos by telephone for play on a local cable channel. For example, CBS's "America on Line" telepoll—conducted in the wake of President Bush's State of the Union speech—garnered over 314,786 calls (Atkin & LaRose, 1994).

As more telcos move to install fiber optics to the "last mile" of line extending to the home, providers will be able to offer a broad range of information and entertainment channels. Bell Atlantic is also developing a personal communications service (PCS) system, which effectively turns pagers into mini-wireless telephones.

Factors to Watch

The landmark Telecom Act of 1996 was passed amidst heady optimism concerning the benefits of deregulation for competition, consumer prices, the construction of an information superhighway, and the resulting job creation in the coming information age.

The act was designed to encourage unprecedented media cross-ownership and competition by enabling cable companies to enter local telephone markets and local and long distance companies to enter each other's markets.

On the second anniversary of its passage, however, the act has encountered heavy criticism in the popular press (Schiller, 1998), as even prominent Congressional supporters now seek hearings to investigate problems with industry conduct that the act was designed to remedy (e.g., service, pricing).

Although it is too early to render a definitive judgment, preliminary returns suggest that the act has not succeeded in that regard. Gene Kimmelman, co-director of Consumers Union, concludes, "It's an abysmal failure so far. The much-ballyhooed opening of markets to competition was a vast exaggeration" (Schiller, 1998, p. A1).

So even without faulty regulation, consumers are paying more for telecommunications services in the new era of deregulation ushered in by the act. In the absence of concerted competition from telcos, cable companies raised rates nearly three times the general inflation rate during the act's first year, while rates for intrastate toll calls rose a hefty 6.1%. Although long distance phone rates increased only an average of 3.7% in 1996, that proportion was closer to 12% for the two-thirds of Americans who are not on any discount calling plan (Mills & Farhi, 1997).

While former FCC Chairman Reed Hundt is dismayed that competition has been delayed by legal wrangling and competitive "détente," he maintains that higher rates may be a necessary first step toward industry competition (Mills & Farhi, 1997). One competitive bright-spot involves a group of smaller companies, such as Teligent, that are building their own advanced, high-speed communications facilities, giving them direct access to their customers and by-passing remnants of the old Bell system (Mandl, 1998).

The ongoing litigation over the FCC denial of RBOC local service applications, mentioned earlier, will have strong ramifications for the structure of the phone industry. It remains to be seen whether long distance companies can compete with the RBOCs, given the latter's advantages in capitalization and switching equipment control.

While politicians might debate the need to maintain regulatory oversight until competition takes hold, few would dispute that competition will eventually be in the offing for telephony. The decade following divestiture saw competitive market forces dramatically reshape the telecommunications landscape. Consumer long distance rates dropped from $0.40 a minute in 1985 to just $0.14 a minute by 1993 (Wynne, 1994). New long distance carriers helped create thousands of new jobs, with millions more likely to accompany the doubling of domestic spending on telecommunications projected by the year 2004—nearly 20% of gross domestic product.

The mergers of NYNEX and Bell Atlantic, as well as Pacific Telesis and SBC, raise the issue of whether the three remaining Baby Bells—U S WEST, BellSouth, and Ameritech—can go it alone. Although their territories range in profitability from the fast-growing south to the depopulated west, analysts expect that there remains plenty of opportunity for all RBOCs in the local phone industry. RBOCs continue to enjoy profit margins nearly double the 20% or so that their counterparts register (Cauley, 1997).

A few days before this chapter went to press, SBC Communications proposed purchasing Ameritech, a move that would make SBC the second largest comunications company in the United States. The progress of this corporate deal will provide a good indication of the likely prospects for other mega-mergers. Considering the fact that the earlier SBC/Pacific Telesis and Bell Atlantic/NYNEX mergers took almost two years to complete, one factor is observing how various state and federal regulators intervene to slow down (or speed up!) this merger.

Although the ongoing appeal of entry restrictions for RBOCs into long distance will be decided in the courts, GTE—as a non-Bell progeny—is free to pursue mergers with long-distance carriers and may well add to the long list of mergers in that area. This new consolidation, combined with obvious telco cross-subsidization concerns, would present a basis for continued vigilance by the Justice Department and FCC.

Telco entry to the video marketplace could create the much-anticipated "revolution" in the broadcast and cable industries. This move can also open up a window of opportunity for various independent programmers that may supply specialty entertainment and information services, much like the specialization in the magazine industry. However, as Mandl (1998) notes, "the federal courts have, at least temporarily, thrown the regulatory environment into turmoil by tossing aside key portions of the act and the regulations it spawned" (p. 18).

In sum, with the myriad avenues for telco entry into other media, the greatest challenge involves preventing the undue consolidation of communications capabilities into ever fewer hands. If we're to maximize innovation, quality, and diversity in tomorrow's information grid, regulators must help channel telephony's vast capital and human resources into allied fields, in a way that augments (rather than depletes) the existing cast of players. Hopefully, the courts will soon clarify the bewildering set of challenges now plaguing implementation of the Telecommunications Act, and do so in a way that safeguards the competitive spirit it was designed to embody.

Bibliography

Ameritech Publishing, Inc. (1995). *Ameritech pages plus Cleveland area white/yellow pages, 1995*. Chicago, IL: Ameritech Publishing.

Ameritech's telco-cable exchange. (1993, June 14). *Broadcasting & Cable*, 70-73.

Atkin, D. (1993). Indecency regulation in the wake of Sable: Implications for telecommunications media. *1992 Free Speech Yearbook 31*, 101-113.

Atkin, D. (1995). Audio information services and the electronic media environment. *The Information Society*, *11*, 75-83.

Atkin, D. (1996). Governmental ambivalence toward telephone regulation. *Communications Law Journal, 1*, 1-11.

Atkin, D., & LaRose, R. (1994). Profiling call-in poll users. *Journal of Broadcasting & Electronic Media, 38* (2), 211-233.

Berniker, M. (1994, December 19). Telcos push for long-distance entry. *Broadcasting & Cable*, 38.

Briggs, A. (1977). The pleasure telephone: A chapter in the prehistory of the media. In I. de Sola Pool (Ed.). *The social impact of the telephone*. Cambridge, MA: MIT Press.

Brooks, J. (1976). *Telephone: The first hundred years.* New York: Harper & Row.

Brown, R. (1993, February 8). U S WEST answers video dialtone call. *Broadcasting & Cable,* 14.

Carterfone. In the matter of use of the Carterfone device in message toll telephone service. (1968). FCC Docket Nos. 16942, 17073; Decision and order, 13 FCC 2d 240.

Cauley, L. (1997, December 10). Genuine competition in local phone service is a long distance off. *Wall Street Journal,* A1, 10.

Chesapeake and Potomac Telephone Co. v. U.S. (1992), Civ. 92-17512-A (E.D.-Va).

Dizard, W. (1989). *The coming information age.* New York: Longman.

Dizard, W. (1994). *Old media, new media.* New York: Longman.

Federal Communications Commission. (1939). *Investigation of telephone industry.* Report of the FCC on the investigation of telephone industry in the United States, H.R. Doc. No. 340, 76th Cong., 1st Sess. 602.

Federal Communications Commission. (1988). *Telephone company-cable television cross ownership rules.* 47 C.F.R. 63.54 - 63.58. Second Report and Order, Recommendation to Congress, and Second Further Notice of Proposed Rulemeking, 7 FCC Rcd. 5781.

Felsenthal, E. (1998, January 13). Court to speed review of ruling in local phone competition case. *Wall Street Journal,* B17.

Gallagher, D. (1992). Was AT&T guilty? *Telecommunications Policy, 16,* 317-326.

Hush-A-Phone Corp. v. AT&T et al. (1955). FCC Docket No. 9189. Decision and Order (1955). 20 FCC 391.

Hush-A-Phone Corp. v. United States. (1956). 238 F. 2d 266 (D.C. Cir.). Decision and Order on Remand (1957). 22 FCC 112.

Jessell, H. A. (1992, July 20). FCC calls for telco TV. *Broadcasting & Cable,* 3, 8.

Keller, J. (1996, February 12). AT&T and MCI explore local alliances. *Wall Street Journal,* A3.

King, J. F. (1994, February 20). Running its own race: Ameritech in the slow lane of the information super-highway. *Cleveland Plain Dealer,* F1.

Kingsbury Commitment. (1913, December 19). Letter from N. C. Kingsbury, AT&T to J. C. McReynolds, Attorney General, Justice Department.

Landler, M. (1995, April 3). The man who would (try to) save New York for NYNEX. *New York Times,* C1, C6.

LaRose, R., & Atkin, D. (1992). Audiotext and the re-invention of the telephone as a mass medium. *Journalism Quarterly, 69,* 413-421.

Mandl, A. (1998, Jan. 26). Telecom competition is coming sooner than you think. *Wall Street Journal,* A18.

Mann-Elkin's Act. (1910). Mann-Elkins Act, Pub. L. No. 218, 36 Stat. 539.

Mehta, S. (1998, Jan. 5). Baby Bells cautious on quick entry to long-distance market after ruling. *Wall Street Journal,* B8.

Microwave Communications, Inc. (MCI) (1969). FCC Docket No. 16509. Decision, 18 FCC 2d 953.

Mills, M., & Farhi, P. (1997, January 27). A year later, still lots of silence: Dial up the Telecommunications Act and get bigger bill and not much else. *The Washington Post National Weekly Edition,* 18.

Naik, G. (1996, February 9). Landmark telecom bill becomes law. *Wall Street Journal,* B3.

Neuendorf, K., Atkin, D., & Jeffres, L. (1998). Understanding adopters of audio information services. *Journal of Broadcasting & Electronic Media, 41,* 111-123.

Pagano, P. (1992, April). Electronic warfare. *Washington Journalism Review,* 20.

Potter, W. (1992, May). Bells start to adopt new services. *Presstime,* 42.

Sable Communications of California, Inc. v. FCC (1989), 109 S. Ct. 2829.

Schiller, Z. (1998, Feb. 9). Local phone competition is still just a promise. *Cleveland Plain Dealer,* A1, 6.

Singer, J. (1993). *Fight for the future: Congress, the Brooks bill and the Baby Bells.* Paper presented to the annual meeting of the Association for Education in Journalism and Mass Communication, Kansas City.

Telecommunications Act of 1996. (1996). 104 Pub. L. 104, 110 Stat. 56, 111 (codified as amended in 47 C.F.R. S. 73.3555).

U.S. v. Western Electric Co., Defendent's Statement. (1980). Civil Action No. 74-1698, 169-170.

U.S. v. Western Electric Co. 767 F. Supp. 308. (D.D.C. 1991).

U.S. v. Western Electric Co. 673 F. Supp. 525 (D.C.C. 1987).

U.S. v. Western Electric Co. and AT&T. (1956). *1956 Consent Decree.* Civil action no. 17-49, 13 RR 2143; 161 USPQ (BNA) 705; 1956 trade cas. (CCH) Section 68246, at p. 71134 (D.C. N.J.).

U.S. v. A.T.& T. (1983). 552 F. Supp. 131, 195 (D.D.C. 1982), aff'd sub nom. *Maryland v. U.S.,* 460 U.S. 1001.

Weinhaus, C. L., & Oettinger, A. G. (1988). *Behind the telephone debates.* Norwood, NJ: Ablex.

Willis-Graham Act. (1921). Pub. L. No. 15, 42 Stat. 27.

Wynne, T. (1994, November 16). An earshocking proposition. *Cleveland Plain Dealer,* B11.

Your new computer: The telephone. (1991, June 3). *Business Week,* 126-131.

18

Broadband Networks

Lon Berquist

T he emergence of new technological capabilities for networks, allowing more efficient means of transmitting voice, video, and data, has led to hopes for a communications revolution offering a variety of enhanced services. The services planned for advanced telecommunications networks include education and distance learning, video on demand, high-definition television, interactive entertainment, interactive program guides and navigators, civic networking, personal communications services, telecommuting, electronic commerce, research support, information services, Internet access, and telemedicine (Cable Television Laboratories, 1995). The attention given by the press and policy makers to the notion of an information superhighway focuses attention on the developing high-speed broadband networks being built throughout the United States. These networks and their advanced services are allowing greater possibilities for seeking and retrieving information, conducting commerce, and communicating.

Background

A number of factors have transformed the public switched telephone network (PSTN), the cable television system, and the infrastructure of the Internet. First, voice transmission has gradually shifted from analog to digital. This, in turn, has necessitated the implementation of more advanced and efficient switching hardware and intelligent software. Finally, the copper transmission medium utilized for traditional networks is rapidly being replaced by optical fiber. Broadband networks are evolving, as users demand more speed, more connections, and the need to reach greater distances.

* Doctoral Candidate, Department of Radio-Television-Film, University of Texas at Austin (Austin, Texas).

Bandwidth, which is actually an analog measure, is a term for describing the transmission capabilities of the network. For digital networks, *data rate* is a more apt description of bandwidth and has to do with the amount of data that can be transmitted in a fixed amount of time. Broadband networks are considered high-speed networks because they transmit data at a higher rate than the 64 Kb/s (Kilobits per second) limit of the narrowband plain old telephone service (POTS).

Bandwidth is determined by a number of factors including the type of transmission media utilized. Increasingly, the infrastructure of high-speed networks consists of fiber optic cable, which has many advantages over typical means of transmission. Unlike broadcasting or wire transmission, such as coaxial cable television or twisted-pair for telephone, fiber transmission is immune to electrical interference. At the same time, fiber is virtually immune to signal jamming and signal stealing through physical tampering. Most important, a single fiber provides the same bandwidth as dozens of cable TV lines or twisted-pair phone wires. A conventional three-inch copper cable for telephony contains 1,200 twisted-pair wires for up to 14,400 phone conversations, while a smaller fiber cable with 72 individual fibers can provide up to 3.5 *million* conversations (Geller, 1991).

For telephone companies, the introduction of fiber into the telephone network, along with a change from analog to digital technologies, allows the transmision of 125,000 times more information than standard copper phone lines (Beacham, 1994). Although standard twisted-pair phone lines offer transmission speeds up to 64 Kb/s and digital phone lines can reach 128 Kb/s, researchers expect fiber optic cable to reach 100 Gb/s (gigabits per second) (Lockton, 1987). This enhanced capability has led many phone companies to consider offering advanced telecom services as they upgrade their network.

Coaxial cable for cable TV provides up to 160 Mb/s (megabits per second) greater bandwidth than phone lines, but it is still far short of fiber's capabilities. With fiber optic cable added to the cable plant, cable TV systems hope to expand programming channels and provide bandwidth-hungry high-definition television, video on demand, interactive services, and enhanced digital audio. However, in order to provide these advanced services, cable systems must upgrade their technology to allow for a two-way infrastructure. Cable, a $20 billion a year business, would like to enhance their technology and enter the $200 *billion* a year business of telephony (Weinschenk, 1995).

In addition to bandwidth, the ability to route signals through switching is a vital part of any network. Because a network is a communications system that connects geographically-dispersed users, the switch must be able to route the signal to the intended receiver. When one makes a call through a typical telephone network, a path is dedicated from the caller to the person being called for the duration of the call. This is an example of a circuit switched network. Data networks, however, need not transmit continuously, since data is made up of "bits" of information. Instead, these bits of data are sent in chunks, or packets, that are forwarded with addresses so all the packets are combined into a whole at the destination (see Figure 18.1). This is a packet switched network, which is the foundation of most data communication networks and related protocols.

Figure 18.1
Packet and Circuit Switching

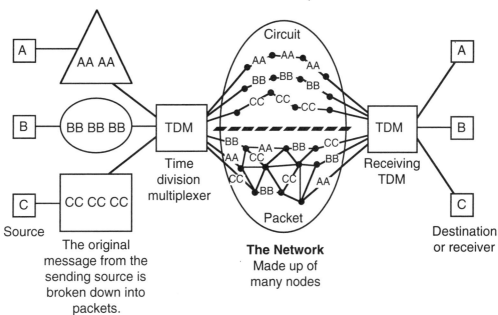

Source

The original message from the sending source is broken down into packets.

The Network
Made up of many nodes

Destination or receiver

Circuit Switching Networks—In a circuit switching network the source establishes a path to the sender and then transmits all the packets along this same path. Circuit switching is used to carry voice and video to ensure the order is maintained when received by the destination site.

Packet Switching Networks—In a packet switching network individual packets must pass through the Nodes on the network to be passed along to the receiver. Throughout its route to the receiver packets traverse any path and do not have to travel together. Packet number three of message one can be received before packet number one. If this does occur the receiver will wait until all the packets arrive and then assemble them into the correct order.

Source: J. Hassay (1996)

The Internet, the network of networks, already utilizes fiber optics in its backbone, but dramatic increases in user traffic threaten to clog the works. According to recent statistics, 62 million people in the United States use the Internet (Weber, 1998), and online traffic has been doubling every 100 days (U.S. Department of Commerce, 1998). Internet access and, more importantly, the World Wide Web have been the applications driving demand for enhanced broadband access. The WWW has significantly increased traffic on the Internet backbone. In 1993, Web traffic made up only 0.1% of Internet network traffic. By 1995, traffic from the Web utilized 26% of the total volume of bytes travelling over the Internet backbone (Staxen & Cavanaugh, 1997). Neither analog 28.8 Kb/s modems nor the more advanced 56 Kb/s modems can meet the ever-expanding needs of Internet users who seek to avoid the notorious "World Wide Wait." The growing number of Internet users and the growing use of bandwidth-hungry applications on the Web have led both cable operators and phone companies to seek new technologies to squeeze every bit of bandwidth out of the last few hundred feet of wire into the home.

As the capabilities of broadband networks are taxed with greater use, new protocols, technologies, and architectures are emerging. Typical packet-switched technologies—such as X.25, frame relay, Internet protocol (IP), and switched multi-megabit data service (SMDS) along with bridges, routers, gateways, and hubs—make up the technologies moving data among local area networks (LANs), wide area networks (WANs), and metropolitan area networks (MANs) (Goldberg, 1997). These technologies primarily serve private business broadband users. As data networking reaches residences through the public networks, such as the telephone and cable television networks, the distinction between public and private networks will blur.

To meet the needs of a growing and diverse number of users on broadband networks, some tout asynchronous transfer mode (ATM) as the ultimate technological solution for dealing with the expected increase in, and variety of, transmitted data (Goldberg, 1997). ATM is a data transmission protocol that handles all traffic types—voice, data, and video—at the same time. It can be utilized from the desktop, the LAN, the WAN, or the backbone. Most importantly, it is compatible with existing networks, allowing seamless network integration. Combined with Synchronous Optical Network (SONET), a transmission technology for fiber optics, ATM networks allow optical carrier transmission signals of up to 2.5 Gb/s with a potential speed of 40 Gb/s (Snyder, 1996).

The Integrated Services Digital Network (ISDN) was the first technology designed to deliver digital voice and data over the circuit switched telephone network. Two types of ISDN lines are available:

(1)   Basic rate interface (BRI) which combines two 64 Kb/s channels to achieve a 128 Kb/s rate.

(2)   Primary rate interface (PRI) with a bandwidth of 1.544 Mb/s.

Although available for over a decade, ISDN has failed to capture a large market share in the United States. Of the 157 million access lines in the United States, less than 750,000 utilize ISDN (ISDN market, 1997).

Recent Developments

xDSL

Despite the lack of success with ISDN, the telephone industry has high hopes for a newer technology similar to ISDN that can run on existing twisted-pair copper wiring at speeds surpassing PRI-ISDN. Digital subscriber line (DSL) technology is rapidly being introduced throughout the United States. DSL is a modem technology able to transmit two-way data communications over ordinary phone lines simultaneously with voice services by using an advanced modulation technique. DSL comes in many variants, hence the more precise description as xDSL.

- *Asymmetrical digital subscriber line (ADSL)*—Originally introduced in the telephone industry's failed attempt at providing video dialtone, ADSL transmits downstream (to the subscriber) at 1.544 Mb/s to 9 Mb/s and upstream (from the residence) at 16 Kb/s to 800 Kb/s. Speed depends on line distance.

- *High-bit-rate digital subscriber line (HDSL)*—Supports two-way transmission of 1.544 Mb/s for up to 12,000 feet. Unlike other xDSL variants, HDSL uses two pairs of copper wire (one of which carries no voice traffic).

- *ISDN-like digital subscriber line (IDSL)*—Uses ISDN transmission technology to deliver 128 Kb/s in both directions.

- *Rate-adaptive digital subscriber line (RADSL)*—A version of ADSL where the modem automatically adjusts to achieve the maximum speed possible on a particular line. Transmission can reach 12 Mb/s downstream and 1 Mb/s upstream.

- *Symmetrical digital subscriber line (SDSL)*—Provides the same amount of bandwidth (1.544 Mb/s) for upstream as for downstream for 10,000 feet.

- *Very-high-bit-rate digital subscriber line (VDSL)*—Deploys speeds up to 56 Mb/s downstream and up to 2.3 Mb/s upstream; however, its range is limited to 1,000 to 6,000 feet. Because of its high speed and short transmission distance, VDSL is likely to be used in the "last 100 feet" for advanced residential broadband networks with fiber-to-the-curb (FTTC) architecture (Makris, 1998; Humphrey & Freeman, 1997).

Of the xDSL technologies, ADSL is currently the only accepted standard by the American National Standards Institute. In early 1998, a consortium of PC manufacturers, telecommunications providers, and networking industry vendors united to form the Universal ADSL Working Group. The goal of the group is to propose to the International Telecommunications Union (ITU) that ADSL be advanced as an interoperable international standard for high-speed communications. However, there is some disagreement on the most appropriate modulation technique for ADSL (Dawson, 1998). ADSL modem vendors Alcatel and Motorola favor the current modulation standard known as "discrete multitone," while Lucent and Fujitsu have built modems utilizing "carrierless amplitude phase" modulation.

Cable Modems

While the telephone companies begin offering broadband network access to businesses and residents though ADSL, cable operators are rolling out cable modems in their attempt to capture bandwidth-hungry customers. Unlike telephone networks, coaxial cable systems are blessed with abundant bandwidth. However, since cable systems lack a two-way infrastructure, they are limited in their ability to provide interactive services.

To remedy this problem, cable systems are rapidly being upgraded to two-way hybrid fiber/coax (HFC) architectures. HFC networks deploy fiber to optical nodes in the customers' neighborhoods, and then transmit telephony, data, or video to the home via coaxial cable. Each node can serve from 500 to 2,000 subscribers. It should be noted that HFC supports the emerging ATM and SONET technologies.

In cable systems, data is shipped downstream via a 6 MHz cable channel at anywhere from 27 Mb/s to 40 Mb/s using a quadrature amplitude modulation technique (Weinschenk, 1996). Individual upstream speeds reach 768 Kb/s and use a different modulation technology, quadrature phase-shift keying. The modem connects to the subscriber's computer via a typical Ethernet LAN cable. For cable systems lacking a two-way architecture, upstream transmission is still possible via a phone return cable Internet technology, which allows speeds up to current analog phone modems, or 128 Kb/s if using ISDN (Stover, 1997).

As with the telephone industry, cable operators have joined together to seek a standard for two-way cable data networks. The Multimedia Cable Network System (MCNS) joint group was recently recognized by the ITU for its cable modem standard—Data Over Cable Service Interface Specifications (Ellis, 1998).

FTTC

Fiber-to-the-curb (FTTC) provides fiber from the central office to an optical network unit (ONU) placed at the curbside of a home. Also termed a switched digital video network, FTTC supports 16 or more homes from the ONU (far less than an HFC system), providing telephone service via twisted pair and video via coaxial cable (McCullough, 1995).

FTTH

For the ultimate in broadband residential service, fiber-to-the-home (FTTH) allows the high speeds of ATM-driven optical fiber directly to a single home. At the house, an optical network unit converts the optical signals to electrical signals, so video, telephony, and data can be distributed to traditional telephone sets, televisions, and data connections (Khasnabish, 1997).

Current Status

Telephone companies have gone beyond the trial stage in offering xDSL service throughout the United States (see Table 18.1). Large cities, particularly those with some high-tech industries or universities, are the initial beneficiaries of xDSL rollouts. Monthly service rates and installation fees vary tremendously among xDSL providers, with monthly rates ranging from $35 to over $1,000, and installation fees ranging from $125 to $1,600.

Table 18.1
U.S. xDSL Service Providers

Provider	Location(s)	Service(s)	Speed Down/Upstream	Monthly Rate	Installation
Ameritech	Ann Arbor, MI Chicago, IL	ADSL	1.5 Mb/s-128 Kb/s	$60	n/a
Ausnet Services	Oregon, Washington, Northern California	ADSL, HDSL, RADSL	2 Mb/s-1 Mb/s, 768 Kb/s-768 Kb/s, 3 Mb/s-2 Mb/s	Usage-based: $350-1st 3 GB, $50-per addl GB, $50-local loop charge	n/a
Concentric Network Corp.	Northern California	IDSL	144-784 Kb/s	$195 $250: local loop	$125
Covad Communications	San Francisco Bay Area (Boston, L.A., N.Y., Seattle, D.C. in 1998)	IDSL	144 Kb/s-1.5 Mb/s, 144 Kb/s-1.5 Mb/s	$90-$195	$325
GTE	Marina Del Ray, CA (16 more states in 1998)	ADSL, RADSL	256 Kb/s-60 Kb/s	$35	n/a
Harvardnet, Inc.	Boston; Portland, ME; Concord, MA; Portsmouth, NH	DSL, RADSL	270 Kb/s-2.5 Mb/s, 128 Kb/s-1.5 Mb/s	$300-$1200	$1,600
Interaccess	Chicago	DSL, RADSL	7 Mb/s-640 Kb/s	$900	$1,000
Northpoint Communications	San Francisco	SDSL	160 Kb/s-1 Mb/s, 160 Kb/s-1 Mb/s	$100-$200	n/a
Onenet Communications	Cincinnnati, Dayton	DSL, RADSL	7 Mb/s-640 Kb/s	$500-$1,500	n/a
Pacific Bell	San Francisco; Austin, TX (other cities to follow)	DSL, ADSL	384 Kb/s or 1.5 Mb/s, 384 Kb/s	$80-$250	$125
SBC Communications	Austin, TX (other cities to follow)	ADSL	384 Kb/s or 1.5 Mb/s, 384 Kb/s	$80-$250	$125
U S WEST Enterprises	Phoenix (40 cities to follow)	RADSL	192 Kb/s-45 Mb/s, 192 Kb/s-5 Mb/s	$60-$840	$200-$300 (45 Mb/s: $970-1st 3 Mb/s, $300-per addl 3 Mb/s, $1,700 ATM port fee)
UUNet Technologies	Northern California, L.A., Boston, N.Y., D.C.	IDSL	128 Kb/s-128 Kb/s	$650-$750, $150-$250: local loop, $3,000: start-up fee	$500
UUNet Technologies	Northern California, L.A., N.Y.	SDSL	768 Kb/s-768 Kb/s	$650-$1,400	$150-$250: local loop

Source: Data Communications (April 1998)

The cable modem market, too, is emerging as a vital option for broadband connectivity (see Table 18.2). Monthly rates for cable data connections range from around $25 to $50 (including modem rental), with installation charges averaging approximately $100.

The three major brand-name services—@Home, Road Runner, and MediaOne Express—have had some early success in gaining subscribers. @Home is a partnership of a number of multiple system operators, such as TCI and Cox, and the venture capital firm Kleiner Perkins Caufield & Byers. Road Runner is Time Warner's cable data service, and Media-One is a venture of the merged U S WEST and Continental Cablevision companies.

Table 18.2
U.S. Cable Modem Market

Cable Modem Service	Subscribers	Monthly Rate	Installation
@Home (TCI, Comcast, Cox, Cablevision, InterMedia, Marcus, Rogers, and Shaw)	50,000	$34.95-44.95	$99.95-175
Road Runner (Time-Warner)	29,000	$24.95-39.95	$99-200
MediaOne Express (Continental, U S WEST)	20,000	$34.95-39.95	$150
PowerLink (Adelphia Communications)	2,400	$34.95-44.95	$99.95
Jones Internet Channel (Jones Intercable)	2,200	$39.95	$99.95
Optimum Online (Cablevision Systems)	600	$45	$150
PeRKInet (Internet Ventures)	500	$49.57	$99
ISP Channel (MediaCity)	500	$49.95	$99
CyberCable (CableVision of Loudoun)	500	n/a	n/a
Internet Commander (Century Communications)	70	n/a	n/a

Sources: Multichannel News (December 1997) and *Austin American-Statesman* (April 1998)

Competition is expected as both cable data services and xDSL services expand operations. In Phoenix, U S WEST plans to offer an FTTC service using VDSL, competing directly with Cox Cable, to provide voice, video, and data (Barthold, 1998). Since cable systems are primarily distributed throughout residential neighborhoods, it appears that telephone companies, with their ubiquitous networks, will have an advantage in capturing businesses, schools, and other institutions. For telecommuters, small office/home office users, and residential users wanting broadband access, competition between cable data services and xDSL providers should force prices down over time.

Factors to Watch

Both xDSL and cable modems are asymmetrical, with far greater downstream bandwidth than upstream capability. This, despite appearing as a limitation, is actually appropriate for Web access, providing substantial download bandwidth for graphics and streaming video or audio, and adequate upstream for Web access commands and searches. Whether or not this will remain adequate in the future remains to be seen.

The Telecommunications Act of 1996 clearly states a desire to make "advanced telecommunications capability" widely available. Section 706 of the act directs the Federal Communications Commission to initiate a notice of inquiry concerning the universal availability of advanced telecommunications. It defines advanced telecommunications capability as "high-speed, switched, broadband telecommunications capability that enables users to originate and receive high-quality voice, data, graphics, and video telecommunications" (Werbach, 1997, p. 79-80). *Originating* advanced telecommunications presumes a greater interactivity than currently available. In order to fulfill this aspect of the Telecom Act, it is likely new technologies will be required.

In early 1998, Vice President Gore announced the cooperation of a number of telecommunications companies in developing the next-generation Internet, called Internet-2, as part of a consortium of universities and research centers (Chandrasekaran, 1998). The Internet-2 project will push the current backbone Internet speeds of 45 Mb/s to a very-high-speed backbone network service (vBNS) of up to 2.4 Gb/s. Just as development of the first generation Internet led to advances in telecommunications capability, it is hoped that the next-generation Internet will lead to even greater advances in broadband technology.

Lee McKnight and W. Russell Neumann (1995) suggest that emerging broadband networks are the "central nervous system" of the new world economy. Cable modem and xDSL technologies are the first step in joining this central nervous system directly to the home.

Bibliography

Barthold, J. (1998, April 27). U S WEST planning to take on Cox with converged Phoenix offering. *Cable World*, *1*, 16.

Beacham, F. (1994, August). Hype, hope, & reality. *Video 18* (5), 36-39, 68.

Cable modem explosion. (1998, April 13). *Austin American-Statesman*, C1.

Cable Television Laboratories, Inc. (1995). *The cable connection: The role of cable television in the national information infrastructure*. A white paper. [Online]. Available: http://cablelabs.com.

Chandrasekaran, R. (1998, April 14). Firm to give research schools super-fast computer services. *Washington Post*, C5.

Dawson, F. (1998, January 26). Technical questions remain on DSL. *Multichannel News*, 43-49.

Ellis, L. (1998, March 30). Cablelabs reports steady progress on several industry standards. *Multichannel News*, 73-74.

Geller, H. (1991). *Fiber optics: An opportunity for a new policy?* Washington DC: The Annenberg Washington Program.

Goldberg, C. J. (1997, July). Charting broadband solutions. *Bandwidth*, 41-48.

Hassay, J. (1996). Broadband Network Technologies. In A. E. Grant. *Communication Technology Update* (5th Ed.). Boston: Focal Press.

Household base for commercially offered cable data services. (1997, December 8). *Multichannel News*, 232.

Humphrey, M., & Freeman, J. (1997). How xDSL supports broadband services to the home. *IEEE Network, 11* (1), 14-23.

ISDN market penetration. (1997, June 1). *America's Network*, 56.

Khasnabish, B. (1997). Broadband to the home (BTTH): Architectures, access methods, and the appetite for it. *IEEE Network, 11* (1), 58-61.

Lockton, J. D. (1987). Information age developments in telecommunications. In W. H. Dutton, J. G. Blumler, & K. L. Kraemer (Eds.). *Wired Cities*. Boston: G. K. Hall.

Makris, J. (1998, April 21). Don't be duped. *Data Communications*, 38-52.

McCullough, D. (1995, July). The raging technology debate: Hybrid fiber/coaxial cable vs. fiber to the curb. *Lightwave*, 36-41.

McKnight, L., & Neuman, W. R. (1995). Technology policy and the NII. In W. J. Drake (Ed.). *The new information infrastructure: Strategies for U.S. policy*. New York: The Twentieth Century Fund Press.

Selected DSL service providers. (1998, April 21). *Data Communications*, 46-49.

Snyder, B. (1996, December 9). Making a higher jump to light speed. *Telephony*, 106, 108.

Staxen, P., & Cavanaugh, K. (1997). Rejuvinated ADSL scores high. In International Engineering Consortium (Ed.). *The local loop: Access technologies, services, and business issues*. Chicago: IEC Publications.

Stover, S. (1997, June). Dose of reality. *Broadband Systems & Design*. 43.

U.S. Department of Commerce. (1998). *The emerging digital economy*. [Online]. Available: http://www.ecommerce.gov/emerging.htm.

Weber, T. E. (1998, April 16). Who, what, where: Putting the Internet in perspective. *Wall Street Journal*, B12.

Weinschenk, C. (1996, December). The great wired hope. *tele.com*, 51-62

Weinschenk, C. (1995). The FSN: What is it good for? *Cable World, 7* (17), 40.

Werbach, K. (1997, March). *The digital tornado: The Internet and telecommunications policy*. OPP working paper 29. Washington, DC: Federal Communications Commission.

19

Satellite Communications

Carolyn A. Lin, Ph.D.*

Artificial satellites have been serving the world's population in nearly every aspect of
their cultural, economic, political, scientific, and military lives during the past three
decades. Since the successful launch of the very first communication satellite—Intelsat
I—in 1965, the satellite was envisioned as the ultimate vehicle for linking the world
together into a global village. As we move to the end of the 20th century, that vision is
being augmented as companion communication technologies in voice, data, and video
transmission—as well as networking systems—are evolving at a rapid pace.

In a multimedia, multichannel communication environment, where wired and wireless technologies compete against each other for the finite pool of consumers, satellite
technology as a broadband and global coverage communication delivery system has
been able to maintain its favorable position in the marketplace. This is because satellite
technology is uniquely suited for delivering a wide variety of signals to accomplish a
great number of different communication tasks. There are three basic categories of non-
military satellite services: fixed satellite service, mobile satellite systems, and scientific
research satellites (commercial and noncommercial).

- *Fixed satellite services* handle hundreds of millions of voice, data, and video
  transmission tasks across all continents between fixed points on the earth's surface.

- *Mobile satellite systems* help connect remote regions, vehicles, ships, and aircraft
  to other parts of the world and/or other mobile or stationary communication
  units, in addition to serving as navigation systems.

* Associate Professor and Graduate Program Director, Department of Communication, Cleveland State
University (Cleveland, Ohio).

- *Scientific research satellites* provide meteorological information, land survey data (e.g., remote sensing), and other scientific applications such as earth science and atmospheric research.

The satellite's functional versatility is embedded within its technical components and its operational characteristics. Looking at the "anatomy" of a satellite, one discovers two modules (Miller, Vucetic & Berry, 1993).

First is the *spacecraft bus or service module*, which consists of five subsystems:

(1)   A *structural subsystem* provides the mechanical base structure, shields the satellite from extreme temperature changes and micro-meteorite damage, and controls the satellite's spin function.

(2)   The *telemetry subsystem* monitors the on-board equipment operations, transmits equipment operation data to the earth control station, and receives the earth control station's commands to perform equipment operation adjustments.

(3)   The *power subsystem* comprises solar panels and back-up batteries that generate power when the satellite passes into the earth's shadow.

(4)   The *thermal control subsystem* helps protect electronic equipment from extreme temperatures due to the intense sunlight or the lack of sun exposure on different sides of the satellite's body.

(5)   The *altitude and orbit control subsystem* is comprised of small rocket thrusters that keep the satellite in the correct orbital position and antennas pointing in the right directions.

The second major component is the *communications payload*, which is made up of transponders. A transponder is capable of:

- Receiving uplinked radio signals from earth satellite transmission stations (antennas).

- Amplifying the radio signals received.

- Sorting the input signals and directing the output signals through its input/output signal multiplexers to the proper downlink antennas for retransmission to earth satellite receiving stations.

The satellite's operational characteristics are literally "out of this world" and reach deep into outer space. Satellites are launched into orbit in outer space via a space shuttle (a reusable launch vehicle) or a rocket launcher (a non-reusable launch vehicle). There are two basic types of orbits: geostationary and non-geostationary orbits. The *geostationary* (or *geosynchronous*) *orbit* refers to a circular or elliptical orbit incline approximately 22,300 miles (36,000 km) above the earth's equator (Jansky & Jeruchim, 1983). Satellites that are launched into this type of orbit at that altitude typically travel around the earth at the same rate that the earth rotates on its axis. Hence, the satellite appears to be "stationary" in its orbital position and in the "line-of-sight" of an earth station, with varying orbital shapes, within a 24-hour period. Therefore, geostationary orbits are most useful for those communication needs that demand no interruption around the clock, such as telephone calls and television signals (see Figure 19.1).

Non-geostationary (or *non-geosynchronous*) *orbits* are located either above or below the typical altitude of a geostationary orbit. Satellites that are launched into a higher orbit travel at slower speeds than the earth's rotation rate; thus, they can move past the earth's horizon to appear for a limited time in the line-of-sight from an earth station. Those satellites that are launched into lower orbits travel at higher speeds than the earth's rotation rate, so they race past the earth to appear more than once every 24 hours in the line-of-sight from an earth station. Non-geostationary orbits are utilized to serve those communication needs that do not require 24-hour input and output, such as scientific land survey data.

Figure 19.1
Satellite Orbits

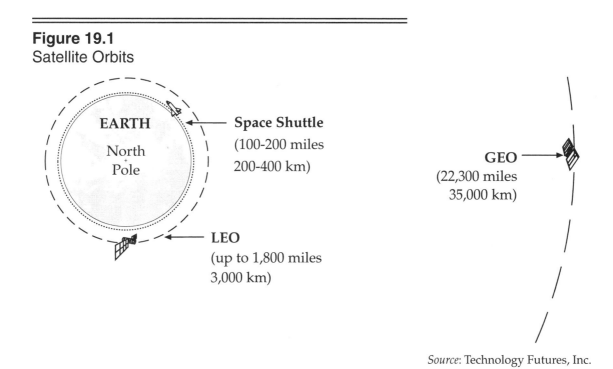

Source: Technology Futures, Inc.

Background

During the early years of satellite technology development, the most notable event was the discovery of the geostationary orbit by the English engineer Arthur C. Clarke in 1945 and the successful launch of the first artificial satellite, Sputnik, by the Soviet Union on October 4, 1957. On January 31, 1958, the first U.S. satellite, Explorer 1, was launched. In July 1958, Congress passed the National Aeronautics and Space Act, which established the National Aeronautics and Space Administration (NASA) as the civilian arm for U.S.

space research and development for peaceful purposes. In August 1960, NASA launched its first communication satellite, Echo I—an inflatable metallic space balloon or "passive satellite," designed only to "reflect" radio signals like a mirror. The next period of satellite technology development was marked by an effort to launch "active satellites," which could receive an uplink signal, amplify it, and retransmit it as a downlink signal to an earth station. In July 1962, AT&T successfully launched the first U.S. active satellite, Telstar I. In February 1963, Hughes Aircraft launched the first U.S. geostationary satellite, Syncom, under a contract with NASA.

In tandem with these experimental developments was the creation of the Communication Satellite Corporation (Comsat), a privately-owned enterprise overseen by the U.S. government, via the passage of the Communications Satellite Act of 1962. Meanwhile, an international satellite consortium, the International Telecommunications Satellite Organization (Intelsat), was formed on August 20, 1964. Intelsat is a non-profit cooperative organization that coordinates and markets satellite services, while member states are both investors and profit sharers of the revenues. The 19 founding member nations, including the United States, 15 western European nations, Australia, Canada, and Japan, agreed to designate Comsat to manage the consortium (Hudson, 1990). The era of global communication satellite services began with the launch of Intelsat-I into geostationary orbit in 1965. Intelsat-I was capable of transmitting 240 simultaneous telephone calls.

Other important issues during this period involved satellite transmission frequency and orbital deployment allocation policy. Both issues were thrust into the management of the World Administrative Radio Conference (or WARC) in 1963. WARC is a technical arm of the International Telecommunications Union (ITU), an international technical organization founded in 1865 to regulate radio spectrum use and allocation. Given that a geostationary satellite's "footprint" can cover more than one-third of the earth's surface, the ITU divides the world into three regions:

- Region 1 covers Europe, Africa, and the former Soviet Union.

- Region 2 encompasses the Americas.

- Region 3 spans Australia and Asia, including China and Japan.

The ITU also designated radio spectrum into different frequency bands, each of which is utilized for certain voice, data, and/or video communication services. Major categories of frequency bands include:

- L-band (0.5 GHz to 1.7 GHz) is used for digital audio broadcast, personal communication services, global positioning systems, and non-geostationary and business communication services.

- C-band (4/6 GHz) is used for telephone signals, broadcast and cable TV signals, and business communication services.

- Ku-band (11 GHz to 12/14 GHz) is used for direct broadcast TV, telephone signals, and business communication services.

- Ka-band (17 GHz to 31 GHz) is used for direct broadcast satellite TV and business communication services.

With the passage of time, the satellite's status as the most efficient technology to connect the world seems firmly established. However, an array of other challenges remains concerning issues related to national sovereignty and the orbital resource as a shared and limited international commodity. At the 1979 WARC, intense confrontations on the issue of "efficient use" versus "equitable access" to the limited geostationary orbital positions by all ITU member nations took place, pitting western nations that owned and operated satellite services against the developing nations that did not. For the developed nations, equitable access to orbital positions suggests a waste of resources, as many developing nations lack the economic means, technical know-how, and practical need to own and operate their own individual satellite services. They consider an "efficient use" (first-come, first-served) system to be more technically and economically feasible. By contrast, developing nations consider equitable access a must if they intend to protect their rights to launch satellites into orbital positions in the future, before developed nations exhaust the orbital slots. This dispute would carry over to the 1990s before it was fully resolved.

International competition for the share of satellite communication markets was not limited to the division between developed and developing nations. In response to U.S. domination of international satellite services, western European nations, for instance, established their own transnational version of NASA in 1964. It was renamed the European Space Agency in 1973, and was responsible for the research and development of the Ariane satellites and launch programs. Modeled after the Intelsat system, Eutelsat was formed in 1977 to serve its western European member nations with the Ariane communication satellites launched by the European Space Agency. A parallel development was the establishment of the Intersputnik satellite consortium by the former Soviet Union, which served all of the eastern European Communist bloc, the former Soviet Union, Mongolia, and Cuba. Other developing nations owned and operated their own national satellite services during this early period as well, including Indonesia's Palap, which was built and launched by the United States in 1976.

The success of Intelsat as an international non-profit organization providing satellite services was repeated in another international satellite consortium—the international maritime satellite system (Inmarsat)—founded in 1979. Inmarsat now provides communication support for more than 5,000 ships and offshore drilling platforms using C-band and L-band frequencies. Inmarsat is also utilized to conduct land mobile communications for emergency relief work for such natural disasters as earthquakes and floods.

Recent Developments

During the 1980s, the satellite industry experienced steady growth in users, types of services offered, and launches of national and regional satellite systems. By the mid-1990s, satellite technology had matured to become an integral part of a global communications network in both technical and commercial respects. The following discussion will highlight a few of these important landmark events, which set the stage for the future of satellite communication in 2000 and beyond.

Due to a new allocation of frequencies in the Ku-band and Ka-band, direct-to-home (DTH) satellite broadcasts or direct broadcast satellite (DBS) services became more economically and technically viable as higher-power satellites can broadcast from these frequencies to smaller size earth receiving stations (or dishes). This development helped stimulate the television-receive-only (or TVRO) dish industry and ignite an era of DBS services around the world (see Chapter 5). The best success story of this particular type of satellite service can be found in Europe and Japan. For the Europeans, direct-to-home satellite services are deemed an economical means to both transport and share television programs among the various nations covered by the footprint of a single satellite, as the pace of cable television development has been slow due to lack of privatization in national television systems. The Japanese utilized this service to overcome poor television reception owing to mountainous terrain throughout the island nation.

Meanwhile, as the economic benefits of satellite communication became apparent for those nations that owned and operated national satellite services during the 1970s, many countries followed suit in the 1980s to become active players as well. In particular, developing countries saw satellite services as a relatively cost-efficient means to achieve their indigenous economic, educational, and social development goals in the vastly-underdeveloped regions outside of their selected urban centers. Following the example of Indonesia (Palap, 1976), India (Insat, 1983), Brazil (Brasilsat, 1985), and Mexico (Morelos, 1985) were among the countries that became national satellite system owners. Arab nations also formed their own satellite communication consortium, Arabsat (1985), to serve their domestic and regional needs. Asiasat (1989), a joint venture between Rupert Murdoch's News Corp. and the Chinese which serves both China and southeast Asia, carries Chinese language television programs targeting the vast Chinese population in the region. In the United States, DTH satellite services have traditionally served remote and rural areas not reached by broadcast and cable television signals. The launch of DirecTV by Hughes Communications in 1994 created an alternative to cable television for urban and suburban consumers as well.

This growth in the number of satellite services available in the marketplace also provided the impetus for free market competition in both domestic and international satellite communications markets. Intelsat has been a near-monopoly for most of its existence. It did endorse two regional consortia, Eutelsat and Arabsat, as well as Inmarsat as "acceptable" competitors because they did not present any significant economic harm to Intelsat's status as a global service provider. This monopoly status was successfully challenged by the Federal Communications Commission (FCC) approval of five private satellite systems (including PamAmSat and Orion) to enter the international satellite communication service market in 1985. As a result, international competition for global satellite services began, when both PamAmSat and Orion launched their satellite services in 1988 (Reese, 1990).

Another aspect of this unstoppable growth trend in the satellite industry involves the expansion of satellite business services. The first satellite system launched for business services was Satellite Business Systems in 1980. Its purpose was to provide corporations with high-speed transmission of conventional voice and data communication, as well as videoconferences to by-pass the public switched network. As earth uplink and downlink station equipment became more affordable, a flurry of corporate satellite communication networks were born (e.g., Sears, Wal-Mart, K-Mart, and Ford). These firms leased satellite transponder time from satellite business service providers to perform such tasks as videoconferencing, data relay, inventory updates, and credit information verification (at the point-of-sale) using very small aperture terminals (VSATs)—earth stations (or receiving antennas) that range from 3 feet to 12 feet in diameter. In many areas, large and small satellite earth stations were clustered in "teleports," allowing a concentration of satellite services and a place to interconnect satellite uplinks and downlinks with terrestrial networks.

As the market grows, digital transmission techniques have become increasingly popular because they allow maximum transmission speed and channel capacity. For instance, voice and video signals can be digitized and transmitted as compressed signals to maximize bandwidth efficiency. Other digital transmission methods, such as time division multiple access (TDMA) and demand assigned multiple access (DAMA), can also be utilized to achieve similar efficiency objectives. TDMA is a transmission method that "assigns" each individual earth station a specific "time slot" to uplink and downlink its signal (for example, two seconds). These individual time slots are arranged in sequential order for all earth stations involved (e.g., earth station nos. 1 through 50). Since such time allotment and sequential order is repeated over time, all stations will be able to complete their signal transmission in these repeated sequential time segments using the same frequency. DAMA is an even more efficient or "intelligent" method. This is because, in addition to the ability to designate time slots for individual earth stations to transmit their signals in a sequential order, this method has the capability to make such assignments based on demand instead of an *a priori* arrangement.

Yet another important technological advance during this period involves launching satellites that have greatly-expanded payload capability and multiband antennas (e.g., C-band, Ku-band, and/or Ka-band receiving and transmitting antennas). These multiband antennas enable more versatile services for earth stations transmitting signals in different frequency bands for different communication purposes, as well as more precise polarized spot beam coverage by the satellite. Such refined precision in spot beam coverage allows for targeted satellite signals to be received by designated earth receiving stations, which can be smaller and more economical due to the strength of the more focused spot beam signal. This, in turn, allows the satellite to better serve greater numbers of individuals and facilitate growth in the business service sector.

Current Status

As of 1998, Intelsat lists 142 member countries. It remains a financially healthy and technologically innovative entity for global communication services. Intelsat's newest satellite package, the Intelsat-VIII series, carries 22,500 two-way voice circuits and three television channels. Employing digital circuit multiplication equipment, it can also include up to 112,500 two-way voice circuits. Private satellite service competitors, such as PamAmSat among other newer entrants, are all sharing the marketplace with relative success in an ever-growing market.

The ITU currently has 187 member nations. It successfully resolved the single-most politically contentious issue ever faced by the organization—balancing equitable access to and efficient use of geostationary orbital positions for all member nations. WARC '95 presented regulations that ensured the rights of those member nations. These regulations enable those who have actual usage needs to obtain a designated volume of orbit/spectrum resources on an efficient "first-come, first-served" basis through international coordination. Additionally, these regulations also guarantee all nations a predetermined orbital position associated with free and equitable use of a certain amounts of frequency spectrum for the future.

Subsequently, during WARC '97, decisions involving frequency allocation for non-geostationary satellite services were resolved, as shown in Table 19.1.

Table 19.1
Frequency Allocation for Non-Geostationary
Satellite Services

Service	Frequency Band	Frequency Allocation
Fixed Satellite Service	Ka-band	18.9/19.3 GHz and 28.7/29.1 GHz
Mobile Satellite Systems	Ka-band	19.3/19.7 GHz and 29.1/29.5 GHz
Low-Speed Mobile Satellite Systems	L-band	454/455 MHz

Source: C. A. Lin

Non-geostationary satellites are launched into low-earth orbits (LEOs) or medium-earth orbits (MEOs) to provide two-way business satellite services. An example of this type of service is the Iridium network illustrated in Figure 19.2, which will use 66 LEO satellites to provide voice communication anywhere on earth. A more ambitious service is Teledesic, which proposes to deliver a 288-satellite constellation to provide high-speed global voice, data, and video communication, as well as "Internet-in-the sky" services. Recent market emergence in cellular radio, personal communication services (PCS), and global positioning systems (GPS) represents another reason why non-geostationary mobile satellite services are on the rise. GPS is a mobile satellite communication service that interfaces with geographic data stored on CD-ROMs. It is a navigational tool used to pinpoint locations in remote regions and disaster areas, and on vehicles, ships, and aircraft; it was originally developed for military use (see Figure 19.3). (For more on the use of satellites for personal communications, see Chapter 22.)

Figure 19.2
Iridium System Overview

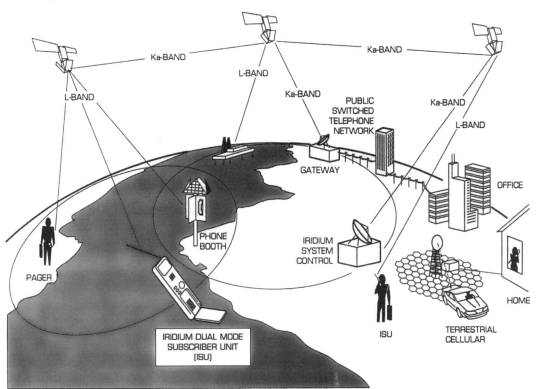

Source: Iridium, Inc.

The increasing interest in launching business satellite services in non-geostationary orbits is an outgrowth of strong market demand and increased congestion in the existing services provided through GEO satellites. The industry has seen an almost exponential market growth around the world as the end of the 20th century approaches. The projected global outlook for this service looks bright beyond 2000.

By contrast, domestic development of DTH satellite services is experiencing only moderate growth. Even though earth receiving satellite dish size can be as small as 18 inches in diameter, a bull market is not on the immediate horizon, as the cable TV industry responds to this competition with improved channel capacity and program lineups. The arrival of digital television broadcasting in 1998 paints an even murkier picture for the future of this industry. Worldwide, direct-to-home satellite services are nonetheless expanding at a respectable rate, especially in those countries that still have relatively limited national television fare due to the lack of privatization of government-owned and/or -operated television systems.

Figure 19.3
Three-Dimensional Triangulation

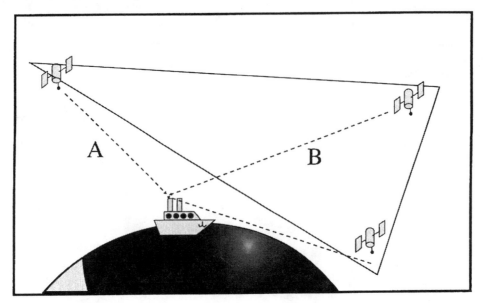

Source: J. Raley

Factors to Watch

Several major developments are expected to take place in the next few years, with implications reaching beyond the millennium. First and foremost is the continuing trend in deregulation for both national and international satellite services. On the international front, European ministries of Posts, Telephone, and Telegraphs and their Japanese counterparts, for instance, should speed up their efforts to deregulate their government control over the telecom industry and to privatize them. The arrival of the European Union will likely hasten those privatization actions, which will include market mergers as well as shake-ups. More important, the stage has been set for the eventual privatization of Intelsat, the largest global satellite communication service (Cole, 1996). Such policies, taken in tandem, should create a vibrant playing field where technological innovations and operational efficiency can help foster a competitive and expanding market.

Increased competition should be expected in the United States as well. In mid-1998, the FCC deregulated COMSAT, allowing it to have greater control over its pricing and packaging practices. This also means that COMSAT will lose its monopoly status as the only domestic satellite company to shuttle signal relays to and from Intelsat (Comsat Corp., 1998).

A related growth event to watch is the Chinese Long March satellite launch program, which suffered launch failures in recent years. A recent successful launch experiment places it on a definite path to emerge as a major player in the commercial satellite launch industry, on par with the European Space Agency's Ariane launch program. Several smaller U.S. commercial launch programs are also developing, attempting to eventually replace NASA's commercial launch program which was discontinued after the 1986 space shuttle Challenger incident. The most promising satellite communication sector involves the business applications via mobile satellite systems as well as fixed satellite services. These services, which operate in non-geostationary orbits (i.e., low- or medium-earth orbits), will compete with existing business services operated in geostationary orbits. The advantages of these new services are that they require a smaller, hence more economical, satellite system and launch vehicles, yet they are versatile enough to provide an entire range of data, voice, and video networking services. Major players entering this market are targeting the wide-open global market for personal communication services, global positioning systems, satellite Internet access, and wireless virtual networks. These include such well-financed ventures as Motorola's Iridium (66 satellites), Loral and Qualcomm's Globalstar (48 satellites), and the much publicized Craig McCaw/Bill Gates joint venture Teledesic (ultimately, up to 800 satellites).

As the global economy continues to become more intertwined, social developments and geopolitical collaboration in developing nations should also follow. Cultural exchanges across national boundaries, instead of the current one-way flow of media products from developed to the developing countries, should also gradually take place. Looking inside this "big picture" of a better world beyond 2000, the satellites' unique role as the chosen vehicle to bring forth and link together our global village is more than apparent and invariably significant.

Bibliography

Berman, H. (1998, March). LEOs and MEOs: Countdown to launch. *Via Satellite*, 46-53.

Christensen, J. (1998, February). WARC '97 results: The effect on the satellite communications industry. *Via Satellite*, 80-96.

Cole, J. (1996, October 17). Intelsat's privatization beams in rivals' cries of foul. *Wall Street Journal*, B4.

Comsat Corp. faces FCC designation as "nondominant." (1998, April 24). *Wall Street Journal*, B10.

Hudson, H. E. (1990). *Communication satellites: Their development and impact*. New York: Free Press.

Jansky, D. M., & Jeruchim, M. C. (1983). *Communication satellites in the geostationary orbit*. Dedham, MA: Artech House.

Miller, M. J., Vucetic, B., & Berry, L. (1993). *Satellite communications: Mobile and fixed services*. Norwell, MA: Kluwer Academic Publishers.

Raley, J. (1994). Global positioning systems. In A. E. Grant (Ed.). *Communication technology update* (3rd Ed.). Boston: Focal Press.

Reese, D. W. E. (1990). *Satellite communications: The first quarter century of service*. New York: Wiley Interscience Publications.

20

Distance Learning Technologies

John F. Long, Ph.D.*

After a decade of moderate growth, distance learning technologies are experiencing rapid expansion. The renewed interest in distance learning is the result of several converging forces. First, it is no longer possible to meet the demand for higher education through the construction of new campuses. In addition, those employed in today's industries find that their skills require constant upgrading in order to remain competitive or prepare for major career changes. This is especially problematic for those making a living in rural areas where retraining can require major lifestyle alterations. Taken in the context of an information-driven economy where most professions rely on a skilled workforce, distance education has emerged as a logical alternative to traditional education (Witherspoon, 1997).

Distance learning systems typically employ an amalgam of communication technologies. The actual system architecture used depends on a number of factors. Costs, geographical location of participants, visual quality requirements, sites supported, interactivity, and pedagogical goals actively define what parameters must be placed on the system design, and which technologies should be utilized for optimal outcomes.

In concert with the articulated social needs for distance education, technological developments have provided a means to accomplish and extend educational directives. The rapid rise of digital telecommunications technology and transformation of media from analog to digital formats have provided a plethora of delivery systems and databases that act as the backbone for contemporary distance education efforts. These resources and technologies are no longer restricted to an elite community, and they are

---

* Department of Communication Design, California State University, Chico (Chico, California).

changing the nature of education and training (Capell, 1995). It is also important to note that the digitization of resources empowers new models of education and an enhanced instructional experience for the traditional classroom.

The component processes that comprise distance education include learning, teaching, communication, and instructional character. It is critical to recognize distance education from the perspective of a systems approach. Anything that happens in one part of the system effects other parts of the system (Moore & Kearsley, 1996). A decision to employ a new technology in a learning system must be grounded in the pedagogical soundness of that overall system. It remains imperative that the learning environment incorporate evolving technologies, as opposed to those technologies driving the learning system.

Background

Historically, distance education has gone through several phases of development. Over a century ago, correspondence study, using mail as the enabling technology, allowed distance learners to obtain degrees and train for professional endeavors. Many of these home study schools are still operating today. For example, International Correspondence School and the American Association for Collegiate Independent Study offer courses to nearly four million students annually. The U.S. military and government are still heavily engaged in this mode of education (Diehl, 1990).

The second development phase could be characterized as an integrated approach to distance education. These systems used multiple media such as correspondence, radio, TV, audiotapes, and telephone conferences. The primary examples of these 1960s and 1970s projects are the University of Wisconsin's AIM Project and Britain's Open University (White 1996). The Articulated Instructional Media (AIM) Project hypothesized that learners could acquire skills from broadcast media and, at the same time, derive an "interactive" experience through correspondence or telephone conferencing. The original conference for AIM became the basis for the British Open University. The British OU model incorporated the public broadcasting resources of the BBC with correspondence materials, offering degree education to any adult. Since 1971, the university has served over a million students.

The emergence of broadcast media and eventually teleconferencing represented another phase of distance education. By the end of 1922, 74 colleges offered classes on radio (Barnouw, 1966). Spurred by the Ford Foundation, educational TV broadcasts from universities began in the 1930s and were omnipresent by the 1950s (Long, 1998). Collegiate consortia developed in 1961 through implementation of the Midwest Program on Airborne Television Instruction. This configuration broadcast educational material to six states from transmitters on DC-6 airplanes (Smith, 1961).

The cause for distance learning took a considerable leap forward in the mid-1960s with the launching of Intelsat I. Offering one television channel or 240 telephone circuits, the satellite provided the hope for distance insensitive relays of educational material.

During the next decade, several universities experimented with the Applications Technology Satellite (ATS). Because of diverse territories and extreme population dispersion, the Universities of Hawaii and Alaska were two of the first to utilize ATS for educational and health training (Rossman, 1992).

As satellite technology improved and increased, its viability in the educational marketplace surged forward. Business, the military, industry, and institutions of higher learning realized the potential that satellites provided. The U.S. military has developed numerous distance education networks using satellites. Using two-way audio and one-way video, the Air Force Technology Network reaches 18,000 students in 69 sites (Moore & Kearsley, 1996). To complement its videoconferencing infrastructure, the military has moved to link up 16 sites for fully-interactive exchange of video, voice, and data.

Business and industry has made substantial use of satellites as well. Nearly half of Fortune 500 companies engage in corporate training using this technology (Irwin, 1992). IBM's Interactive Satellite Educational Network broadcasts courses to over 50 locations. The system is configured as a one-way video, two-way audio system with digital keypads for testing responses. Wang Laboratories uses a similar system for international training as does Ford Motor Company and Aetna Life & Casualty. Currently, there are over 14,000 receive sites for corporate education (Dominick, Sherman & Copeland, 1996).

Because of expertise in the academic fields and corresponding needs in private industry, a virtual university, similar in concept to the British Open University, was founded in 1985. The National Technological University (NTU) offers advanced degrees by distributing digitally-compressed video by satellite. Offering more than 400 courses, its clients are corporations, such as AT&T, General Electric, and Kodak, whose employees require advanced training. Because of its digitally-compressed signal, NTU was able to substantially reduce transmission costs and increase channel capacity (Witherspoon, 1997).

Other higher educational consortia have developed to defray institutional costs and to share expertise. The Adult Learning Satellite Service operated by the Public Broadcasting Service (PBS) offers courses to public and private entities also using satellite delivered compressed video. Through generous funding from Congress, the Star Schools Program initiated regional satellite consortia designed to enhance educational opportunities for children in kindergarten through grade 12 (K-12). Most programs were advanced to educate audiences unserved by traditional methods.

The next phase of distance learning technologies utilizes computer networking and multimedia. The Internet has provided the opportunity to offer courses and complete degree programs directly through a student's computer. Unlike satellite conferencing, the student does not have to attend a site to receive instructions and interact with the instructor. Computer networking also provides asynchronous learning alternatives through the use of electronic mail, bulletin boards systems, and the like. Groups such as the Open University, University Online, and CalCampus transform regular university classes into digital format. These classes can then be replayed live or delayed through a Web browser. Graphics, digital images, and computer simulations can be added to the presentation (Open University, 1997). Most programs using the Internet have faced technological limitations due to modem speeds. The narrow bandwidths cannot yet simulate virtual classrooms as in satellite teleconferencing.

In a pilot program funded by GTE, Coast Community College and California State University, Dominquez Hills used CODEC technology and T-1 lines to share courses. The system provided for a high-quality interactive environment, but with a high price. After funding the T-1 line at $18,000 annually, it was replaced by a lower-cost ISDN (Integrated Services Digital Network) system with little loss of quality (Coast Community College, 1996).

Some educational requirements demand high-quality video. In a cooperative venture, U S WEST Communications prepared a 100-mile asynchronous transfer mode (ATM) network to train resident radiologists on medical techniques. This system provided remote participants with high-resolution image data streams, audio, and video directly to hospital sites (Schnepf, Du & Ritenour, 1995).

Recent Developments

The convergence of monetary, faculty, institutional, and technological resources continues to characterize distance learning development. Large consortia are engaged in cooperative efforts to integrate industry, higher education, and government initiatives. For example the Western Governors University (WGU) has brought together 17 western states and Guam to create a virtual university. Western institutions and companies (such as IBM, 3M, and Microsoft) will market their courses on the World Wide Web. This interactive "smart catalog" and technological nerve center of WGU identifies the media delivery system for each class. The principal means of delivery will be satellites using compressed digital video and interactive Web-based designs (Blumenstyk, 1998).

A similar design emerged with the California Virtual University. In this model, over 300 California public and private higher education institutions will provide access to the expected 500,000 new students during the next decade. At the center of this cooperative is an interactive Internet catalog of technology-mediated distance learning offerings, combined with a Web site and Intranet tools available to all campus Web sites (Design Team, 1997). Technologies used by each campus vary considerably. Some deploy ATM, ISDN, and Internet, whereas others use only broadcast television. The fastest growing method for course delivery, either as a complete course or combined with other instructional methods, is interactive Internet design (Office of the Chancellor, 1998).

Now serving students, teachers, and adult learners in all 50 states, Star Schools networks reach over one million students and 50,000 students annually. For the past 10 years, these networks have used interactive satellite conferencing. During 1998 to 1999, the STEP-Star network, previously using C-band analog technology, will expand to DBS using the Ka-band. (For a discussion of satellite bands, see Chapter 19). This expansion inexpensively increases the network scope, allowing access directly to homes. In addition, current satellite learning programs in K-12 and higher education integrate Web and Internet resources and create virtual online classrooms that expand the scope of the satellite delivered activities. This distributed learning model allows for greater flexibility by incorporating other technologies into the classroom (Krebs, 1997).

Distributed learning environments using convergent technologies are the mainstays of the International University. This virtual university offers business degrees combining the Internet, videotape, telephone, and print media. Instructors and students interact online using a virtual classroom. By 1999, Ka-band satellites will deliver multimedia directly to homes, enabling places without high bandwidth connections to access an array of materials and resources. NTU has similarly made plans to enter the distributed learning market. Its Home Learner System, beginning in late 1998, will engage engineering students with high-speed Internet connections via Ka-band directly to the students' homes (Krebs, 1998)

Clearly, a robust, high-bandwidth network that allows for seamless movement of large quantities of digital information to distributed environments has recently become a requirement for system development. The struggle to keep up with this rapidly-growing demand has caused some systems to have several different configurations available, or to prioritize information according to importance (3Com, 1998).

Because of its enormous traffic, commercial use, and small bandwidth, the Internet has severe distance learning drawbacks. In response, at a meeting in Chicago in October 1996, representatives of 34 universities agreed to endorse the creation of Internet-2. President Clinton adopted the central goals as the next-generation Internet (NGI) initiative and committed his administration to a system 1,000 times faster than today's. A very elaborate Internet-2 partnership was developed that included industry, universities, and government resources. Currently, over 100 universities are involved in research and development on collaborative projects.

Internet-2 uses very high-performance backbone network service (vBNS) (622 Mb/s), a system developed through the cooperative efforts of MCI and National Science Foundation (NSF). This technology is a combination of ATM and SONET, and 53 universities are currently using it in concert with the NSF funding (vBNS, 1998). Most research to date has resulted in collaborative simulation efforts to understand earthquakes, the ice age, molecular modeling, and other projects engaging the physical sciences.

Current Status

In order to facilitate development and use of Internet-2 infrastructure, NSF and private industry have funded various phases for state or regional internetworking. Current system architecture requires a gigaPOP (network point-of-presence transmitting gigabytes). In contemporary design and experimentation, gigaPOPs are shared by geographically proximate institutions, and they provide high-bandwidth desktop-to-desktop communication between institutions. One prototype, VITALnet, interconnects Duke, UNC-Chapel Hill, North Carolina State, and MCNC (an Internet service provider) at 2.4 Gb/s. System advocates recognize its value in distance collaborative learning and telemedicine (NC Giganet, 1997). The California Research and Education Network–2 (Cal-REN–2) has initiated a venture involving major research campuses in two major high-speed clusters or aggregation points (APs) to form distributed gigaPOPs. APs connect to each other using OC-12 lines, and OC-3 or DS-3 lines connect to regional universities. The

OC-12 lines then connect APs to the national Internet-2 and vBNS for national collaborative efforts.

Figure 20.1
CalRen-2 GigaPOP Architecture

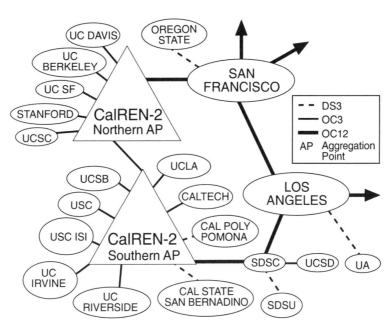

Source: J. F. Long

The NSF Bay Networks funded project is designed to establish a distributed knowledge system to support statewide and national research and educational applications (Bay Networks, 1997). The NSF has funded other regional university systems in Virginia, Washington, and Oregon for specialized high-speed applications for network-based experiments with videoconferencing, distance education, and cooperative seminar series. By late 1999, 71 institutions will be engaged in applied connectivity research (Oregon State, 1997).

With universal high-bandwidth networking capability on the horizon, the issue of instructional software that would serve as the basis of distributed instruction becomes an issue. Harnessing information in a comprehensible fashion must be addressed. In groundbreaking experimentation, Sun Microsystems has developed Distributed Tutored Video Instruction (DTVI) in concert with "Kansas," a distributed classroom. The DTVI is a 3×3 matrix of seven students, an instructor, and videotape of class content filling the nine squares of the matrix on the computer screen. This form of desktop videoconferencing allows for "social presence" and interaction among visually present learners. "Kansas" allows these participants to move about in 2-D space, to make or break connections with others to collaborate, and have independent discussions. The "Kansas" window establishes the instructional feasibility of a network-based virtual space (SunLabs, 1998).

Figure 20.2

"Kansas" DTVI System

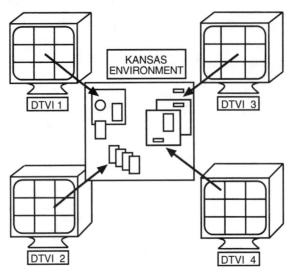

Source: J. F. Long

The issue of platform standards for high-bandwidth distance learning will likely emerge from the efforts of EDUCOM's National Learning Infrastructure Initiative. Termed "instructional management system" or IMS, this organization's standards and services will create an environment where distributed instructional environments will be managed by software. IMS standards will reflect the adoption of the HTML, VRL, and HTTP standards relative to the World Wide Web. Learning software created by corporations or universities will embody IMS protocol, thus permitting cross-platform collaborative learning and research. Currently, California State University, University of Michigan, and University of North Carolina have become the primary participants in this standards development effort (Internet-2, 1998).

Factors to Watch

The immediate future of distance learning will be characterized technologically by further development of Ka-band distributed collaborative systems to geographically diverse areas. A parallel development is likely to occur with continuous investment into Internet-2, and plausibly Internet-3. And, considering historic convergence trends, each technology will likely hybridize to offer high-bandwidth access universally.

One potential application is tele-immersion. This would permit individuals at different locations to share a single virtual environment. These systems would emulate the "Kansas" classroom described earlier, but with a third dimension and facilitated communication.

The virtual laboratory is similar to tele-immersion. This environment would create a distributed problem-solving environment wherein researchers interact and direct simulations of virtual projects. Sometimes called "collabortoriums," these projects allow for simultaneous exploration and manipulation of physical properties from multiple disciplines. NSF is planning to support software infrastructure for virtual laboratories in the next few years (NGI, 1998).

Bibliography

3Com. (1998). *3Com takes lead in race to introduce policy based networking.* [Online]. Available: http://www.3com.com/pressbox.

Barnouw, E. (1966). A *tower in Babel: A history of broadcasting in the United States.* New York: Oxford Press.

Bay Networks. (1997). *Adaptive networking strategy.* [Online]. Available: http://www.Baynetworks.com.

Blumenstyk, G. (1998, February 6). Western Governors U. takes shape as a new model for higher education. *Chronicle of Higher Education. 44* (8), 21-24.

Capell, P. (1995). *Report on distance learning technologies.* Pittsbugh, PA: Carnegie Mellon Software Engineering Institute.

Coastline Community College. (1996). [Online]. Available: http://www.cccd.edu.

Design Team. (1997). *California academic plan.* California Virtual University, Draft 3.0.

Diehl, G. (1990, January). Recent research activities of the USAF extension course institute (ECI). *Research in Distance Education, 2* (1), 16-19.

Dominick, J., Sherman, B., & Copeland, G. (1996). *Broadcasting/cable and beyond: An introduction to modern electronic media* (3rd Ed.). New York: McGraw-Hill.

Geary, S. (1997). Educators take a trip on the Internet. *Teleconference, 16* (8), 21.

Internet-2. (1998). *Learningware and the Instructional Management System.* [Online]. Available: http://www.internet2.edu/html/learningware.html.

Irwin, S. (1992). *The business television directory.* Washington, DC: Warren Publishing.

Krebs, A. (1997, November). Star achools: Approaching a decade of accomplishments. *Via Satellite, 12* (11), 56-65.

Krebs, A. (1998, February). The room–size world revisited: Global distance learning. *Via Satellite, 13* (2), 55-65.

Long, J. (1998). Public television: The struggle for an alternative. In J. Walker & D. Ferguson (Eds.). *The broadcast television industry.* Needham Heights, MA: Allyn & Bacon.

Moore, M., & Kearsley, G. (1996). *Distance education: A systems view.* Belmont, CA: Wadsworth.

NCGiganet. (1997). *Background information.* [Online]. Available: http://www.ncgni.org/background.html.

NGI. (1998). *Netamorphosis demonstration.* [Online]. Available: http://www.ngi.gov/events/netamorphosis/press.html.

Office of the Chancellor. (1998). *Survey of technology delivered distance education in the California State University.* Working Paper, CSU, Office of the Chancellor.

Open University. (1997). [Online]. Available: http://www.online.edu.

Oregon State. (1997). *Now a part of the new national high-speed network—vBNS.* [Online]. Available: http://www.osu.orst.edu.

Rossman, P. (1992). *The emerging worldwide electronic university.* Westport, CT: Greenwood Press.

Schnepf, J., Du, D., & Ritenour, R. (1995). Building future medical education environments over ATM networks. *Communications of the ACM, 38* (2), 54-70.

Smith, M. (1961). *Using television in the classroom.* New York: McGraw Hill.

SunLabs. (1998). *Kansas: A distributed classroom.* [Online]. Available: http://www.sunlabs.com/researcg/distancelearning/kansas.html.

vBNS. (1998). *vBNS University case studies.* [Online]. Available: http://www.vbns.net/press/case.

Wedenmeyer, C. (1981). *Learning at the back door: Reflections on non-traditional learning in the lifespan.* Madison, WI: University of Wisconsin Press.

White, J. (1996). Britain's open road. *The Distance Educator, 2* (2), 8-11.

Witherspoon, J. (1997). *Distance education: A planner's casebook.* Boulder, CO: Western Interstate Commission for Higher Education.

21

Cable Distributed Telephony and Data Services

Donald R. Martin, Ph.D.*

F or several years, cable companies have been expanding their system capacities to enable them to distribute more video channels, as well as offer new services requiring bidirectional capability. Cable systems with this capability can provide the necessary point-to-point connectivity for telephony, data transmission, and other services. The cable industry considers the development of these value-added services essential to its ability to effectively compete in the deregulated telecommunications marketplace.

Background

Historically, cable has been a point-to-multipoint medium, distributing broadcast, satellite-originated, and locally-originated video services to subscribers. Many cable systems have also offered audio services originating from both radio stations and satellite-delivered audio networks. As the market for pay-per-view services was developed, cable operators began rebuilding their systems to provide upstream data to manage these new services. While some systems have been engineered to provide routine transactional services, pay-per-view has been one of the most common uses of a cable company's two-way capabilities. However, pay-per-view is essentially a point-to-multipoint service for

* Associate Professor of Communication, School of Communication, San Diego State University (San Diego, California).

distributing subscriber-ordered television signals. Even the next generation of video service, video on demand, will be a medium for distributing programming designed for mass consumption.

In the early to mid-1990s cable operators began to seriously pursue the development of telephony and data services. This activity was stimulated by several significant judicial and legislative events:

(1)   A series of judicial decisions since the 1984 AT&T divestiture granted telephone companies more access into the video distribution business. This competitive threat stimulated the cable industry to consider diversifying by developing new services for their subscribers.

(2)   The Telecommunications Act of 1996 was written to encourage the telephony and cable industries to enter each other's markets (Freed, 1997). In anticipation of this deregulated environment, cable companies began to develop point-to-point services to offset the anticipated drop in video distribution revenues.

Since upgraded cable systems often did not use all of their available bandwidth for video distribution, system operators began to seek ways to generate revenue from this unused network capacity.

Through the first half of the 1990s, cable operators launched a number of demonstration projects to test the feasibility of offering residential telephony service over cable networks. These trials were geographically dispersed, and often had relatively few sample subscribers. Generally, the projects indicated that switched telephony could be offered on cable despite a number of technical issues that needed to be resolved before the technology could be widely deployed. While there was significant progress, there was no widespread deployment of cable telephony during this period.

By the mid-1990s, however, a number of cable operators also began testing the feasibility of providing data services on their systems. These technical and marketing tests confirmed that both information providers and subscribers would be interested in data services delivered at much greater speeds via a cable network than were possible with conventional computer modems through twisted-pair telephone lines. However, during this period, data transport did not become a regular service provided by cable companies for several reasons:

• Many cable companies had not finished upgrading their systems to transport data and other "auxiliary" services. In fact, even in late 1997, only about 20% of cable operators were capable of doing two-way data transfer (Quick, 1997).

• Industry standards had not been developed for cable modems, and equipment manufacturers and cable operators were unwilling to commit to equipment that might become prematurely obsolete when standards were established.

• The World Wide Web had not developed to the point that residential subscribers realized the value of faster Internet services.

Recent Developments

Recently, there have been a number of significant developments in cable delivery of wired-switched telephony and data transmission services. These developments have been driven as much by economic, regulatory, and consumer behavior as they have been by technological innovation.

Wired Switched Telephony

Residential cable telephony continues to be demonstrated in several franchises. However, many major companies have been reluctant to invest in extensive deployment of telephony service. This industry-wide caution appears to be based on uncertainty about the new competitiveness of the local telephony business, the remaining technical obstacles associated with cable telephony, and a reassessment of priorities based on anticipated subscriber interest in cable delivered online data services. Some of the uncertainty has been fueled by an assessment of the consequences that the Telecommunications Act of 1996 may have on the local exchange business. While the act encourages competition within the local exchange carrier business, this competitive atmosphere may fragment the market, making cable's entry a problematic investment. Of special concern is potential competition from resellers who can quickly enter the market and begin to offer service with a minimum of capital investment. Cable's reluctance to embrace telephony has concerned telecommunications policy makers such as Vice President Al Gore, who has urged the industry to revisit this decision (Hearn, 1997).

While this caution has slowed cable's entry into telephony, it has not been halted. Many multiple system operators (MSOs) continue to build the necessary infrastructure for telephony. However, recent rollouts of tariffed local exchange services are primarily focused on the more lucrative business segment of the telephony market. For example, in Long Island, New York, Cablevision Lightpath Inc., the telephone subsidiary of Cablevision Systems Corporation, offers regular telephone service to over 900 business customers (Cablevision Lightpath, 1997). The company has recently expanded its service into Connecticut, and claims that subscribing companies are saving 30% on their monthly telephone bills.

It appears that, after a cable company establishes telephony in a market for business subscribers, it begins to use the available plant and personnel to initiate residential service. Cablevision Lightpath, for example, has initiated limited residential service in its Long Island franchise and plans to offer similar service in Connecticut (Gibbons, 1997).

Some cable companies are also offering limited telephony services to subscribers in multi-family dwelling complexes such as apartments and condominiums. These services allow a cable operator to offer subscribers an array of cable, telephony, and security services without having to be certified as local exchange carriers. These residential shared services (RSS) also ensure that other competitors will not infringe on a cable company's video distribution business in multi-family dwellings within its franchises. Time Warner, for example, has RSS contracts with buildings representing over 25,000 residential units in Texas, Florida, and New York. Jones Communications offers services similar to RSS in its suburban Washington, D.C. franchises (Gibbons, 1997).

Some cable companies are also exploring the possibilities of offering Internet protocol (IP) telephony services on their systems. IP telephony is currently a non-switched system used by personal computer owners seeking a low-cost alternative to regular long distance telephone service. However, this emerging technology could become more sophisticated when used over hybrid fiber/coax (HFC) cable systems via cable modems (Gibbons, 1997).

Wired Data Services

In contrast to their cautious approach toward telephony, cable operators are enthusiastically entering the data transmission business. Their enthusiasm can be attributed to the incredible growth in the number of subscribers to the Internet and a realization that the data transmission speed of a transitional telephone modem is relatively slow. Many cable modems are designed to deliver downstream data at over 10 Mb/s, almost 300 times faster than the speed of popular 33.6 Kb/s telephone modems. It should be noted, however, that when many users are simultaneously using their cable modems, they may experience a decrease in transmission speeds.

A number of telephone companies are currently offering residential ISDN (Integrated Services Digital Network) services. ISDN has been relatively expensive, and it does not approach the transmission speeds offered by cable modems. Until telephone companies can extend their fiber networks or develop asymmetrical digital subscriber line (ADSL) services over the existing copper, they will not be competitive with high-speed data services provided via cable. At the present time, ADSL appears to be the telephone industry's most promising alternative to cable data services (Vittore, 1996).

A major difficulty in rolling out data transmission as a commercial venture has been developing cable modems which could be mass-produced at reasonable costs. In the last several years, prices have declined; as of mid-1998, a cable modem could be purchased for approximately $250.

New online services are being developed to take advantage of this high-speed connectivity. These services feature full-motion graphics and video segments which can only be delivered at high-speed data rates. Two competing cable services account for most of this segment of the high-speed data market. Founded by Tele-Communications Inc. (TCI), Comcast, and Cox Communications, @Home was launched in 1996. It is also now offered by a number of other MSOs such as Cablevision (Lesly, 1997). In early 1997, the Excalibur Group, a joint venture of Time Warner Cable and Time, Inc., launched a competing high-speed service called Road Runner. By early 1998, Road Runner had over 35,000 subscribers in 17 Time-Warner cable systems in the United States. In 1997, Cablevision introduced a similar service called Optimum Online, but this service is much smaller than either @Home or Road Runner (Internet access, 1997).

The costs associated with residential cable modem service vary. However, most companies charge their subscribers an additional $40 per month for unlimited data service

(Cohen, 1998). This price generally includes cable modem rental and online service via a high-speed provider such as @Home or Road Runner. Installation of a cable modem in a subscriber's residence is fairly expensive and labor intensive. It usually requires the placement of an Ethernet card inside a subscriber's computer, as well as modem set-up and cable connection. Therefore, most companies are currently charging $100 to $150 for installation (Finnie, 1997).

These complex installations and their high attendant costs have been identified as one of the major impediments to the widespread deployment of cable modems (Hettrick, 1997). In an attempt to address this problem, late in 1997, a consortium of cable industry operators and equipment manufacturers adopted interoperability specifications for cable modems. Future cable modems based on these specifications will be "plug-and-play" devices, which consumers will be able to purchase at retail outlets and install themselves without technical assistance (Goldberg, 1997).

In the past several years, the cable industry has also been developing high-speed data services targeted for businesses; they are generally designed to serve small and medium-sized businesses not large enough to justify their own networks. Depending upon the required application, the cable operator may provide the business with cable modems tied into its HFC system, or it may provide the business with discreet fiber connectivity for even faster applications. Online services are also being developed for business customers. One of these services, @Work, is the companion to the @Home residential service.

Current Status

While a few cable companies are currently offering wired telephony, the cable industry's national penetration of the local exchange carrier market is relatively insignificant. Lingering technical issues and a revised regulatory landscape have persuaded MSOs to move cautiously when considering deploying telephony services. Cox Communications, however, has been positioning itself as a telephone provider by offering discounts on long distance services provided by Frontier Communications.

On the other hand, cable is aggressively pursuing data transmission opportunities. In early 1998, there were 100,000 cable modem subscribers nationwide (Cable modems, 1998). A consumer need for connectivity to online services has stimulated a robust market for high-speed modems and the development of new online services for cable modem customers. In fact, one of the problems currently faced by cable operators is their inability to keep up with the overwhelming demand and install the modems in a reasonable amount of time.

Factors to Watch

Several factors will significantly determine when wired telephony and high-speed data services will be available from cable companies.

Wired Telephony

Despite the delayed schedules in deploying telephony, look for some cable companies to begin offering telephony to businesses and to some residential customers in selected markets. For example, Cox Communications, Cablevision, and other companies plan to deploy new local telephony services on some of their franchises in 1998 (Hearn, 1997). In general, watch for cable companies to carefully reassess their competitive positions and commit to local telephony service in selected markets where revised predictions suggest that such services are economically viable.

Data Services

Forrester Research estimates that seven million households will subscribe to a cable modem service by 2001 (Snyder, 1997). However, look for some lowering of prices to occur before high levels of penetration will be reached. A Yankee Group study suggested that 66% of U.S. households with Internet service want faster access, and 52% are interested in cable modem services. However, when confronted with the current cable modem pricing structure, only 4% of the respondents continued to express an interest in cable data services (Cable modems, 1997).

The longer it takes for telephone companies and other competitors to build the requisite infrastructure to enter the high-speed online business, the more established cable data services will become. Therefore, the timetable of the competition's roll-out may directly affect cable's success in retaining a significant market share of this business.

Bibliography

Cable modems: Speed increases. (1998, January). *Television Business International*, 89.

Cablevision Lightpath and Bell Atlantic forge comprehensive interconnection agreement. (1997, October 9). *Business Wire*.

Cohen, S. (1998, January 5). Cable modems battle DSL. *Electronic News*, 44, 56.

Finnie, S. (1997, April 22). Home, home on the Web. *PC Magazine*, 16, 64.

Freed, L. (1997, March 27). Future home connectivity. *PC Magazine*, 16, 220.

Gibbons, K. (1997, March 17). Operators plow ahead with telephony. *Multichannel News*, 18, 8.

Goldberg, L. (1997, March 3). Interoperability specs for cable modems released amid industry fanfare. *Electronic Design*, 45, 34.

Hearn, T. (1997, February 10). Phone retreat surprises Gore. *Multichannel News*, 18, 5.

Hettrick, S. (1997, December 10). Do online right, cablers told. *The Hollywood Reporter*.

Internet access: Cablevision introduces high-speed cable modem Internet service to Connecticut with launch of Optimum Online. (1997, October 20). *Edge*.

Lesly, E. (1997, October 20). Cablevision loses its tunnel vision. *Business Week*, 106.

Quick, G. (1997, November 18). 3Com prepares low-cost cable modems for home. *TechWeb News*.

Snyder, B. (1997, October 20). Interactive: Time Warner kicks off ad campaigns for Road Runner: Integrated spots promote cable modem service. *Advertising Age*, 54.

Vittore, V. (1996, July 29). Telephone industry vendors push high-speed data option for telco customers. *Cable World*, 1.

22

Wireless Telephony: Cellular, PCS, and MSS

Michael D. Milnes*

They can be found in movie theaters—both on-screen and in the audience. They are found in briefcases, purses, pockets, and backpacks. They are used for business, social, safety, crime, and disaster relief purposes. They are used while driving, walking, shopping, and writing term papers. They are cellular phones. Currently, nearly 56 million U.S. subscribers (CTIA, 1998a) make use of this technology that keeps people on the go and in touch with their employers, colleagues, customers, and families.

The advances in wireless communication technology that make cellular phones possible flowed from improvements in computer processing, battery technology, miniaturization, and new digital signal processing and transmission methods. Cellular phones are now so popular that, in some densely-populated areas, it can be troublesome to place a call during rush hour because the system is overloaded (OTA, 1995).

Wireless telephony can be broken down into four different technologies:

- Traditional cellular phones.

- Personal communication systems (PCS).

- Specialized mobile radio (SMR).

- Mobile satellite service (MSS).

* Graduate Student, Duquense University (Pittsburgh, Pennsylvania) and Senior Public Relations Specialist, Westinghouse Savannah River Company (Aiken, South Carolina.

Cellular phones and PCS are the most commonly-used technologies, and typically, hand-held phones are carried with the user. SMR is a dispatch radio service used by mobile transportation companies such as taxicab companies, trucking operations, and delivery services. In SMR, mobile radio units interconnect with the public switched telephone network (PSTN), which allows the mobile radio to function as a telephone (FCC, 1997c). MSS is defined by the FCC as "radio communication service (1) between mobile earth stations and one or more space stations; or (2) between mobile earth stations by means of one or more space stations" (FCC, 1997c). Mobile satellite services can be provided by geostationary satellites (GEOs) or low earth orbiting satellite systems (LEOs) (Schwartz, 1996).

Originally, all wireless telephony used analog radio signals, which mimic the wave pattern of the human voice to carry conversations and data. However, digital technologies, which can fit more transmissions into the same spectrum, have been developed and are being introduced into the market. Digital radio signals transmit messages by converting the speech or data into the binary language of computers. There are three major types of digital technologies competing for the cellular market: group standard mobile (GSM), time division multiple access (TDMA), and code division multiple access (CDMA) (Schwartz, 1996).

The most common wireless telephony technology is cellular telephony. The system uses a *cellular* arrangement of many low-power transmitters. The calling service area is divided into clusters, with seven cells per cluster that range in size from a single building to an area with a 20-mile radius. As a customer approaches the boundary of one cell, the network computers recognize that the cell strength is becoming weak and transfer (or "hand off") the call to the next cell without noticeable interruption (CTIA, 1998b).

Each cell has its own radio telephone and control equipment. Every cell is assigned a set of voice channels and a control channel, with adjacent cells given different channels to prevent interference. The control channel transmits data about the incoming call to the cellular phone from the mobile telephone switching office (MTSO) and informs the controller when a call is being made from a cellular phone (see Figure 22.1). The MTSO also uses the control channel to prescribe the voice channel for the call. The 25 MHz assigned to each cellular system presently incorporates 395 voice channels and 21 control channels (FCC, 1997b).

Current cellular architecture makes efficient use of the spectrum and increases system capacity. In the past, there was a single-tower system, in which each channel was used by only one customer at a time. However, a cellular system allows a channel used in one cell to be reused by a different user in another cell, as long as there is enough distance between the cells to prevent interference.

The "heart" of a cellular system is the MTSO, which is connected by microwave or landline links to all the base stations. The MTSO is connected via a high-speed digital link to the PSTN. The caller's voice signal is sent from the phone over-the-air to a base station (commonly called a tower), and back to the MTSO. The call is then either sent through a landline network or to another tower to be forwarded to the receiving phone. The MTSO manages the assignment of radio channels to users. When a caller dials a number and presses the "send" button on the cell phone, the MTSO determines channel availability and assigns the channel. During the call, the MTSO monitors the signal strength to see if it should initiate a hand-off to a nearby cell (OTA, 1995).

Figure 22.1
Diagram of a Cellular System

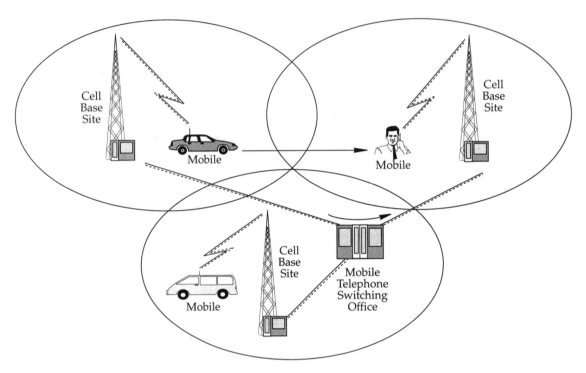

Source: U.S. Department of Commerce

Finally, there is PCS technology, which is broken down into broadband and narrowband. Broadband PCS is defined by the Federal Communications Commission (FCC, 1998a) as "radio communications that encompass a variety of mobile and/or portable radio services that provide services to individuals and businesses and can be integrated with a variety of competing networks." The PCS market is predicted to encompass small, lightweight, multifunction phones; portable facsimile and other imaging devices; new types of multifunction cordless phones; and advanced devices with two-way data capabilities.

Narrowband PCS is defined by the FCC as a family of mobile or portable radio services that may be used to provide wireless telephony, data, advanced paging, and other services to individuals and business, and which may be integrated with a variety of competing networks. Narrowband PCS could also be used for pagers equipped with a small keyboard to allow the subscriber to both receive and send complete messages through microwave signals (FCC, 1998c).

In practice, PCS is defined less as a particular service and more as a label for the operators that are using:

- The 2 GHz allocation of 1,850 MHz to 1,990 MHz for three channels of broadband PCS in each area.

- The 901-902 MHz, 930-931 MHz, and 940-941 MHz bands for three channels of narrowband PCS in each area.

At first, it was believed that PCS providers would offer a service somewhat different from that of established cellular services. PCS was supposed to be lower in cost than cellular, but would offer fewer functions. However, today's PCS services look much like those of traditional cellular providers (FCC, 1998a). The main impact of PCS is to provide price competition to established carriers and bring digital technology to the market faster. In addition, PCS also provides more services such as paging, caller ID, and voice mail.

Background

Wireless telephony can trace its roots back to Morse code, wireless telegraphy, and Guglielmo Marconi. Marconi sent the first wireless telegraph message across the English Channel in 1899. The early development of wireless was marked by the politics of the age. The first applications that became profitable for the Marconi company were ship-to-shore services for the British, Italian, and German navies in the time period leading up to World War I. Marconi's services were not allowed to compete with Great Britain's government-administered telephone and telegraph services. As a result, mobile telephony was delayed for several decades (Regli, 1997). In 1906, Reginald Aubrey Fessenden sent the first transmission of human voice by radio from Massachusetts to ships in the Atlantic Ocean. The "wireless" was used by the military in World War I to communicate over short distances (Mottershead, 1997).

Mobile telephone service began in 1946 in St. Louis and, within a year, it was being offered in 25 cities. As implied by the name, mobile telephone combined telephone protocol with radio wireless technology. The FCC allocated only a small portion of spectrum to mobile telephony; therefore, systems could only support a limited number of users. Demand for service quickly surpassed capacity and led to poor service at busy times of the day. Often, users would have to try several times before their call was placed. In some cities, carriers had to restrict the number of subscribers in order to maintain a reasonable level of service. For example, in 1978, the mobile telephone service in New York served only 525 customers and had a waiting list of 3,700. Even with restrictions on the number of subscribers, over half of the calls attempted did not go through (OTA, 1995).

Cellular technology, proposed in 1947 and developed at Bell Labs during the 1970s, made it possible for callers to move about more freely. This technology also vastly increased the capacity of mobile systems. The development of software-controlled switching technology and microprocessor-controlled radio technology made cellular telephony practical (Zysman, 1995).

In 1981, the FCC adopted rules creating a commercial cellular radio telephone service. The FCC set aside 50 MHz of spectrum in the 800 MHz frequency band for two competing cellular systems in each market, one for the local wireline telephone company and the other for non-wireline companies. The FCC designated 306 metropolitan statistical areas (MSAs) and, from the remaining counties not included in the MSAs, the FCC created 428 rural service areas (RSAs) for a total of 734 cellular markets. The FCC used comparative hearings to select the licensees in the top 30 markets where there was more than one applicant. In the remaining markets, the FCC used a lottery system to select the licensees (FCC, 1997a).

In 1994, the demand for mobile telephony led the FCC to allocate a large amount of additional spectrum to PCS providers. PCS was envisioned to reflect the future of mobile communications systems targeted to users with pocket phones rather than car phones. This new spectrum is shared by up to six additional wireless operators in each market.

Recent Developments

Although cellular phone service is a prime example of a consumer-driven market, there are other key factors in the development of this market which include:

- New standards.

- New frequency reuse concepts.

- Digital compression.

- New frequency allocations.

- The Telecommunications Act of 1996 with its resulting increase in competition.

In the United States, one of the biggest by-products of increased competition in the cellular market is the need for more antennas. The placement and the number of these towers are causing increased controversy in local zoning meetings throughout the country. Section 704 of the Telecommunications Act of 1996 governs federal, state, and local government policy on the placement of wireless antennas. The FCC claims that the new law preserves local zoning authority, but, in the same sentence, it states that the law "clarifies when the exercise of local zoning authority may be preempted by the FCC." In addition, Section 704 requires the federal government to take steps to help licensees in spectrum-based services, such as PCS and cellular, get access to preferred sites for their facilities (FCC, 1996).

The FCC has projected that the number of antennas will increase from fewer than 40,000 in 1997 to 125,000 in 2006. The growing opposition to antennas has led to an effort to conceal the towers. Towers are disguised as 100-foot-tall pines, mock bell towers, windmills, and even a three-story saguaro cactus. Antennas can be found on existing structures such as water towers, church steeples, and the Green Monster left-field wall in Boston's Fenway Park (Revkin, 1998).

While the growth of the wireless market has created antenna-siting issues, it has also generated opportunities for companies that construct the antenna sites and maintain antennas and the accompanying equipment. Consumers can now obtain cellular phone and PCS service contracts while shopping at Radio Shack or Wal-Mart. Whenever there is a consumer product, advertising accompanies it. Combined media spending among the top three wireless phone manufacturers—Qualcomm, Nokia, and Ericsson—increased by roughly 100% from 1996 to the first four months of 1997. Four wireless carriers—Air-Touch Communications, Omnipoint, Nextel, and PrimeCo Personal Communications—raised their media spending by 45% to $131 million from 1996 to the first 10 months of 1997 (Snyder, B., 1998).

Unfortunately, wireless technology also brings opportunities for the unethical portion of our population. The cellular industry estimates that carriers lose more than $400 million per year to cellular fraud (FCC, 1997a). Every wireless phone has a mobile identification number (MIN) and a manufacturer's electronic serial number (ESN) used to validate a legitimate customer. The MIN is meant to be changeable, but the ESN is fixed and unchangeable so it can act as a "fingerprint" for each phone.

In late 1990, "tumbling ESN" fraud developed. This type of fraud alters a wireless phone so it would tumble through a series of ESNs and make a caller appear to be another new customer each time a call was made. Wireless companies moved to pre-call validation to eliminate this fraud. As this diminished, ESN cloning developed. In this fraud, the criminal puts a computer chip into the phone that can be programmed with both the ESN and MIN of a legitimate user. The criminal then obtains valid combinations through various methods and makes calls until the cloned numbers are detected. The wireless industry has also taken steps to prevent this fraud (CTIA, 1998c).

Fraud is not the only illegal use of wireless technology taking place. The mere interception of some telephone-related radio transmissions may constitute a criminal violation. However, the move to digital technology is making interception of calls more difficult.

New ventures with wireless phones are not the sole domain of the criminal element. A woman in the tiny village of Bora in Bangladesh has 20 to 30 people lined up outside her house every morning to use her cell phone. Noa Jahan Begum received a loan from the Gremlin Bank, and she established a business that allows the residents of her village to communicate with the world (Kennard, 1998).

Wireless technology is used throughout the world from remote villages to congested urban areas. In Japan, the Handy Phone or PHS (a cheaper, stripped-down version of the traditional cellular phone) was introduced primarily for use in the urban areas of Tokyo and Osaka. The service has been extremely successful and spurred the sales of regular cell phones. There are now more cell phones per square mile in Japan than in the United States. As a result, Japan is quickly exhausting existing radio spectrum. In March 1997, Motorola signed a contract reportedly worth over a billion dollars to create new digital networks in Japan using CDMA technology (Snyder, D., 1997).

CDMA, developed by Qualcomm, encodes each phone conversation, creating a different binary language for each call. A code is assigned to each transmission which permits a microprocessor to pick the appropriate transmission from the airwaves and translate it for the user. CDMA does not use specific channels on set frequencies. Transmissions can occur over a range of frequencies, which allows a particular signal to be spread across several frequencies and avoid interference in clogged parts of the assigned spectrum (Regli, 1997).

Although CDMA is the choice of Motorola in Japan, it is not the only form of digital transmission. A prevalent alternative to CDMA is TDMA, which divides a single channel into time blocks. Each user's phone transmits a short burst of data, waits as other users assigned to the channel transmit their data, sends another burst, and so on. At the receiver, the bursts are reassembled into a continuous signal and converted to speech. In the U.S. TDMA system, three users share the same channel that the analog Advanced Mobile Phone Service (AMPS) system assigns to a single user (OTA, 1995). TMDA takes advantage of the frequent pauses and brief moments of silence in phone conversations to intersperse other information into a single transmission.

While the main digital adoption battle in the United States appears to be between TDMA and CDMA, much of the world has adopted another transmission standard—GSM. Similar to TDMA, GSM employs a form of time division access. Because GSM was established as a pan-European system for digital transmission, subscribers can travel throughout Europe and still use the same wireless phone.

Until the early 1980s, the FCC dictated the standards for the wireless industry in the United States. For example, all cellular carriers were required to use AMPS. However, the FCC is now leaving it to the industry to determine what digital standard should be adopted. Various industry groups have tried to determine a standard for digital transmission, but so far have been unsuccessful.

Therefore, each manufacturer of cellular and PCS devices is free to incorporate a different transmission technology. The goal of these manufacturers is to convince cellular carriers to switch to their form of digital. Some carriers have aligned themselves with specific manufacturers, while others have remained on the sideline in order to assess the success of various systems. As Brian Regli (1997) explains in *Wireless: Strategically Liberalizing the Telecommunications Market*, in order to make a service possible, it is necessary to find some sort of technological common ground. At the same time, defining standards through political means is not the best way to ensure continuing technological innovation. Pressures for innovation in the market push technical advances faster than a politician's exhortation.

Why are digital transmission standards so important? Because callers want the ability to make calls when they are outside their home geographical area (roaming). Because there is no existing common technology in the PCS band, PCS carriers may offer phones that incorporate multiple PCS technologies. Also, the technology choices of the larger PCS carriers and alliances may begin the process of reducing the number of contending technologies from seven to two or three. Another option would be to expand the area that they are licensed to serve, reducing the need to coordinate with other carriers. One strategy is to build nationwide coverage through consolidation or establishing contiguous license areas through FCC auctions.

Although the FCC claims that the auction of PCS licenses helped to kick off an entirely new industry, others claim that the auctions had more to do with obtaining billions of dollars for the Treasury than for advancing PCS services and the efficient allocation of the spectrum. Table 22.1 gives a breakdown of each of the PCS auctions, the type of licensing scheme offered, the number of licenses the FCC offered in the auction, and perhaps most important, the amount of money the auctions generated for the federal government. During the ascendancy of the idealized version of the PCS model—a wireless phone with added features such as two-way paging, voice mail, call forwarding, and others—competition was the most crucial element. But as PCS services have become a reality, many feel that the emerging players look a great deal like the cellular players of old (Regli, 1997).

This allocation of the PCS frequencies should help to expedite the shift from analog to digital PCS services. Existing digital cellular systems should also expand. In addition, spaced-based systems (Iridium, Globalstar, and Teledesic) will seek PCS allocations and other new space-based frequency allocations.

Current Status

As of mid-1998, the Cellular Telecommunications Industry Association (CTIA) estimates that there are 55,168,754 wireless telephony subscribers in the United States. The spectacular rise of wireless telephony beginning in the 1980s was "underpredicted" by analysts, government officials, equipment suppliers, and service providers. The sustained market growth for wireless service has exceeded 25% annually since 1984, often by sizable margins (see Table 22.2). This growth suggests an almost pure example of a customer-demand-driven market. In today's technologically advanced society, a new market demand perceived as essential and desirable by a broad segment of the business and consumer population will attract significant financial resources (Pelton, 1995).

Who are the major players in the wireless telephony market? When talking about the telephone world, AT&T's name usually is mentioned early and often. In the case of wireless telephony, AT&T Wireless Services, Inc. is ranked number one. Table 22.3 ranks the top 10 U.S. cellular, PCS, and enhanced specialized mobile radio licensees.

Table 22.1
FCC PCS Auction Summary

Auction	Licensing Scheme	Number of Licenses Auctioned	Net Winning Bids (in Billions)
Nationwide Narrowband PCS	Nationwide	10	$617.0
Regional Narrowband PCS	Regional	30	$394.8
A&B Block PCS	MTA	99	$7,034.2
C Block PCS	BTA	18	$904.6
D, E, and F Block PCS	BTA	1,479	$2,523.4
Total		**2,129**	**$20,744.3**

Note: 18 of C-Block's 493 licenses were reauctioned in the C-Block reauction. BTA (basic trading area) and MTA (major trading area) are based on the Rand McNally *1992 Commercial Atlas & Marketing Guide*. The FCC has divided the 120 MHz of spectrum allocated to broadband PCS into six frequency blocks (A through F). Blocks A, B, and C are 30 MHz each. Blocks D, E, and F contain 10 MHz each.

Source: FCC (1998b)

Table 22.2
Annualized Wireless Industry Data Survey Results Reflecting Domestic U.S. Commercially-Operational Cellular, EMSR, and PCS Providers—June 1985 to June 1997

Date	Estimated Total Subscribers	Annualized Total Service Revenue (in 000s)	Number of Cell Sites	Cumulative Capital Investment (in 000s)	Average Local Monthly Bill
1985	203,600	$354,316	599	$588,751	n/a
1986	500,000	$666,782	1,194	$1,140,163	n/a
1987	883,778	$941,981	1,732	$1,724,348	n/a
1988	1,608,697	$1,558,080	2,789	$2,589,589	$95.00
1989	2,691,793	$2,479,936	3,577	$3,675,473	$85.52
1990	4,368,686	$4,060,494	4,768	$5,211,765	$83.94
1991	6,380,053	$5,075,963	6,685	$7,429,739	$74.56
1992	8,892.535	$6,688,302	8,901	$9,276,139	$68.51
1993	13,067,318	$9,688,302	8,901	$9,276,139	$67.31
1994	19,283,306	$12,591,936	14,740	$16,107,920	$58.65
1995	28,154,415	$16,460,516	19,833	$21,709,286	$52.45
1996	38,195,466	$21,525,861	24,802	$26,707,046	$48.84
1997	48,705,553	$26,707,04	38,650	$37,454,294	$43.86

Source: CTIA (1998a)

Table 22.3
Top 10 Licensees Ranked by Total Population
Across Markets

Licensees	Type of License	Number of Markets	POPs
AT&T Wireless Services	PCS A/B/D/E Blocks	240	242,904,066
Sprint PCS	PCS A/B/D/E Blocks	189	234,042,364
Nextel Communications	EMSR	353	165,286,639
NextWave Personal Communications	PCS C/D/E/F Blocks	95	152,878,416
Omnipoint Communications	PCS A/C/D/E/F Blocks	127	126,599,013
AT&T Wireless	Cellular	109	71,742,921
Geotek	ESMR	25	69,045,270
AirTouch	Cellular	107	64,290,313
Western Wireless	PCS A/B/D/E Blocks	107	58,056,142
Bell Atlantic Mobile	Cellular	78	56,858,587

Source: Wireless Week

What is the make-up of the U.S. wireless market? According to Peter D. Hart of Research Associates, cellular subscription is evenly divided between men and women. However, there are approximately twice as many male PCS subscribers (66%) as women (34%). Further details on the wireless market in the United States are shown in Table 22.4.

The introduction of PCS into the market has increased the number of competitors and decreased the average wireless monthly bill. PCS prices are averaging about 20% below analog cellular in 42 of the top 50 cities with at least one PCS competitor. The average discount is about the same in 21 cities that have two new PCS competitors. The Yankee Group estimates that the average per-minute price, which was $0.45 for the typical user prior to PCS introduction, will settle at about $0.20 per minute by 1999. As costs drop, wireless will become part of the overall telephony pie and begin to displace landline traffic (Yankee Group, 1997).

With four or more wireless competitors (including Nextel) in half of the major U.S. markets, prices have dropped by an average of 25%. However, carriers want to resist price wars. Mark Lowenstein with Yankee Group suggests that enhanced wireless services (such as voice-activated dialing and electronic personal assistants) will be a method for the carriers to differentiate themselves. The Yankee Group predicts enhanced wireless service revenues will grow from $98 million in 1997 to $773 million by the year 2002 (Yankee Group, 1998b).

Table 22.4
Wireless Market Make-Up

	Cellular Subscribers	PCS Subscribers
Gender		
Male	50%	66%
Female	50%	34%
Age		
18-34	31%	36%
35-49	34%	45%
50 and over	35%	19%
Type of Use		
Business	25%	30%
Personal	58%	49%

Source: CTIA (March 1997)

Although a great deal of money is being spent on PCS and cellular, the most significant wager is in mobile satellite technology and the development and deployment of LEO and medium-earth orbit (MEO) systems. These systems allow calls to be made from anywhere in the world—from the back of a camel in the Sahara Desert to the beaches of Fiji. It is estimated that the industry is committing more than $50 billion to establish telecommunications systems in space. Motorola's Iridium LLC is investing almost $5 billion in its 66 LEO satellites and 15 to 20 earth-based gateways. Globalstar (Loral/Qualcomm) is spending $2.5 billion for its network of 48 satellites (Bruno, 1998).

In February 1998, Motorola put five Iridium satellites into coordinated orbit with the 46 already in place. The networked satellites will orbit the earth on six different planes of 11 satellites. Motorola CEO Christopher Galvin maintains that Iridium's 66 satellite constellation will be complete by mid-year 1998 with scheduled commercial service activation taking place in the fall of 1998 (PR Newswire, 1998). Iridium is estimating the costs of handsets to be $3,000, and the cost of calls averaging from $1.75 to $3.00 per minute.

The system will use a combination of TDMA and frequency division multiple access (FDMA) schemes (Bruno, 1998).

The Globalstar system will use six satellites in each of eight different planes. Backed by Qualcomm (developer of CDMA), it is little wonder that the Globalstar system will use CDMA. Globalstar estimates that its handset costs will be $750, and airtime will cost $.30 per minute. The first four Globalstar satellites were placed in orbit on February 14, 1998. Globalstar expects to launch 44 satellites into orbit by the end of 1998 and initiate commercial service in early 1999 (Globalstar, 1998).

Currently, satellite telephony's only viable customer appears to be the business person who travels around the globe. In addition, LEOs have problems serving urban areas due to the "canyons" created by skyscrapers. Users may experience signal shadowing from buildings. Within buildings, if users are not within sight of the satellite, they may not be able to use their phones (Bruno, 1998).

Factors to Watch

If satellite telephony has limiting factors, what is the future of satellites? Broadband service may be the savior of the satellite services. Motorola has proposed a hybrid LEO/GEO system, the Celestri System, formed from an integrated family of communication satellites, ground stations, and terrestrial equipment designed to provide a range of multimedia, video, and data services. By networking LEO satellites with GEO satellites, the Celestri System will provide regional broadcast capabilities with real-time interactivity. Earth-based control equipment would include terrestrial-based network interfaces to telecommunications infrastructures, the Internet, corporate and personal networks, entertainment networks, and residences. Motorola claims that the system would provide users instant access to a broadband network and true bandwidth-on-demand (Motorola, 1998).

The Teledesic LEO network is being called the "Internet in the sky" and is backed by Microsoft's Bill Gates and cellular phone pioneer Craig McCaw. Teledesic will require 288 satellites, none of which were launched as of early 1998 (Rohde, 1997). The Teledesic network will be a high-capacity broadband network that combines global coverage of LEO satellites, the flexibility of the Internet, and fiber-like quality of service. Teledesic will offer bandwidth on demand ranging from 64 Kb/s to 155 Mb/s. The big disadvantage is that broadband satellite data services are four or five years away (Bruno, 1998).

However, wireless data transmission is here now. One-third of the current U.S. workforce (43 million) is mobile. These mobile workers need easy access to huge resources of legacy enterprise data and the ability to use simple messaging technology. The Yankee Group estimates that 33% of large U.S. corporations will provide field service and sales personnel with wireless Intranet access by 2000 (Yankee Group, 1998a).

Wireless data will enhance the cellular phone as a communication device, but what will be the next application? Perhaps the future will not be one function, but a group of complimentary applications. These functions may include wireless fax, e-mail, and Internet. Short messaging services will combine a phone and a pager to allow simple transactions such as banking and reservations. Finally, one of the biggest reasons people adopted cell phones was the security they provided in case of a car breakdown or the ability to contact others when delayed in traffic. Wireless data will provide users access to navigation and traffic systems, roadside assistance, and interactive maps. Although the market is young, it promises to bring many exciting innovations (CITA, 1998d).

Bibliography

Bruno, C. (1998). Trouble in the skies? *NetworkWorld*. [Online]. Available: http//www.nwfusion.com/netresources/0119.sat4.html.

Cellular Telecommunications Industry Association. (1998a). *CTIA's annualized wireless industry data survey results*. [Online]. Available: http://www.wow-com.com/professional/reference/graphs/gdtable.cfm.

Cellular Telecommunications Industry Association. (1998b). *How wireless works*. [Online]. Available: http://www.wow-com/consumer/works/index.cfm.

Cellular Telecommunications Industry Association. (1998c). *Wireless fraud*. [Online]. Available: http://www.wow-com.com/professional/fraud/background.cfm.

Cellular Telecommunications Industry Association. (1998d). *The world with wireless*. [Online]. Available: http://www.wow-com.com/professional/wirelessapps/white.cfm.

Cellular Telecommunications Industry Association. (1997). *Wireless statistics*. [Online]. Available: http://www.wow-com.com/professional/reference/usdemog.cfm.

Emmett, A. (1997). Technologist of the year. *America's Network*. [Online]. Available: http://www.americasnetwork.com/issues/97issues/970115/011597_cover.html.

Federal Communications Commission. (1998a). *Broadband PCS fact sheet*. [Online]. Available: http://www.fcc.gov/wtb/pcs/bbfctsh.html.

Federal Communications Commission. (1998b). *FCC auction summary*. [Online]. Available: http://www.fcc.gov/wtb/auctions/mta.html.

Federal Communications Commission. (1998c). *Narrowband PCS fact sheet*. [Online]. Available: http://www.fcc.gov/wtb/pcs/nbfctsh.html.

Federal Communications Commission. (1997a). *Cellular fraud*. [Online]. Available: http://www.fcc.gov/wtb/cellular/celfrd.html.

Federal Communications Commission. (1997b). *Cellular radio telephone service fact sheet*. [Online]. Available: http://www.fcc.gov/wtb/cellular/celfctsh.html.

Federal Communications Commission. (1997c). *Specialized mobile radio service*. [Online]. Available: http://www.fcc.gov/wtb/specrdsv.html.

Federal Communications Commission. (1996). *New national wireless tower siting policies*. [Online]. Available: http://www.fcc.gov/siting.html.

Globalstar. (1998). *Launch schedule*. [Online]. Available: http://www.globalstar.com/intro.htm.

Kennard, W. (1998). *Wireless 98 speech in Atlanta*. [Online]. Available: http://www.wow-com.com/professional/convention/atlanta1.cfm.

Mason, C. (1997). The truth about CDMA. *America's Network*. [Online]. Available: http://www.americasnetwork.com/issues/97issues/971201/120197_cdma.html.

Mottershead, A. (1997). Radio. *Groller Electronic Encyclopedia*. New York: Groller Electronic Publishing.

Motorola. (1998). *Facts about the Celestri System*. [Online]. Available: http://www.mot.com/GSS/SSTG/projects/celestri/ASD_product.html.

Office of Technology Assessment. (1995). *Wireless technologies and the national information infrastructure*. Washington, DC: U.S. Government Printing Office (OTA-ITC-622).

Pelton, J. (1995). *Wireless and satellite telecommunications*. Upper Saddle River, NJ: Prentice-Hall.

PR Newswire. (1997). *Iridium completes the year with a successful launch; Global system enters 1998 with more than two-thirds of satellites in orbit.* [Online]. Available: http://news.wirelessdesignonline.com/PRNewsWire/prn19971220-74862.html.

Regli, B. J. W. (1997). Wireless: *Strategically liberalizing the telecommunications market.* Mahwah, NJ: Lawrence Erlbaum Associates.

Revkin, A. (1998, January 11). It's a tree! A cactus! Wireless carriers resort to hiding antennas. *New York Times*, 21.

Rohde, D. (1997). Satellites: Are they for you? *Network World.* [On-line] Available: http://www.nwfusion.details.html.

Schwartz, R. E. (1996). *Wireless communications in developing countries.* Boston: Artech House.

Snyder, B. (1998, February 2). Ad budgets boom as wireless phones get into branding. *Advertising Age, 69* (6), 12.

Snyder, D. (1997). *Marketplace.* [Online]. Available: http://www.usc.edu/search97cgi/s97-cgi?

Wireless Week. (1998). *Top 200 licensees cellular, PCS, ESMR carriers ranked by POPS.* [Online]. Available: http://www.wirelesswek.com/industry/top200.htm.

Yankee Group. (1997). *PCS is driving down U.S. wireless pricing.* [Online]. Available: http://www.yankeegroup.com/press_release/pcs_pricing.html.

Yankee Group. (1998a). *Synergies between the Internet and mobile computing will accelerate wireless data adoption.* [Online]. Available: http://www.yankeegroup.com/press_release/impact.html.

Yankee Group. (1998a). *Wireless competition will force carriers to deliver enhanced wireless services.* [Online]. Available: http://www.yankeegroup.com/press_release/wireless_force.html.

Zysman, G. (1995). Personal communications and instant networks. *AT&T Technology, 10* (2), 2-7.

23

Personal Communication: Pagers, Palmtops & PDAs

Philip J. Auter, Ph.D*

A man sits on the deck of a beachside bar on a Wednesday afternoon, sipping his drink and soaking in the beauty of white beaches and azure water. He is not on vacation. He is not at home, nor at the office. He is working—telecommuting.

Checking the datebook on his handheld personal digital assistant (PDA), he sees that his book chapter revisions are due in two days. Pressing another button on the device, he looks up the editor's name, address, and contact numbers. The editor needs to be contacted today to clarify a few possible changes.

E-mail is the easiest way to reach her on Wednesdays, so the author presses another button on the PDA, opening up the e-mail software. Using a digital stylus, he "writes" on the screen of his PDA, composing a quick message to the editor.

The persistent beeping of his pager interrupts his thoughts. He's received two digital messages. One is the number of his voice mailbox (which he has forwarded to his office phone number). The other is his home number—his son is calling.

Heading over to the payphone across the deck, he checks in with his voice mail and his home. He then snaps a credit-card-sized modem onto the bottom of the PDA and, because there is no phone jack available, he adds his acoustic coupler. He attaches the device—which resembles an antique modem—to the handset of a payphone. Depositing more coins, he dials the local phone number to his Internet service provider. His message is uploaded, and his incoming mail down-

---

* Assistant Professor of Communication, Department of Communication, University of South Alabama (Mobile, Alabama).

loaded. He also takes a moment to check CNN Interactive's Website—through his PDA—for the latest news.

Returning to his table, he taps out the latest revisions to his chapter on his palmtop computer. The WinCE operating system and the stripped-down word processor on the small device exchange files easily with his desktop computers at home and at work.

The persistent beeping of his pager again interrupts the sounds of ocean waves and seagulls. He's received another message. This alphanumeric page is actually a forwarded e-mail message from the editor. She has received the author's message and would like to schedule a conference call for tomorrow morning at 9 A.M. He adds that entry to his PDA datebook, packs up his toys, and heads home. It's been a productive day.

Whether this sounds like a dream—or a nightmare—it's not a vision of the future. All the technology and services described above are available today—although they rarely work together this well. Pagers, palmtops, and PDAs, along with cell phones and wireless modems, are helping to make communication truly personal, portable, and available to almost anyone.

Background

"Personal communication" devices have developed along four lines: cellular telephones (discussed in Chapter 22), paging devices, palmtop computers, and personal digital assistants. Each has its own unique history.

Paging

The concept of paging was born in 1949 when hospitalized radio engineer, Charles Neergard, was annoyed at the public address system used to notify doctors that they had messages. He felt that there must be a quieter way to inform physicians and, shortly thereafter, the first internal radio-paging network was developed. In the 1970s, voice circuitry was added to pagers. Digital displays became common in the early 1980s, and alphanumeric paging technology soon followed (Hon, 1996; Rose, 1995).

All paging systems work on the same basic concept. A caller dials a phone number, which is sent to a paging terminal that transmits a tone, voice, numeric, or alpha message. The paging terminal converts the message into a pager code and relays it to transmitters throughout the coverage area. The transmitters send out the code to all pagers in the coverage area tuned to a specific frequency, but only the pager with the proper address (cap code) will be alerted (Rose, 1995; Motorola, 1996).

PDAs

Personal digital assistants are small, calculator-like devices that offer a set of specific information storage and processing functions. Their evolution started with organizers with 32 kilobytes or more of memory that provided a limited set of functions. Companies such as Sharp used the technology to produce address books, calendars, to-do lists, and calculators. These first-generation PDAs are still available today at remarkably low prices.

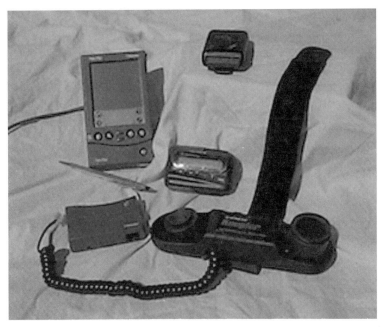

The Wired Roadwarrior's Travel Kit (clockwise from top left): PalmPilot PDA, Motorola numeric pager, Konexx acoustic coupler, PalmPilot modem, Cross Pen with Digital Writer refill, and Motorola alphanumeric pager (center). Photo courtesy of Philip Auter.

Apple's Newton Message Pad, introduced in 1993, was one of the first PDAs and, for a while, held over 75% of the small PDA market. Although large and expensive by today's standards, it incorporated some of the current PDA units' features, including handwriting recognition and synchronization of data to a desktop computer (Bergwerk, et al., 1995; Hamm, 1996). Current PDAs are smaller, less expensive, and more powerful, and they are becoming more and more indistinguishable from palmtops, pagers, and cell phones.

Palmtops

The ever-shrinking laptop computer evolved into a sub-notebook and finally a palmtop computer. Most palmtops can be differentiated from PDAs in that they have a tiny keyboard and no handwriting recognition. Operating systems and user interfaces have varied, but an attempt has been made to mimic the Windows-based desktop computer operating system (OS). Although the WinPad OS was doomed to failure, Microsoft's newer WinCE operating system has been a strong entrant in the field. Sporting stripped-down versions of word processors, spreadsheets, and Internet connectivity, today's palmtops look and act more like pint-sized desktop computers than PDAs (Hamm, 1996).

Recent Developments

Paging

A number of developments have occurred in paging which have generated more interest in advanced services and resulted in greater penetration of all paging services.

Alphanumeric paging has become the fastest-growing segment of the industry, in part because you no longer have to buy a dedicated "box" to send alpha pages. Today, many commercial and shareware programs allow anyone with a PC (or a Macintosh) and a modem to send a text message to an alpha pager.

Many paging companies have added enhanced services such as e-mail-to-pager gateways and information services that send news updates and stock quotes directly to alphanumeric pagers.

Two-way paging, a system that allows someone in the field to respond to a page directly from the pager, has finally begun to work well. Another innovation that has shown greater success is "flex" technology. Fewer pages are lost with "flex"-equipped units, because a message is continually rebroadcast from the tower until the pager confirms that it has been received. Messages sent while your pager is off or out of the service area are thus received once the device is active and within its reception range.

Palmtops and PDAs

Certainly one of the biggest advancements in PDAs and palmtops is the ability to "easily" synchronize data with a desktop PC. Early attempts to do this via infrared light beams were cumbersome and expensive. Today, PDAs such as the PalmPilot are placed in a cradle that connects to a dataport on any PC or Mac. After an initial configuration, the device can synchronize data with the press of a single button.

New operating system software can be credited for this development, along with other advances in PDAs and palmtops. The open architecture of the PalmPilot OS has enabled the development of many commercial and shareware applications that stretch this device's flexibility to the limit. Games, calculators (of many types), drawing applications, and Internet access are just a few of the functions that make this PDA seem more like a pocket-sized desktop computer.

Not to be outdone, Microsoft has finally developed a stable palmtop/PDA operating system with their WinCE. Although it operates in an entirely different way from the desktop versions of Windows (programs that are tens of megabytes larger), this stripped-down OS mimics the Windows operating environment—making life simpler for some users. Publishers of many of the more popular desktop programs are developing palmtop versions of their software for these portable devices.

Another major advancement in PDAs is more stable—and forgiving—handwriting recognition. Here again, the PalmPilot from 3Com has received rave reviews. Ironically, one of the first PDAs with handwriting recognition, the costly Apple Newton and its younger relative, the eMate, have finally "died" after a lingering illness (Apple, 1998). "Writing" on a PDA is accomplished by using a digital stylus—a not-too-sharp, pointed object—onto an electronic screen. Regular pens, pencils, and similar items are not recommended. However, forward-looking pen companies like Cross have developed inserts for their traditional pens, effectively turning any pen into a digital stylus.

Connectivity has also advanced dramatically for PDAs and palmtops. Streamlined Internet software, credit-card-sized modems, and new access options have allowed PDA and palmtop users almost seamless access to e-mail and the Web. Wireless cell modems make it possible for almost universal access for the financially well off. For those of us on a more limited budget, tiny modems allow access wherever a dataport can be found. And acoustic couplers (reminiscent of early modems) can help you establish a link through almost any telephone handset—even a payphone (e.g., Konexx).

Most of the PDA and palmtop offerings are becoming indistinguishable from each other and from other electronic devices. As the technology advances, pagers, PDAs, palmtops, and cell phones are moving toward one hybrid communication unit.

Current Status

Paging

Paging is one of the fastest-growing segments of the telecommunications industry, with more than 42 million units in service in 1996. By 2000, industry analysts predict that the number of subscribers will grow to almost 70 million. One out of every seven people will be carrying a wireless device (PageMart, 1997).

The greatest demand growth is in the sector of alphanumeric paging, in part because these devices allow for the forwarding of electronic mail. Of the 600+ paging companies in the United States, 95% of the business is controlled by the top 30 companies (PageMart, 1997).

Palmtops and PDAs

Although initial interest in palmtops and PDAs has been low, advances in hardware and software have increased growth. The current market for both types of units is estimated to be two million units—with a projected growth to five million units by 2000 (3Com, 1997).

Due in part to the flexibility of its OS, the PalmPilot quickly grabbed 51% of the standard handheld computer market (which includes palmtop devices) just one year after the product began shipping (3Com, 1997). Although it is quite popular with consumers, and has been licensed for devices such as IBM's WorkPad (a legal PalmPilot clone), the "Palm OS" faces a tough challenge against Microsoft's newest competing OS, WinCE.

Factors to Watch

Pagers, palmtops, PDAs, and cell phones will continue to shrink in size, offer more features, and, most important, merge into one unit. Market size will increase as manufacturers target low-end versions of their products to new market segments. Parents are already providing young children with their own digital pagers, and toy manufacturers have started developing "digital diaries"—low-end PDAs targeted toward kids (e.g., Tiger Toys "Dear Diary"). In the near future, the paging industry may also finally work the bugs out of two-way paging.

As palmtop computers and PDAs become more similar, look for the Microsoft WinCE and 3Com PalmPilot operating systems to battle for market dominance. Some companies will be in search of the "incredible shrinking PDA"—attempting to make functional devices that are as small as possible. One company, Rolodex Electronics, already has a PDA/organizer that is built into the architecture of a PCMCIA card. Inserting the card into the desktop computer's PC card slot and synchronizing it transfers data from a desktop PC quickly and easily. The credit-card-sized organizer can carry thousands of addresses, appointments, and "to-do's." Some even predict that a more advanced PDA will appear in the form of a watch in the near future (Andress, 1997). Watch pagers already exist.

As most PDAs and palmtops include more and more features, they will consume more power, possibly straining the life of even the new "high energy" batteries (Eveready Battery Company, 1998). Features standard in today's laptop and desktop machines, including color, will be incorporated into many future palmtops and PDAs. And if researchers at MIT's Wearable Computing Project have their way, we'll be wearing our computers in the not too distant future. Perhaps that would be the truest form of personal computing.

Bibliography

3Com Corporation. (May, 1997). *Dataquest and PC Data Research studies place U.S. Robotics' PalmPilot at lead of thriving handheld computer market.* [Online]. Available: http://palmpilot.3com.com/pr/pressrel_study.html.

Andress, S. R. (1997). *Digital concepts—PalmPilot 2001c, Electra, Communica, and Exec.* [Online]. Available: http://www.pe.net/~scotta/digitalconcepts/.

Apple Inc. (February, 1998). Apple discontinues development of Newton OS. *Apple media & analyst information press releases.* [Online]. Available: http://www.apple.com/pr/library/1998/feb/27newton.html.

Bergwerk, J., Chan, T., Haddad, R., Lee, R., & Sudyatmiko, R. (1995). *Personal digital assistants.* [Online]. Available: http://ieem.stanford.edu/ie275/PDA/home.html.

Eveready Battery Company Inc. (March, 1998). *Energizer announces superior high-drain battery.* [Online]. Available: http://www.eveready.com/news/highrate.html.

Hamm, S. (1996). PDA vets are getting ready for round two. *PC Week.* [Online]. Available: http://www.zdnet.com/pcweek/news/1118/18vets.html.

Hon, A. S. (1993, 1996). An introduction to paging: What it is and how it works. *Motorola Electronics.* [Online]. Available: http://www.mot.com/MIMS/MSPG/Special/explain_paging/ptoc.html.

Motorola Electronics Pte Ltd. (1996). *What is paging?* [Online]. Available: http://www.mot.com/MIMS/MSPG/Special/explain_paging/overv.html.

PageMart. (1997). *Paging facts & industry links.* [Online]. Available: http://www.pagemart.com/f_l.html.

Rose, D. S. (1995). A very brief history of paging. *AirMedia, Inc.* (formerly Ex Machina). [Online]. Available: http://www.gowireless.net/histpage.htm.

24

Teleconferencing

Kyle Nicholas*

The terms teleconferencing and videoconferencing are often used interchangeably in trade journals and marketing promotions. A teleconference can describe any kind of two-way multiparty communication that uses a telephone line link. However, in this chapter, we define a teleconference as when two or more individuals "meet" via a two-way, interactive audio/visual link. For most users, teleconferencing acts and feels like two-way television, joining them across distance in real time. Teleconferencing is made possible by sending audio and visual signals from cameras through the air or down phone lines, generally with a computer in between to compress and filter the signal for greater speed and clarity. Teleconferences facilitate large groups such as corporate conventions, small groups such as work groups, and one-on-one conferences. Users employ different hardware and software depending on the scale of the conference, the level of interactivity required, and the locations of participating conferees.

Large group teleconferences are generally held in convention halls or theaters, where hundreds of spectators can watch a speaker, and the speaker gets a view of the crowd. The system resembles broadcast television, and generally utilizes satellite or fiber optic transmission networks. Such conferences are generally less interactive; often, only an audio link is provided for the crowd to respond to the speaker. Small group conferences may be held in specially-built studios or in small rooms utilizing "roll around" monitors with attached video cameras. By using omnidirectional microphones and voice tracking cameras, small group teleconferences allow a great deal of interactivity, much as in a workgroup or small seminar.

Both small groups and one-on-one teleconferences may utilize desktop computer technology. Desktop systems utilize a small video camera and microphone attached to the computer. Incoming video is visible in a frame on the user's monitor. Information is

* Doctoral Candidate and Researcher, Texas Telecommunications Policy Institute, University of Texas at Austin (Austin, Texas).

usually transmitted via a dedicated local area network (LAN) or wide area network (WAN) using an all-digital transmission protocol, such as the Integrated Services Digital Network (ISDN). However, some new applications utilize standard telephone lines, and others communicate via the Internet. If a group is using a desktop videoconference system, users will be connected through a "bridge," which can either be a hardware switch or software that allows a single user's transmission to reach the desk of each member of the group. For a desktop teleconference participant, the process is much like a conference phone call, only with video, graphics, and other computer applications included. Desktop video is the fastest growing segment of teleconferencing technologies.

Background

In 1935, the German Post Office linked four locations via a studio-based audio/visual system. Shortly thereafter, an Iowa school district employed audio teleconferencing to teach homebound students. By 1958, AT&T had linked 10 U.S. Department of Defense sites through a fast-connecting audio system. A few years later, the first private video teleconferencing system was installed by First National City Bank (Olgren & Parker, 1983). Despite these pioneering efforts, and at least two concentrated marketing efforts by AT&T, teleconferencing failed to catch on with consumers.

AT&T demonstrated a two-way videophone at the 1934 World's Fair and again at the 1964 fair. Participants in these demonstrations were impressed with the technology, but considered it more novelty than necessity (Noll, 1992). The technology continued to improve, and color video images approaching broadcast quality were available by the early 1980s. High-speed phone lines and the emergence of digital ISDN lines, combined with improving compression techniques, enabled more reliable and functional teleconferences. Innovators in medicine, education, and industry experimented with new applications (Olgren & Parker, 1983.)

By the late 1980s, however, teleconferencing was still waiting for widespread acceptance. One important reason was the cost; the special studios required for a teleconferencing were prohibitively expensive. Another critical factor inhibiting teleconferencing acceptance may be human behavior. Users have had to learn how to "stage" conferences, adapting their normal meeting behavior to the technology. Because of the technical demands of computerized video compression, users must move and speak more slowly to avoid a fuzzy or jerky picture. Users also had to become accustomed to seeing themselves on camera without becoming distracted. Finally, much business gets done outside of meetings, and much learning is accomplished outside the classroom so, for many users, teleconferencing still needs to be supplemented by face-to-face or group meetings (Noll, 1992).

Despite the factors mentioned above, teleconferencing does appear to be coming into its own (King, 1996). Adoption of teleconferencing in the business sector is being driven by an emerging need for routine dissemination of multimedia information (Cukor & Coppock, 1995). Consolidation of firms into giant global corporations and the fact that even small startup companies need to compete in the global marketplace have resulted in an increased need for communication and collaboration. While many companies still employ LANs for teleconferencing, improvements in the CODEC—the computer processor which utilizes software to compress and decompress digital information in teleconferencing—allow small companies to teleconference via the public phone network with only somewhat diminished results. Desktop teleconferencing is improving due, in part, to higher computer processing speeds and faster modems. In a 1996 survey, 19% of U.S. businesses indicated that they already employ desktop teleconferencing, and 47% said they planned to implement it by 1999 (Korostoff, 1996). Network security is apparently not a limiting factor; according to a recent survey, corporate users are not concerned with teleconference hacking (Cukor & Coppock, 1995).

The key inhibiting factor has been a lack of interoperability. Vendors have employed proprietary algorithms and packaged hardware and software so that one system will not "communicate" with another. This has led to two problems:

(1)   Proprietary systems can quickly become obsolete, as new innovations are not "backward compatible" with old systems.

(2)   Consumers are accustomed to technological transparency in communication media. Phones and faxes "talk" to each other regardless of brand. Teleconference users expect the same from conferencing technologies (Videoconferencing, 1996.)

Recent Developments

Standards

The development and acceptance of technological standards for compression and transmission of audio and video signals is key to teleconferencing. The International Telecommunications Union (ITU) has adopted several related standards, and vendors are rapidly building compliant software and hardware. The three most important standards for the implementation and adoption of teleconferencing are known as H.320, H.324, and H.323.

H.320—The ITU's H.320 standard covers direct connections between conference sites. The connection is usually made using a dedicated, circuit switched network (see Chapter 18), such as the ISDN employed by the majority of large group videoconferencers (ITU, 1996). International teleconferencing may get a boost from the adoption of this standard, particularly in Europe and the Pacific Rim where ISDN is widespread. The H.320 standard will be particularly important to organizations with an ISDN network already installed, especially in studio and arena configurations.

H.324—Currently, 6% of Americans (approximately 8.2 million people) telecommute on a regular basis, according to the Kensington Telecommuting Survey. The number is expected to grow to 18 million by 2005 (ITCA, 1998). The H.324 standard, ratified in 1996, has been developed to facilitate desktop teleconferencing in the home via twisted copper pair lines. With the advent of H.324, vendors can exploit the home teleconferencing and videophone markets, including the market for telecommuters. Several vendors, including giants Intel and Compaq, now include teleconferencing capabilities as part of a standard personal computer package (Korostoff, 1996). With the introduction of home videoconferencing packages as low as $260, small non-profit groups and individual users can afford to try teleconferencing, dramatically increasing the number of video-ready endpoints (Davis, 1997). Standards are particularly important in this market because they allow individuals to shop for price and features, without worrying about compatibility with other systems.

H.323—The Internet utilizes packet switching—digitized messages are broken up into small chunks and routed through various pathways to their final destination, a method that makes teleconferencing over the Internet particularly problematic (Labriola, 1997). The third ITU standard, H.323, applies to networks using Internet protocol (IP), generally LANs and WANs. Data conferencing via the Internet is already appearing as an embedded capability in Internet browsers. Some data conferencing software, such as Microsoft's NetMeeting, can be downloaded free. H.323-compliant videoconferencing capabilities are being embedded into the latest versions of Microsoft Explorer and Netscape Communicator Web browsers (Davis, 1997).

The move to desktop teleconferencing for organizations is also enabled by more sophisticated bridging systems. IP multicast software using H.323 allows a source to send a data stream (video, audio, and text) that several destinations can receive by subscribing to a designated address (Krapf, 1997). Multicast means that users do not have to view all audio/visual activity in a conference. For instance, in a virtual classroom, a student doesn't necessarily want to watch all the other students. She may receive video only from the teacher, although the teacher may monitor video inputs from all students. Multicast reduces desktop clutter and frees up valuable bandwidth on the network.

Some vendors have introduced teleconferencing products designed for home users (generally referred to as videophones) compatible with H.323. However, few of the products utilizing Internet connections from the home support frame rates of more than three to five frames per second, compared with 30 frames per second for full-motion video. Additionally, the Internet is constantly plagued with frustrating delays that can deteriorate video quality (Labriola, 1997).

Emerging social and organizational factors are driving the diffusion of teleconferencing. Collaboration and productivity are the watchwords for enterprises in the 1990s, and teleconferencing is increasingly accepted as a tool that enables both. Enterprise networking—connecting the computers in an organization with the resources of vendors, suppliers, and collaborators—has gained increasing acceptance as an organizational model (Bracker, 1995). Teleconferencing facilitates enterprise networking to the extent that it allows users to share information efficiently and collaborate effectively (Morrison & Liu Sheng, 1992). The introduction of high-speed modems in general purpose computers and the implementation of higher-speed LANs and WANs in the organizational setting enable faster, more "life-like" transmission of audio/visual signals. The widespread use of general-purpose video cameras means that many people have had the experience of appearing on video screens, and e-mail is a part of the daily life of millions. These experiences may make users more comfortable with desktop communications and video imagery in general.

Current Status

The teleconferencing industry is dominated by PictureTel, which accounts for about 53% of the market with approximately 50,000 endpoints (Dataquest at PictureTel, 1998). VTel and Compression Labs Inc. (CLI) merged in 1997, creating the number two player with 25% of the market and about 23,000 endpoints (VTel, 1998). ViaVideo was absorbed by Polycom in early 1998; the company is now poised to offer video, audio, and data conferencing suites, a first in the industry (Polycom, 1998). Other vendors with significant market share include Nortel, Objective Communications, Databeam, and US Robotics.

Teleconferencing products comprise a $5 billion industry (ITCA, 1997). Hardware sales totaled $1 billion in 1997 and are predicted to grow to $5 billion by 2001. In 1997, vendors selling H.323-compliant products aimed at organizations using LANs garnered approximately $500 billion in revenues. That figure is predicted to be about $1.8 billion in 1999. Sales of circuit-switched (H.320 compliant) systems totaled about $250 million in 1997 and are expected to exceed one-half billion dollars in revenues by 1999. The market for products aimed at the copper-based phone line market (H.324) is predicted to grow from $250 million in 1997 to about $400 million in 1999 (Forward Concepts, 1997). The majority of this growth will be in the desktop market, driven primarily by professionals and individual consumers, a group sometimes referred to as "prosumers." Corporate purchases will drop from about one-third of the total market to about 25% in 1999 (Forward Concepts, 1997). Predicting growth in the teleconferencing industry has been particularly problematic in years past, however, and these figures may be colored by the wishful thinking of an industry that has yet to develop its "killer app." To date, the exponential growth regularly forecast has not materialized. Additionally, gross revenues for the industry will be tempered by steadily declining prices for teleconferencing products.

There has been significant movement from hardware to software solutions for teleconferencing. Although a recent comparison notes hardware solutions maintain a slight edge in quality, software is often less expensive, more adaptable, and easier to implement (Hibner, et al., 1997). Prices for ISDN desktop systems using H.320 are well under $2,000. Prices for systems using H.324 are under $400, and some systems bundled with new PCs are far less than that (Forward Concepts, 1997). Polycom now sells the full-featured View-station for a base price of less than $6,000 (see photo).

A significant factor in adoption may be the cost of ISDN lines, still the preferred transmission medium. ISDN prices in the United States vary widely by provider and region. Prices for the first year of business use (installation plus monthly charges) range from $470 plus usage charges for PacBell customers on the West Coast to a $706 flat rate for Southwestern Bell customers in Texas. Bell Atlantic customers on the East Coast would pay about $617 plus usage for the same service. Residential prices are somewhat lower (ISDN, 1998).

Polycom's Viewstation System consists of a set-top box, remote, and microphone unit. It sells for less than $6,000 and features high-quality, full-motion video, an automatic-tracking camera, and the ability to display computer presentations. Photo courtesy of Polycom.

Costs associated with ISDN lines remain a considerable part of the teleconferencing price equation, especially when multiple lines are required for teleconferences. But for many businesses, ISDN enhances communication speeds and efficiencies across the board, including accelerated Internet access and faster processing throughout the LAN. The portion of ISDN costs accorded specifically to teleconferencing will vary from business-to-business or home-to-home. One of the interesting ways ISDN prices fluctuate is the contrast between Southwestern Bell's flat rate pricing compared with usage charges imposed by some other carriers. Bell Atlantic, for instance, charges $0.01 per channel per minute in local, non-toll areas. (Generally, at least two channels are needed for teleconferencing.) These charges vary for intraLATA (local toll) calls, and long distance charges vary by carrier. ISDN charges are in addition to normal service charges.

Factors to Watch

Internet service providers (ISPs) are likely to offer teleconferencing capabilities as value-added services for customers (Borthick, 1997). Innovations in bridging hardware and software, as well as remote diagnostic capabilities, will extend the reach and scope of teleconferencing networks. New software will allow corporate users to teleconference through their Intranet firewall onto the Internet, facilitating packet-switched teleconferencing. Teleconferencing over the Internet can dramatically reduce phone charges, and Internet phone use is already reshaping the long distance phone companies. As Internet teleconferencing advances, revenue streams for phone companies will be affected. However, cultural factors, such as language barriers and cultural norms, continue to inhibit the effectiveness of international teleconferencing (Cukor & Coppick, 1995). The bundling of teleconferencing software and hardware with home computers is an important innovation on the horizon. An estimated 300 million people have computers equipped for Internet access. This represents a potential explosion in teleconferencing, with broad implications for users and the industry.

Bibliography

Borthick, S. (1997, November/December). DVC market: Are we there yet? *Desktop Videoconferencing Magazine*. [Online]. Available: http://www.bcr.com/dvcmag/novdec/dvc6p10.htm.

Bracker, Jr., W., & Sarch, R. (1995). *Cases in network implementation: Enterprise networking*. New York: Van Nostrand Rheinhold.

Cukor, P., & Coppock, K. (1995, April). International videoconferencing: A user survey. *Telecommunications*, 33-34.

Davis, A. (1997, September/October). Videoconferencing markets and strategies. *Desktop Videoconferencing Magazine*. [Online]. Available: http://www.bcr.com/dvcmag/septoct/dvc5p23.htm.

Forward Concepts. (1997). *Teleconferencing markets and strategies: From novelty to necessity*. [Online]. Available: http://www.fwdconcepts.com/studindx.htm.

Hibner, D., Noufer, M., Ahuja, Y., Reed, G., Mangia, K., & Kovac, R. (1997). *Desktop videoconferencing systems get better*. Applied Research Institute, Center for Information and Communication Sciences, Ball State University. [Online]. Available: http://nelsonpub.com/cn/c12video.htm.

Integrated Services Digital Network. (1998). *Prices quoted by PacBell*. [Online]. Available: http://www.pacbell.com/products/index.html.

International Teleconferencing Association. (1997). *Teleconferencing revenues passed $5 billion in North America in 1996*. News release. [Online]. Available: http://www.itca.org/web/memberservices/pressitca/970627.html.

International Telecommunications Union. (1997). *Summary of H.320, H.323 and H.324 standards*. [Online]. Available: http://www.itu.int/itudoc/itu-t/rec/h/s_h320_e_43422.html; http://www.itu.int/itudoc/itu-t/rec/h/s_h323_e_43422.html; and http://www.itu.int/itudoc/itu-t/rec/h/s_h324_e_43422.html.

King, R. (1996, December). A work (still) in progress. *tele.com*, 85-88.

Korostoff, K. (1996, September 16). Desktop video is coming, really! *Computerworld*, 35.

Krapf, E. (1997, November/December). The role of multicast in videoconferencing over IP networks. *Desktop Videoconferencing Magazine*. [Online]. Available: http://www.bcr.com/dvcmag/novdec/dvc6p32.htm.

Labriola, D. (1997, December 16). Videoconferencing on the net. *PC Week Online*. [Online]. Available: http://www.zdnet.com/pcmag/pclabs/nettools/1622/tools/tools1.htm.

Lapolla, S. (1997, March 17). Vendors bolster Web-based videoconferencing efforts. *PC Week*, 123.

Morrison, J., & Liu Sheng, O. (1992). Communication technologies and collaboration systems: Common domains, problems, and solutions. *Information and Management, 23*, 93-112.

Noll, A. M. (1992). Anatomy of a failure: Picturephone revisted. *Telecommunications Policy*, 307-317.

Olgren, C., & Parker, L. (1983). *Teleconferencing technology and applications*. Dedham, MA: Artech House.

PictureTel quoting Dataquest. (1998). [Online]. Available: http://www.picturetel.com/coinf.htm#numbers.

Polycom. (1997). [Online]. Available: http://www.polycom.com

Videoconferencing: Ready for prime time? (1996, May). *Managing Office Technology*, 52-53.

VTel. (1997). [Online]. Available: http://www.vtel.com.

V

CONCLUSIONS

25

Trends in Selected U.S. Communications Media

Dan Brown[*]

T his chapter updates two attempts (Brown & Bryant, 1989; Brown, 1996) to chart the progress of various communications media from their outsets. The philosophy of reporting in all three efforts focuses on media units and penetration (i.e., percentage of marketplace use such as households) rather than on dollar expenditures. The example of box office receipts from motion pictures offers a notable exception. Even here, however, the data include unit attendance figures with the dollar amounts. Considering the dynamic value of the dollar, more meaningful media consumption trends emerge from examining changes in non-monetary units.

Another premise of this chapter holds that government sources should provide as much of the data as possible. Researching the growth or sales figures of various media over time quickly reveals conflict in both dollar figures and units shipped or consumed. Such conflicts exist even among government reports. For example, marked differences exist between the statistical abstract tables and industrial outlook tables in reporting motion picture box office receipts. Even so, emphasizing government sources provides at least some consistency to the data-gathering process, although these sources frequently draw from private sources for their own reports. Readers should use caution in interpreting data for individual years and instead emphasize the trends over several years.

* Associate Dean of Arts & Sciences, East Tennessee State University (Johnson City, Tennesseee).

Print Media

The American print industry consists of approximately 70,000 firms that generated $210 billion in 1997 domestic revenues. World trade in print industry products reached $40 billion in 1997, and the United States led all nations in exporting about $4.9 billion of print materials in that year. That figure represented an increase of approximately 12% over 1996 totals. Challenges from new electronic media appear likely to exceed the threat from previous print media competition from radio, television, and motion pictures. As catalogs, instruction sets, and other traditional print media move to CD-ROM and Internet distribution, print industries may lose business (U.S. Department of Commerce, 1998, p. 25-1).

Both the number of American newspaper firms and newspaper circulation (Figure 25.1 and Table 25.1) decreased slightly in recent years. Early 1990s trends in the number of firms fluctuated somewhat, but circulation steadily declined throughout the decade. As large newspaper chains bought smaller ones, only 77 independent American newspapers with circulation exceeding 30,000 remained in business in 1996 (U.S. Department of Commerce, 1998, p. 25-4). With the dwindling number of owners, however, profits remained strong, and circulation actually increased in the 25 largest markets.

Figure 25.1
Newspaper Firms and Daily Newspaper Circulation

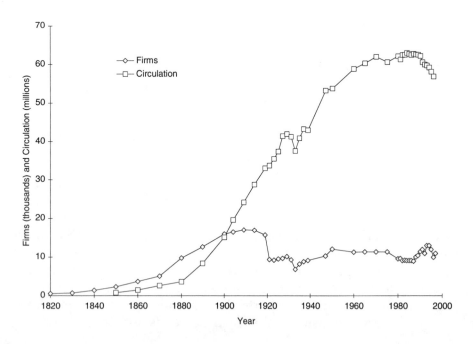

Table 25.1
Newspaper Firms and Daily Newspaper
Circulation

Year	Firms	Circulation (millions)	Year	Firms	Circulation (millions)
1704	1		1931	9,299	41.3
1710	1		1933	6,884	37.6
1720	3		1935	8,266	40.9
1730	7		1937	8,826	43.3
1740	12		1939	9,173	43.0
1750	14		1947	10,282	53.3
1760	18		1950	12,115	53.8
1770	30		1960	11,315	58.9
1780	39		1965	11,383	60.4
1790	92		1970	11,383	62.1
1800	235		1975	11,400	60.7
1810	371		1980	9,620	62.2
1820	512		1981	9,676	61.4
1830	715		1982	9,183	62.5
1840	1,404		1983	9,205	62.6
1850	2,302	0.8	1984	9,151	63.1
1860	3,725	1.5	1985	9,134	62.8
1870	5,091	2.6	1986	9,144	62.5
1880	9,810	3.6	1987	9,031	62.8
1890	12,652	8.4	1988	10,088	62.7
1900	15,904	15.1	1989	10,457	62.6
1904	16,459	19.6	1990	11,471	62.3
1909	17,023	24.2	1991	11,689	60.7
1914	16,944	28.8	1992	11,339	60.1
1919	15,697	33.0	1993	12,597	59.8
1921	9,419	33.7	1994	12,513	59.3
1923	9,248	35.5	1995	12,246	58.2
1925	9,569	37.4	1996	10,466	57.0
1927	9,693	41.4	1997	10,616	
1929	10,176	42.0			

Note: The data from 1704 through 1900 are from Lee (1973). The data from 1904 through 1947 are from U.S. Bureau of the Census (1976). The number data between 1947 and 1986 are from U.S. Bureau of the Census (1986). The data from 1987 through 1988 are from U.S. Bureau of the Census (1995c). The data after 1988 are from U.S. Bureau of the Census (1997).

Except for a dip in 1996, the number of periodical titles (Table 25.2 and Figure 25.2) remained fairly stable. Many American firms rapidly expanded international operations in the mid-1990s. Concern about paper and distribution costs, together with new marketing opportunities, sparked higher per-copy prices among periodicals and considerable corporate interest in new media. Single copy newsstand sales of magazines fell below 20% of magazines sold. In 1995, half of the 10 largest American magazines reduced circulation to focus on target demographic groups and minimize marketing costs.

The American book publishing industry accounts for about 30% of worldwide demand for books, and 90% of the American industry product goes to domestic trade (U.S. Department of Commerce, 1998, p. 25-10). During the 1990s, the number of book titles printed (Table 25.3 and Figure 25.3) experienced sharp growth after a period of steady increase, with the 1995 total reflecting nearly one-third more titles than the 1990 output.

Table 25.2
Published Periodical Titles

Year	Titles	Year	Titles
1904	1,493	1975	9,657
1909	1,194	1980	10,236
1914	1,379	1981	10,873
1921	4,796	1982	10,688
1923	3,747	1983	10,952
1925	3,829	1984	10,809
1927	4,496	1985	11,090
1929	4,659	1986	11,328
1931	5,157	1987	11,593
1933	4,887	1988	11,229
1935	3,459	1989	11,556
1937	4,019	1990	11,092
1939	4,202	1991	11,239
1947	4,985	1992	11,143
1954	3,427	1993	11,863
1958	4,455	1994	12,136
1960	8,422	1995	11,179
1965	8,990	1996	9,843
1970	9,573	1997	11,408

Note: Data from 1904 through 1958 are from U.S. Bureau of the Census (1976). The data from 1960 through 1985 are from U.S. Bureau of the Census (1986). Data from 1987 through 1988 are from U.S. Bureau of the Census (1995c).

Figure 25.2
Published Periodical Titles

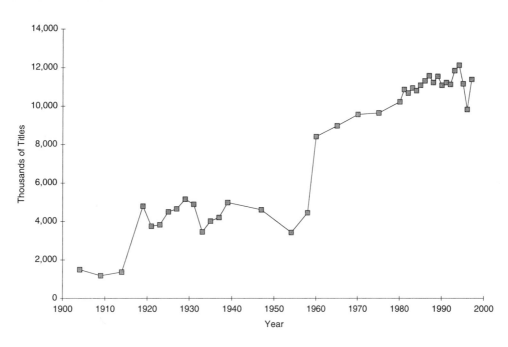

Figure 25.3
Published Book Titles

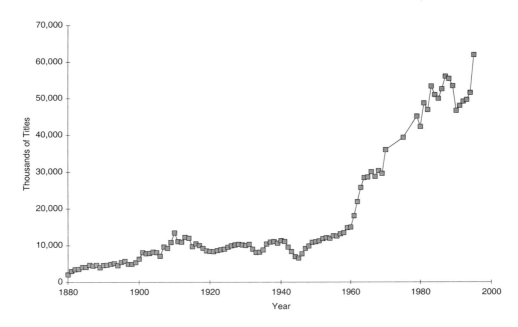

Table 25.3
Published Book Titles

Year	Total Titles	Year	Total Titles	Year	Total Titles	Year	Total Titles
1880	2,076	1911	11,123	1942	9,525	1980	42,377
1881	2,991	1912	10,903	1943	8,325	1981	48,793
1882	3,472	1913	12,230	1944	6,970	1982	46,935
1883	3,481	1914	12,010	1945	6,548	1983	53,380
1884	4,088	1915	9,734	1946	7,735	1984	51,058
1885	4,030	1916	10,445	1947	9,182	1985	50,070
1886	4,676	1917	10,060	1948	9,897	1986	52,637
1887	4,437	1918	9,237	1949	10,892	1987	56,057
1888	4,631	1919	8,594	1950	11,022	1988	55,483
1889	4,014	1920	8,422	1951	11,255	1989	53,446
1890	4,559	1921	8,329	1952	11,840	1990	46,738
1891	4,665	1922	8,638	1953	12,050	1991	48,146
1892	4,862	1923	8,863	1954	11,901	1992	49,276
1893	5,134	1924	9,012	1955	12,589	1993	49,756
1894	4,484	1925	9,574	1956	12,538	1994	51,663
1895	5,469	1926	9,925	1957	13,142	1995	62,039
1896	5,703	1927	10,153	1958	13,462		
1897	4,928	1928	10,354	1959	14,876		
1898	4,886	1929	10,187	1960	15,012		
1899	5,321	1930	10,027	1961	18,060		
1900	6,356	1931	10,307	1962	21,904		
1901	8,141	1932	9,035	1963	25,784		
1902	7,833	1933	8,092	1964	28,451		
1903	7,865	1934	8,198	1965	28,595		
1904	8,291	1935	8,766	1966	30,050		
1905	8,112	1936	10,436	1967	28,762		
1906	7,139	1937	10,912	1968	30,387		
1907	9,620	1938	11,067	1969	29,579		
1908	9,254	1939	10,640	1970	36,071		
1909	10,901	1940	11,328	1975	39,372		
1910	13,470	1941	11,112	1979	45,182		

Note: The data for 1880-1919 include pamphlets; 1920-1928, pamphlets included in total only; thereafter, pamphlets excluded entirely. Beginning in 1959, the definition "book" changed, rendering data on prior years not strictly comparable with subsequent years. Beginning in 1967, the counting methods were revised, rendering prior years not strictly comparable with subsequent years. The data from 1904 through 1947 are from U.S. Bureau of the Census (1976). The data from 1975 through 1983 are from U.S. Bureau of the Census (1984). The data from 1984 are from U.S. Bureau of the Census (1985). The data from 1985 and 1989 through 1992 are from U.S. Bureau of the Census (1995c). The data from 1986 and 1987 are from U.S. Bureau of the Census (1990). The data from 1988 are from U.S. Bureau of the Census (1992). The data from 1985 and 1989 are from U.S. Bureau of the Census (1995c). The data from 1989 are from U.S. Bureau of the Census (1997).

Telephone

The Telecommunications Act of 1996 cleared the way for competition between providers of electronic communications. Although few telephone companies have entered the video services sector, cable companies have begun offering telephone services in many communities (FCC, 1998a). The telephone remains one of the most ubiquitous of American communications devices (Table 25.4 and Figure 25.4), and the 1990s have seen explosive growth in wireless telephones. The nine years of the cellular industry, beginning in 1983, brought just more than 10 million subscribers, but the 1993-1996 period brought an additional 30 million subscribers (U.S. Department of Commerce, 1998, p. 31-14).

The number of cellular phones sold increased more than sevenfold from 1990 to 1996. With increasing sales, the cost of using cellular phones is also dropping. Typical monthly bills fell from more than $100 in the mid-1980s when cellular calling began, to less than $44 by early 1998 (Wald, 1998).

Figure 25.4
Telephone Penetration and Cellular
Telephone Subscribers

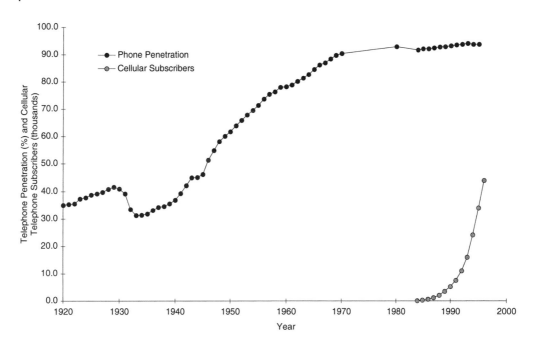

Table 25.4
Telephone Penetration and Cellular
Telephone Subscribers

Year	Households with Telephones (%)	Year	Households with Telephones (%)	Cellular Subscribers (thousands)
1920	35.0	1956	73.8	
1921	35.3	1957	75.5	
1922	35.6	1958	76.4	
1923	37.3	1959	78.0	
1924	37.8	1960	78.3	
1925	38.7	1961	78.9	
1926	39.2	1962	80.2	
1927	39.7	1963	81.4	
1928	40.8	1964	82.8	
1929	41.6	1965	84.6	
1930	40.9	1966	86.3	
1931	39.2	1967	87.1	
1932	33.5	1968	88.5	
1933	31.3	1969	89.8	
1934	31.4	1970	90.5	
1935	31.8	1975		
1936	33.1	1979		
1937	34.3	1980	93.0	
1938	34.6	1981		
1939	35.6	1982		
1940	36.9	1983		1
1941	39.3	1984	91.8	100
1942	42.2	1985	92.2	350
1943	45.0	1986	92.2	682
1944	45.1	1987	92.5	1,231
1945	46.2	1988	92.9	2,069
1946	51.4	1989	93.0	3,509
1947	54.9	1990	93.3	5,283
1948	58.2	1991	93.6	7,557
1949	60.2	1992	93.9	11,033
1950	61.8	1993	94.2	16,009
1951	64.0	1994	93.9	24,134
1952	66.0	1995	93.9	33,786
1953	68.0	1996		44,043
1954	69.6			
1955	71.5			

Note: 1950-1982 data applied to principal earners filing reports with FCC; earlier data applies to Bell and independent companies. Beginning in 1959, data includes figures from Alaska and Hawaii. The data for 1986 and 1987 are estimates. The data to 1970 are from U.S. Bureau of the Census (1976). The data from 1970 through 1982 are from U.S. Bureau of the Census (1986). The data after 1982 are from U.S. Department of Commerce (1987). The data from 1986 and 1987 are from U.S. Bureau of the Census (1992, 1993). The data after 1987 are from U.S. Bureau of the Census (1997).

Motion Pictures

Table 25.5 and Figure 25.5 show motion picture box office receipts and average weekly attendance. During the 1990s, average weekly attendance reflected little change. In fact, the stability of attendance reflects little change for more than three decades. However, box office receipts increased by nearly one-third during the 1990s, increasing every year after 1991.

The 1996 box office take of $5.9 billion represented a 7.3% increase over the comparable measurement from 1995. Box office revenues were projected to reach $6.2 billion in 1997, "$6.5 billion in 1998 and $7.6 billion in 2002" (U.S. Department of Commerce, 1998, p. 32-3).

Figure 25.5
Motion Picture Attendance and
Box Office Receipts

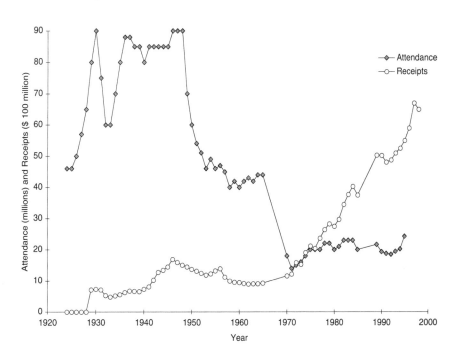

Table 25.5
Motion Picture Attendance and
Box Office Receipts

Year	Average Weekly Attendance*	Receipts ($ million)	Year	Average Weekly Attendance*	Receipts ($ million)
1922	40		1961	42.0	921
1923	43		1962	43.0	903
1924	46	1.71	1963	42.0	904
1925	46		1964	44.0	913
1926	50		1965	44.0	927
1927	57		1966		964
1928	65		1967		989
1929	80	720	1968		1,045
1930	90	732	1969		1,099
1931	75	719	1970	18.0	1,162
1932	60	527	1971	14.0	1,214
1933	60	482	1972	15.0	1,583
1934	70	518	1973	16.0	1,524
1935	80	556	1974	18.0	1,909
1936	88	626	1975	20.0	2,115
1937	88	676	1976	20.0	2,036
1938	85	663	1977	20.0	2,372
1939	85	659	1978	22.0	2,643
1940	80	735	1979	22.0	2,821
1941	85	809	1980	20.0	2,749
1942	85	1,022	1981	21.0	2,966
1943	85	1,275	1982	23.0	3,453
1944	85	1,341	1983	23.0	3,766
1945	85	1,450	1984	23.0	4,030
1946	90	1,692	1985	20.3	3,749
1947	90	1,594	1986	19.6	3,780
1948	90	1,506	1987	20.9	4,250
1949	70	1,451	1988	20.9	4,460
1950	60	1,376	1989	21.8	5,030
1951	54	1,310	1990	22.8	5,020
1952	51	1,246	1991	21.9	4,800
1953	46	1,187	1992	22.6	4,870
1954	49	1,228	1993	23.9	5,200
1955	46	1,326	1994	24.8	5,400
1956	47	1,394	1995	24.3	5,500
1957	45	1,126	1996		5,900
1958	40	992	1997		6,700
1959	42.0	958	1998		6,500
1960	40.0	951			

* In millions.

Note: The data to 1970 are from U.S. Bureau of the Census (1976). The data from 1970, 1975, and 1979 through 1985 are from U.S. Bureau of the Census (1986). The box office data from 1971 are from U.S. Bureau of the Census (1975). The box office receipts data from 1972 through 1988 are from U.S. Department of Commerce (1988), and the data for 1988 through 1992 are from U.S. Department of Commerce (1994). The 1991 attendance came from U.S. Bureau of the Census (1996). The data for 1993 through 1998 are from U.S. Department of Commerce (1998).

Recording

Overall recorded music shipments steadily declined through the 1990s, as Table 25.6 shows. Figure 25.6 illustrates the trends in various music types other than compact discs, and Figure 25.7 tracks overall music sales of non-CD types. The Recording Industry Association of America reported the first declines in CD sales (Table 25.7 and Figure 25.8) since CDs were introduced in 1983, with total CD sales falling by 3.3% in 1997 from 1996 totals (First-ever, 1998).

Reports in popular media (Brull, 1998) documented the popularity of downloading music directly to consumers using the Internet. Brull wrote that one Website downloaded 4,000 copies of independent label-recorded singles at $0.99 each between July 1997 and March 1998. He mentioned the existence of more than 30,000 music Websites.

Table 25.6
Recorded Music Unit Shipmentst

Year	Singles	LPs/EPs	Cassettes	8-tracks	Total
1973	228.0	280.0	15.0	91.0	614.0
1974	204.0	276.0	15.3	96.7	592.0
1975	164.0	257.0	16.2	94.6	531.8
1976	190.0	273.0	21.8	106.1	590.9
1977	190.0	344.0	36.9	127.3	698.2
1978	190.0	341.3	61.3	133.6	726.2
1979	195.5	318.3	82.8	104.7	701.1
1980	164.3	322.8	110.2	86.4	683.7
1981	154.7	295.2	137.0	48.5	635.4
1982	137.2	243.9	182.3	14.3	577.7
1983	125.0	210.0	237.0	6.0	578.0
1984	132.0	205.0	332.0	6.0	675.0
1985	121.0	167.0	339.0	4.0	631.0
1986	93.9	125.2	344.5		658.0
1987	82.0	107.0	410.0		683.3
1988	65.6	72.4	450.1		588.1
1989	36.6	34.6	446.2		517.4
1990	27.6	11.7	442.2		481.5
1991	22.0	4.8	360.1		386.9
1992	19.8	2.3	336.4		358.5
1993	15.1	1.2	339.5		355.8
1994	11.7	1.9	345.4		359.0
1995	10.2	2.2	272.6		285.0
1996	10.1	2.9	225.3		238.3

Note: The data for all years prior to 1983 are from U.S. Department of Commerce (1986). The data from 1983 through 1985 are from U.S. Bureau of the Census (1986). The data for 1986 through 1994 are from U.S. Bureau of the Census (1995c). The data after 1994 are from U.S. Bureau of the Census (1997).

Figure 25.6
Recorded Music Unit Shipments

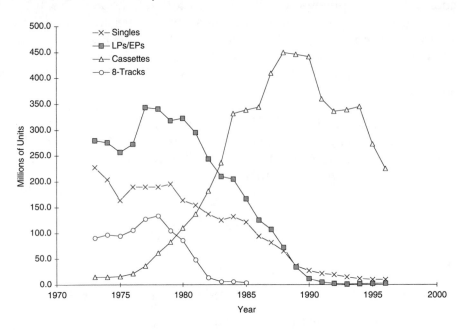

Figure 25.7
Total Recorded Music Sales

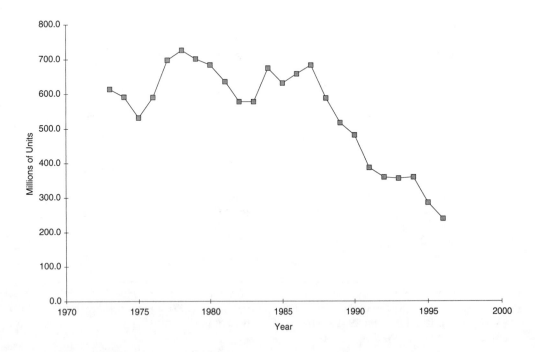

Table 25.7

Compact Discs and Players

Year	Disc Shipments (millions)	Player Shipments (thousands)	Penetration (%)
1983	1	45	0
1984	6	200	1.60
1985	23	850	6.40
1986	53	1,500	7.00
1987	102.1	1,384	
1988	149.7	4,800	7-8
1989	207.2	4,280	12-16
1990	286.5	6,900	19
1991	333.3	9,260	28
1992	407.5		35
1993	495.4	21,500	42
1994	662.1	26,500	
1995	722.9		
1996	778.9		
1997	753.2		

Note: The data for discs through 1985 are from U.S. Bureau of the Census (1986), and after 1986 are from U.S. Bureau of the Census (1995c). The data for players through 1985 are from U.S. Bureau of the Census (1986), for 1986 from U.S. Bureau of the Census (1988), and 1986-1991 from respective issues of U.S. Department of Commerce (1986, 1987, 1988, 1989, 1991, and 1992). The data from 1993 and 1994 are from Trachtenberg (1995). The data for penetration for 1984 and 1985 are from Graham (1986). The data for penetration for 1986 are from Lewyn (1987) and from 1992-1993 from U.S. Bureau of the Census (1995c). Data from 1995-1996 are from U.S. Bureau of the Census (1997). The 1997 figure was calculated based on a reported 3.3% drop (First-ever, 1998).

Figure 25.8
Compact Discs and Players

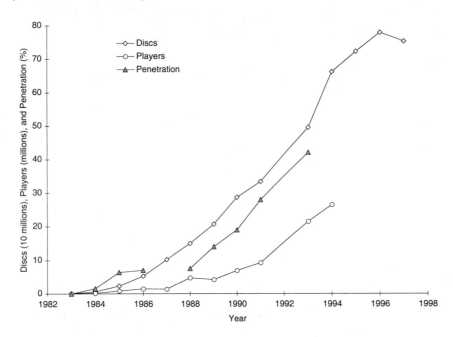

Radio

As of November 1997, 10,470 commercial radio stations held licenses in the United States. This number included 5,656 (54%) FM stations and 4,819 (46%) AM stations. All the growth in stations since passage of the Telecom Act has been in FM stations. Since the Telecommunications Act of 1996 eliminated the national multiple radio ownership rule and eased the local ownership rule, the number of commercial broadcasting radio stations grew by about 2.5%. Although the number of radio stations increased, the number of radio owners declined by 11.7% since March 1996. The decrease in owners occurred mostly from mergers between existing owners. The result of these mergers has been to change the ranking and composition of the top radio station owners (FCC, 1998b), which captured larger shares of total radio advertising revenues in respective radio markets. However, no loss of variety of radio station formats has occurred (FCC, 1998b). Table 25.8 and Figure 25.9 show the penetration of radio.

Table 25.8
Radio Households and Penetration

Year	HHs with Sets (000s)	% Penetration	Year	HHs with Sets (000s)	% Penetration	Year	HHs with Sets (000s)	% Penetration
1922	60		1947	35,900	91.8	1972	67,200	
1923	400		1948	37,623		1973	69,400	
1924	1,250		1949	39,300	93.4	1974	70,800	
1925	2,750		1950	40,700		1975	72,600	98.6
1926	4,500		1951	41,900		1976	74,000	
1927	6,750		1952	42,800		1977	75,800	
1928	8,000		1953	44,800		1978	77,800	
1929	10,250		1954	45,100		1979	79,300	
1930	13,750	40.3	1955	45,900	95.9	1980	79,968	99.0
1931	16,700		1956	46,800	95.7	1981	81,600	99.0
1932	18,450		1957	47,600	95.8	1982	82,691	99.0
1933	19,250		1958	48,500	96.1	1983	83,078	99.0
1934	20,400		1959	49,450	96.1	1984	84,553	99.0
1935	21,456		1960	50,193	95.1	1985	85,921	99.0
1936	22,869		1961	50,695	94.7	1986		99.0
1937	24,500		1962	51,305	93.7	1987		99.0
1938	26,667		1963	52,300	94.6	1988	91,100	99.0
1939	27,500		1964	54,000	96.2	1989	92,800	99.0
1940	28,500	80.3	1965	55,200	96.1	1990	94,400	99.0
1941	29,300		1966	57,200	97.6	1991	95,500	99.0
1942	30,600		1967	57,500	97.1	1992	96,600	99.0
1943	30,800		1968	58,500	96.2	1993	97,300	99.0
1944	32,500		1969	60,600	97.4	1994	98,000	99.0
1945	33,100		1970	62,000	97.8	1995	98,000	99.0
1946	33,998		1971	65,400				

Note: Authorization of new radio stations and production of radio sets for commercial use was stopped from April 1942 until October 1945. 1959 is the first year for which Alaska and Hawaii are included in the figures. The data prior to 1970 are from U.S. Bureau of the Census (1976). The households with sets data from 1970-1972 are from U.S. Bureau of the Census (1972). The households with sets data from 1973 and 1974 are from U.S. Bureau of the Census (1975). The households with sets data from 1975 through 1977 are from U.S. Bureau of the Census (1978). The households with sets data from 1978 and 1979 are from U.S. Bureau of the Census (1981). All data from 1988 through 1994 are from U.S. Bureau of the Census (1995c). The data from 1995 are from U.S. Bureau of the Census (1997).

Figure 25.9
Radio and TV Households

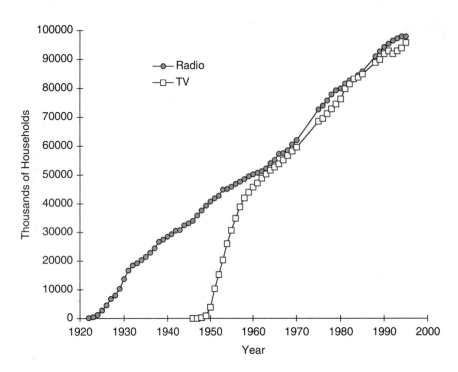

Television

As radio remains strong, so does television. Documenting the steady public attitude toward television as a household necessity, Table 25.9 shows the steady rate of penetration and continuing growth in the number of households with TV sets. Figure 25.9 charts the close parallel growth in households with both television and radio. Despite growth in competing multichannel video media, the four largest broadcast television networks still attract 59% of prime-time viewers. The number of broadcast television stations increased to 1,561 in 1997 over 1,550 in 1996, and they provide the only means of television reception for 23% of television households (FCC, 1998a).

As discussed in Chapter 7, the Federal Communications Commission has adopted rules for digital television broadcasting which will soon allow broadcast stations to provide multiple channels. FCC rules permit broadcasting two channels of high-definition television (HDTV), multiple streams of standard signals (SDTV), or a combination of these types. HDTV broadcasting is expected to be in the top 10 markets in 1998, with a transition to digital broadcasting for the remainder of the country to be completed by 2006 (FCC, 1998a).

Table 25.9
Television Households and Penetration

Year	HHs with Sets (000s)	% Penetration	Year	HHs with Sets (000s)	% Penetration
1946	8		1973		
1947	14		1974		
1948	172		1975	68,500	97
1949	940		1976	69,600	
1950	3,875	9	1977	71,200	
1951	10,320		1978	72,900	
1952	15,300		1979	74,500	
1953	20,400		1980	76,300	98
1954	26,000		1981	79,900	98
1955	30,700		1982	81,500	98
1956	34.90		1983	83,300	98
1957	38,900		1984	83,800	98
1958	41,924		1985	84,900	98
1959	43,950		1986	85,900	98
1960	45,750	87	1987	87,400	98
1961	47,200		1988	89,000	98
1962	48,855		1989	90,000	98
1963	50,300		1990	92,000	98
1964	51,600		1991	93,000	98
1965	52,700		1992	92,000	98
1966	53,850		1993	93,000	98
1967	55,130		1994	94,000	98
1968	56,670		1995	95,900	98
1969	58,250		1996	97,000	99
1970	59,550	95	1997	97,000	99
1972					

Note: 1959 is the first year for which Alaska and Hawaii are included in the figures. The data dealing with households with television to 1971 are from U.S. Bureau of the Census (1976). The data dealing with households with television from 1980 through 1984 are from U.S. Bureau of the Census (1985). The data about penetration for all other pre-1987 years and all data for 1985 and 1986 are from U.S. Bureau of the Census (1986), and data for 1987-1994 are from U.S. Bureau of the Census (1995c). The data for 1995 are from FCC (1995). The data for 1996-1997 are from FCC (1998a).

Cable Television

Table 25.10 shows the number of cable systems, subscribers, and proportion of homes using cable. Figure 25.10 portrays the growth of subscribers and penetration. Between the mid-1990s and the present, the cable industry increased most measures of reach and capacity, including homes with cable available, subscribers, penetration, premium channel subscriptions, proportion of overall television viewing, and channel capacity (FCC, 1998a). New cable offerings such as digital video, cable modems for Internet access, and telephone services emerged in 1997.

Table 25.10
Cable TV Systems, Subscribers, and
Penetration

Year	Systems	Subscribers (thousands)	Penetration (Percent)
1952	70	14	
1955	400	150	
1960	640	650	
1965	1,325	1,275	
1967	1,770	2,100	
1968	2,000	2,800	
1969	2,260	3,600	
1970	2,490	4,500	
1971	2,639	5,300	
1972	2,841	6,000	
1973	2,991	7,300	
1974	3,158	8,700	
1975	3,506	9,800	
1976	3,681	10,800	
1977	3,832	11,900	
1978	3,875	13,000	
1979	4,150	14,100	
1980	4,225	16,000	
1981	4,375	18,300	25.3
1982	4,825	21,000	29.0
1983	5,600	25,000	37.2
1984	6,200	30,000	41.2
1985	6,600	31,275	44.6
1986	7,600	36,933	46.8
1987	7,900	41,100	48.7
1988	8,500	41,100	49.4
1989	9,050	48,600	52.8
1990	9,575	52,600	56.4
1991	10,704	54,900	58.9
1992	11,075	55,800	60.2
1993	11,217	56,400	61.4
1994	11,230	57,200	62.4
1995	11,126	61,700	63.4
1996	11,119	63,500	67.8
1997	10,750	64,200	68.2

Note: The systems and subscribers data are from U.S. Bureau of the Census (1986). The penetration data through 1986 are from U.S. Bureau of the Census (1986). The penetration data through 1986 are from U.S. Bureau of the Census (1986) except for 1970, 1980, and 1987 through 1995. Of the latter, 1987 data are from U.S. Bureau of the Census (1988), and data from 1970, 1980, and 1988-1995 are from U.S. Bureau of the Census (1995c). The subscriber and penetration data for 1995-1997 are from FCC (1998a). The number of systems for 1996 is from U.S. Bureau of the Census (1997), and the number of 1997 systems is from FCC (1998a).

Figure 25.10
Cable TV Subscribers and Penetration

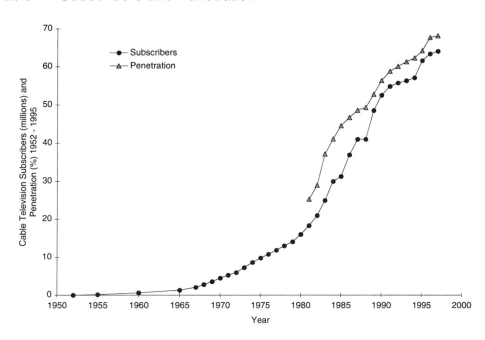

Of 97 million homes with televisions in 1997, cable access passed 94.2 million (97.1%). This proportion grew slightly more than 1% since 1995. The number of homes subscribing to cable television grew to 64.2 million homes, or 66.2% of all homes with televisions, by the end of June 1997 (FCC, 1998a).

During 1996, the number of homes subscribing to premium cable services increased by 5.6% to 31.5 million homes from 29.8 million in 1995. Many homes subscribed to more than one premium cable service, bringing the total number of separate pay subscriptions to 57.2 million units in 1997, a 5% increase over 1996 (FCC, 1998a). The channel capacity of typical cable systems reached 53 channels by late 1996, an increase from 47 channels a year earlier. In October 1996, 1,724 (16.4%) cable systems offered 54 or more channels, and that figure increased to 1,886 (19%) systems a year later. By then, 58.4% of cable subscribers could receive 54 or more channels (FCC, 1998a).

Broadcast television has steadily lost viewer share to cable over the last 10 years. The average cable share for all viewing times grew from 11.5 in the 1987-1988 broadcast year to 36.25 during 1996-1997. During the same period, audience share from all viewing sources for broadcast television programming fell from 87.7 to 66.5. Viewing shares for premium cable channels remained unchanged at 6.92 during that time.

Direct Broadcast Satellite and Other Cable TV Competitors

Competitors for the cable television industry include a variety of technologies, such as direct broadcast satellite (DBS), home satellite dish services, multichannel multipoint distribution services (MMDS), and satellite master antenna television (SMATV). These technologies and cable TV make up a group known as multichannel video programming distributors (MVPD). Table 25.11 and Figure 25.11 display trends for the non-cable categories.

Cable television accounted for 87% of MVPD subscribers in June 1997, down from 89% in September 1996. Non-cable MVPD proportions reached the levels forecast for the year 2000 by the U.S. Department of Commerce (1998, p. 32-5) in the same year that the forecast was published. During the 12 months beginning in fall 1996, non-cable MVPD subscribers grew from 8.1 million to 9.5 million, an increase of almost 20% (FCC, 1998a). The factors accounting for the shift included increased marketing and consumer awareness, along with technological improvements in cable alternatives (U.S. Department of Commerce, 1998). The U.S. Department of Commerce predicted that about eight million DBS subscribers would exist by the year 2000. However, that figure also seemed low, considering that, between 1994 and 1998, consumers purchased eight million home satellite dishes (Gross, 1998).

Figure 25.11
Multichannel Video Program Distribution

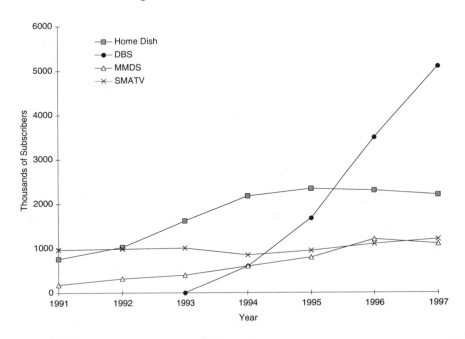

Table 25.11

Multichannel Video Program Distribution

Year	Satellite Dishes Sold	Home Satellite Dish	DBS	MMDS	SMATV
1980	4				
1981	20				
1982	60				
1983	225				
1984	525				
1985	700				
1986	280				
1987	400				
1988	330				
1989	335				
1990	380				
1991	146	764		180	965
1992	153	1,023		323	984
1993	201	1,612	<70	397	1004
1994	436	2,178	602	600	850
1995	222	2,341	1,675	800	950
1996	3,223	2,300	3,500	1,200	1,100
1997	3,740	2,200	5,100	1,100	1,200

Note: For dish sales, the 1986 figure is estimated, and the data from 1980 through 1983 are from Lewyn (1986). The dish sales for 1984 through 1986 are from U.S. Department of Commerce (1987), and the dish sales for 1987 through 1993 are from U.S. Bureau of the Census (1995c). All data are from FCC (1995). Dish sales for 1995 exclude sales after August of that year. Subscriber data for 1996 and 1997 are from FCC (1998a).

Videocassette Recorders

Table 25.12 shows the shipments and household penetration of videocassette recorders (VCRs) by year, and Figure 25.12 illustrates the penetration of VCRs. The FCC sees VCRs as competition for premium cable television and pay-per-view services (FCC, 1995). Cassette rentals and sales accounted for $4.5 billion or 45% of American domestic motion picture studio revenues in 1996, forming the largest single revenue source for the studios. These sales accounted for 62.5% of the $7.2 billion spent in 1996 by consumers for VCR competition from cable television, satellite, and other multichannel pay television services. By 1997, 88% of American homes owned at least one VCR (FCC, 1998a).

Table 25.12

VCR Shipments and Penetration

Year	VCRs (000s)	VCR Penetration (%)
1978	402	
1979	478	
1980	804	1.1
1981	1,330	1.8
1982	2,030	3.1
1983	4,020	5.5
1984	7,143	10.6
1985	18,000	20.8
1986		36.0
1987	43,000	48.7
1988	51,000	58.0
1989	58,000	64.6
1990	63,000	68.6
1991	67,000	71.9
1992	69,000	75.0
1993	72,000	77.1
1994	74,000	79.0
1995	77,000	81.0
1996		88.0

Note: The data from 1978 are from U.S. Bureau of the Census (1982). The data from 1979 through 1984 are from U.S. Bureau of the Census (1984). The data from 1985 are from U.S. Bureau of the Census (1986). The penetration data are from U.S. Bureau of the Census (1986). The VCR sales data for 1985 and 1988-1994 are from the U.S. Bureau of the Census (1995c), and data from 1987 are from the U.S. Bureau of the Census (1990). The VCR penetration data for 1988-1994 are from U.S. Bureau of the Census (1995c), and the data for 1995 and 1996 are from FCC (1998a).

When laserdisc players entered the American market in 1981, some observers predicted that the superior audio and video quality of the medium would provide significant competition for VCRs as home movie viewers. Although the laserdisc found followers, it never challenged the levels of penetration achieved by VCRs. Sales of laserdiscs declined by 35% during the first nine months of 1996, compared with the same period in 1995. By 1997, sales of laserdiscs plummeted (Combos, 1997) as digital versatile discs (DVD) entered the market in February.

This new medium is an enhanced CD that stores between two and eight hours of high-definition video with a storage capacity far exceeding that of muscial CDs. The devices were delayed in coming to the market because of disputes over copyright considerations (U.S. Department of Commerce, 1998). The Consumer Electronics Manufacturers Association (CEMA) reported 1997 sales of 350,000 DVD players. DVD unit sales were expected by CEMA to reach 750,000 players in 1998 (CE sales, 1998).

Figure 25.12
VCR Shipments and Penetration

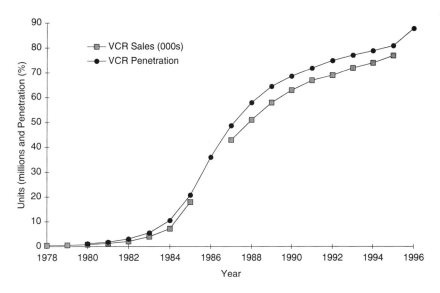

Personal Computers

The previously-mentioned media in this chapter now link with personal computers to provide communications, entertainment, and information to families and businesses. Shipments of personal computers, including desktops, servers, and notebooks, surpassed 25 million units in 1996, a 14% increase over the figures from the previous year. The U.S. Department of Commerce forecast shipments reaching 60 million units by 2002, accompanied by "substantially more pervasive" home and educational use (U.S. Department of Commerce, 1998, p. 27-11).

The target market of home computer buyers shifted to lower income homes in the late 1990s, as more affluent families and businesses reached saturation. Computer manufacturers and sellers shifted focus to buyers who were interested in machines with lower prices and less technology-oriented computer functions. For example, advertising of computers in mass media outlets featured ease-of-use, aesthetics, and ability to perform common home tasks, such as telephone services, Internet access, television viewing, and playing games. Table 25.13 and Figure 25.13 trace the steady rise in sales and penetration of personal computers.

Table 25.13
Personal Computer Shipments and Home Use

Year	PCs Shipped	Homes with PCs	% of Homes with PCs*
1978	212,000		
1979	246,000		
1980	371,000		
1981	1,110,000	750,000	1%
1982	3,530,000	3,000,000	4%
1983	6,900,000	7,640,000	9%
1984	7,610,000	11,990,000	14%
1985	6,750,000	14,960,000	18%
1986	7,040,000	17,380,000	20%
1987	8,340,000	19,970,000	23%
1988	9,500,000	22,380,000	25%
1989	9,329,970		21%
1990	9,848,593		22%
1991	10,903,000	25,000,000	25%
1992	12,544,374	25,000,000	27%
1993	14,775,000	31,000,000	30%
1994	18,605,000	32,010,000	33%
1995	22,582,900	35,890,000	37%
1996	25,000,000		

* The percentages of households with computers before 1988 are calculated by dividing the number of homes with computers by the number of homes with TVs from Table 25.9. The figures after 1988 are from Electronic Industries Association (1995).

Note: Shipments for 1978 are from U.S. Bureau of the Census (1983); from 1979 and 1980 are from U.S. Bureau of the Census (1984). The data for 1981-1988 are from U.S. Bureau of the Census (1992). Shipments for 1989-1990 are from U.S. Bureau of the Census (1993), 1991-1993 data are from U.S. Bureau of the Census (1995c), and 1994-1995 data are from U.S. Bureau of the Census (1997). The computer shipments for 1996 are from U.S. Department of Commerce

Figure 25.13
Personal Computer Shipments and Home Use

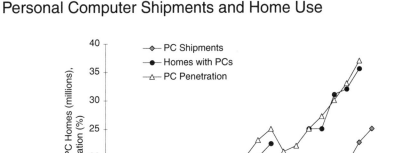

Synthesis

This chapter documents, mostly from government records, trends involving traditional media. During the 1990s, media forms showing declines included newspaper circulation and number of newspaper firm owners. Motion picture attendance and box office receipts showed signs of leveling off. Most forms of recorded music showed declines, including the first decline in CD sales since their introduction. A modest rebound in unit sales of LPs failed to stop the decade-long slide in overall sales of non-CD recorded music. Radio households remained steady, and television households increased slightly. Cable television penetration increased slightly.

Media showing increases in recent years included the number of published periodical titles, which reversed a two-year decline. Published book titles continued to increase sharply. Although household telephones remained steady in unit shipments, dramatic increases occurred in wireless phones. Explosive growth occurred in sales of small satellite dish DBS systems, propelling an increase in penetration of overall multichannel video media. VCR penetration increased its rate of growth, but VCRs have encountered a formidable digital competitor in DVD. Home computers as a mass market commodity began to offer commodity characteristics formerly associated with older media.

Americans in 1994, for the first time, bought more personal computers than television sets (Farynski, 1995), and Americans in 1998 used computers and television sets to accomplish similar tasks. The merging of computers and telecommunications as "infotainment," a combination of information and entertainment, presided over a trend toward increased household penetration of a variety of digital devices. These recent trends suggest a bright future for digital media. For example, digital wireless telephones have begun serious competition with cellular devices, and HDTV appears poised to enter the American mass market. The growing variety of consumer media choices seems certain to quickly influence consumption patterns of both traditional and new media.

Bibliography

Associated Press. (1993, July 15). *Johnson City Press*, 14.

Brown, D. (1996). A statistical update of selected American communications media. In A. E. Grant (Ed.), *Communication technology update* (5th Ed.). Boston: Focal Press, 327-355.

Brown, D., & Bryant, J. (1989). An annotated statistical abstract of communications media in the United States. In J. Salvaggio & J. Bryant (Eds.), *Media use in the information age: Emerging patterns of adoption and consumer use*. Hillsdale, NJ: Lawrence Erlbaum Associates, 259-302.

Brull, S. V. (1998, March 2). Net nightmare for the music biz. *Business Week*, 89-90.

CE sales rose 5.2% in 1997—CEMA. (1998). *Television Digest*, 37, 14 (3).

Combos, PTV, and VCR decks lead upbeat July. (1997, August 18). *Television Digest*, 37 (33), 14 (2).

Electronic Industries Association. (1995, August 24). Computer sales. *Johnson City Press*, 15.

Farynski, J. (1995, September 21). *High quality video on the Internet*. [Online]. Available: http://www.interaus.net/1995/9/video.html.

Federal Communications Commission. (1995, December 11). *Annual assessment of the status of competition in the market for the delivery of video programming*. CS Docket No. 95-61. Washington, DC: U.S. Government Printing Office.

Federal Communications Commission. (1998a, January 13). *In the matter of assessment of the status of competition in markets for the delivery of video programming* (Fourth Annual Report). CS Docket No. 97-141. Washington, DC. [Online]. Available: http://www.fcc.gov/Bureaus/Cable/Reports/fcc97423.txt.

Federal Communications Commission. (1998b, March 13). *Review of the radio industry, 1997*. MM Docket No. 98-35. Washington, DC: U.S. Government Printing Office.

First-ever decline hits CD album shipments—RIAA. (1998, February 23). *Television Digest*, 38 (8), 14 (2).

Graham, J. (1986, July 7). Sales of CD players are soaring. *USA Today*, 1D.

Gross, N. (1998, February 16). On the digital frontier: A shoot-out with no winners. *Business Week*, 92-93.

Lee, A. (1973). *The daily newspaper in America*. New York: Octagon Books.

Lewyn, M. (1986, July 16). Sending a new signal. *USA Today*, 4B.

Lewyn, M. (1987, May 29). Video CDs to debut tomorrow. *USA Today*, 1B, 2B.

National Association of Broadcasters (1996). *NAB radio: Radio fast facts*. [Online]. Available: http://www.nab.org/radio/default.html-ssi.

Trachtenberg, J. A., & Shapiro, E. (1996, February 26). Record store shakeout rocks music industry. *Wall Street Journal*, B1, B8.

U.S. Bureau of the Census. (1972). *Statistical abstract of the United States: 1972* (93rd Ed.). Washington, DC: U.S. Government Printing Office.

U.S. Bureau of the Census. (1975). *Statistical abstract of the United States: 1975* (96th Ed.). Washington, DC: U.S. Government Printing Office.

U.S. Bureau of the Census. (1976). *Statistical history of the United States: From colonial times to the present*. New York: Basic Books.

U.S. Bureau of the Census. (1978). *Statistical abstract of the United States: 1978* (99th Ed.). Washington, DC: U.S. Government Printing Office.

U.S. Bureau of the Census. (1981). *Statistical abstract of the United States: 1981* (102nd Ed.). Washington, DC: U.S. Government Printing Office.

U.S. Bureau of the Census. (1982). *Statistical abstract of the United States: 1982-1983* (103rd Ed.). Washington, DC: U.S. Government Printing Office.

U.S. Bureau of the Census. (1983). *Statistical abstract of the United States: 1984* (104th Ed.). Washington, DC: U.S. Government Printing Office.

U.S. Bureau of the Census. (1984). *Statistical abstract of the United States: 1985* (105th Ed.). Washington, DC: U.S. Government Printing Office.

U.S. Bureau of the Census. (1985). *Statistical abstract of the United States: 1986* (106th Ed.). Washington, DC: U.S. Government Printing Office.

U.S. Bureau of the Census. (1986). *Statistical abstract of the United States: 1987* (107th Ed.). Washington, DC: U.S. Government Printing Office.

U.S. Bureau of the Census. (1988). *Statistical abstract of the United States: 1989* (109th Ed.). Washington, DC: U.S. Government Printing Office.

U S. Bureau of the Census. (1990). *Statistical abstract of the United States: 1991* (111th Ed.). Washington, DC: U.S. Government Printing Office.

U.S. Bureau of the Census. (1992). *Statistical abstract of the United States: 1993* (113th Ed.). Washington, DC: U.S. Government Printing Office.

U.S. Bureau of the Census. (1993). *Statistical abstract of the United States: 1994* (114th Ed.). Washington, DC: U.S. Government Printing Office.

U.S. Bureau of the Census. (1995a). *Consumer electronics 1992*. [Online]. Available: http://www.census.gov/ftp/pub/industry/ma36m92.txt.

U.S. Bureau of the Census. (1995b). *Consumer electronics annual 1994.* [Online]. Available: http://www.census.gov/ftp/pub/industry/ma36m94.txt.

U.S. Bureau of the Census. (1995c). *Statistical abstract of the United States: 1996* (116th Ed.). Washington, DC: U.S. Government Printing Office.

U.S. Bureau of the Census. (1996). *Statistical abstract of the United States: 1997* (117th Ed.). Washington, DC: U.S. Government Printing Office.

U.S. Department of Commerce. (1986). *U.S. industrial outlook 1986*. Washington, DC: U.S. Department of Commerce, U.S. Bureau of Economic Analysis, and U.S. Bureau of Labor Statistics.

U.S. Department of Commerce. (1987). *U.S. industrial outlook 1987*. Washington, DC: U.S. Department of Commerce, U.S. Bureau of Economic Analysis, and U.S. Bureau of Labor Statistics.

U.S. Department of Commerce. (1988). *U.S. industrial outlook 1988*. Washington, DC: U.S. Department of Commerce, U.S. Bureau of Economic Analysis, and U.S. Bureau of Labor Statistics.

U.S. Department of Commerce. (1989). *U.S. industrial outlook 1989*. Washington, DC: U.S. Department of Commerce, U.S. Bureau of Economic Analysis, and U.S. Bureau of Labor Statistics.

U.S. Department of Commerce. (1991). *U.S. industrial outlook 1991*. Washington, DC: U.S. Department of Commerce, U.S. Bureau of Economic Analysis, and U.S. Bureau of Labor Statistics.

U.S. Department of Commerce. (1992). *U.S. industrial outlook 1992*. Washington, DC: U.S. Department of Commerce, U. S. Bureau of Economic Analysis, and U.S. Bureau of Labor Statistics.

U.S. Department of Commerce. (1994). *U.S. industrial outlook 1994*. Washington, DC: U.S. Department of Commerce, U. S. Bureau of Economic Analysis, and U.S. Bureau of Labor Statistics.

U.S. Department of Commerce. (1998). *U.S. industry and trade outlook 1998*. New York: McGraw-Hill.

Wald, M. L. (1998, March 15). Untying cellular phones from those annual contracts. *New York Times on the Web*. [Online]. Available: http://www.nytimes.com/library/tech/98/03/biztech/articles/15phones.html.

26

Conclusions

August E. Grant, Ph.D.

This is the sixth book in the *Communication Technology Update* series. The pace of change in communication technologies has been remarkable since the first edition was published in 1992. Some of the most important issues in that first edition are still important today, including regulatory changes, organizational battles, and standards issues.

On the other hand, many aspects of the communication technology environment are dramatically different. The Internet, which was not discussed in the first edition, is discussed in more than half of the chapters in this edition. Regulatory changes have fueled the pace of mergers and acquisitions to a level that was virtually unforeseen in 1992.

The most important lesson from looking back at previous editions of this *Update* is how much the landscape of communication technologies has changed. Table 26.1 provides a picture of how that landscape is different, indicating technologies that have been added and deleted from the table of contents since 1992.

The lesson from Table 26.1 is that we should expect similar, dramatic changes in the communication technology landscape over the next six years as well. In the same manner that the Internet and World Wide Web have emerged since 1992 to become one of the most prominent technologies, perhaps another technology will emerge in the coming six years to have the same remarkable impact.

One of the most important issues is how to analyze emerging technologies. Ironically, when predicting the likely success of any innovation, the safest prediction is always failure. Of all new products, services, and technologies introduced, only a fraction reach the market, with few of these lasting more than a few years. The ones that last, such as the telephone, radio, television, and (perhaps!) the Internet have the potential to bring about profound changes in society.

Table 26.1

Changes in *Communication Technology Update*
Table of Contents Since 1992

Additions	Deletions
Internet	Low-Power Television
Switched Broadband Networks	Television Shopping
Virtual Reality	Videotext and Teletext
Personal Communications Services	Videodisc
Multimedia	Satellite News Gathering
	Desktop Publishing
	Digital Video Interactive/ Compact Disk-Interactive
	Digital Compact Cassette

Source: A. E. Grant

The first prediction, therefore, is that many of the newest technologies discussed in this book will fail. Some will fail because the technology could never live up to its promise, while others will succumb to marketplace competition. Some superior technologies will fail because of inferior marketing or distribution. Finally, some will fail because they were introduced at the wrong time: too early, too late, or during economic recession.

External factors such as the general state of the economy are critical to any new technology. The level of investment in development of a technology must be matched by the investment in marketing the technology. Growth of many of the technologies discussed in this book has been fueled by a strong economy, which included a great deal of investment in research and development of new technologies.

One key question is whether—or when—a general economic downturn might occur. During times of economic recession, many companies are wary of making investments in technology development, preferring to postpone the expense until conditions improve. By that time, many technologies lose their windows of opportunity.

Competitive factors will become more important than ever in determining the success of new communication technologies. Many industries such as telephony and cable television, which have operated as virtual local monopolies, will face increasing competition from both inside and outside their industries. The primary beneficiary of competition will be the consumer, who will see prices for services fall to the cost of providing the service.

As consumers face an array of options for any communication technology, their choices will most likely be affected by the choices they have made in the past. They will want to do business with familiar company names and products. Brand identity will increase in importance for all communication technologies. One example of the importance of brand identity can be found in the new television distribution technologies such as direct broadcast satellite (DBS) and wireless cable. Technologically, these services have almost nothing in common with cable television, but the software (programming) is almost exactly the same.

As people make decisions to adopt new communications technologies, the thing they will look for first is the delivery of information they are already receiving from another source. Then, they will look for a relative advantage in price, quality, etc. As important as these factors are for early adoption, the next step is even more important. Once a technology is able to gain a foothold in the market, it can begin offering new services or features not found in the older technology. Consumers like both novelty and familiarity—but not too much of either.

Among all the technologies reported in this book, the one that is continuing to grow the fastest (for the fourth consecutive year) as of mid-1998 is the Internet. People began using the Internet for interpersonal communication via e-mail, but the service has evolved to provide a wealth of information retrieval tools, most notably the World Wide Web. The key question is how long can the Internet retain its frontier spirit, a direct by-product of its non-commercial roots?

New commercial applications of the Internet are introduced daily, with most of these making heavy use of the Internet's ability to transmit digital images, audio, and video as easily as text. The result is an exponential increase in data traffic on the network. As long as the capacity of the Internet can keep up with the increase in traffic, the Internet will remain the site of the boldest innovation among all communication technologies.

On the other hand, capacity limits may well lead to a "tragedy of the commons." This term refers to the use of a common grazing area by a village to support all the animals in the village. The field is only able to support a limited number of animals. The problem is that the prosperity of individuals is based upon the number of animals they raise, so each attempts to raise as many animals as possible. Eventually, the field becomes over-grazed, and everyone suffers. The Internet is today's common field. If demand increases to the point at which the flow of data slows, users will be forced to migrate to different media (if they can afford it).

The Internet is clearly both the site of the fastest developments in the field of communication technology and the ideal subject for research on a variety of communication and social issues. A few years from now, we will know whether the Internet is a fad (the CB radio of the 1990s) or not; in either case, research on the Internet should yield lessons that can be applied to a wide range of communication technologies.

Analysis of the other communication technologies discussed in this book indicates the importance of regulatory policy in shaping a technology. There is no technological reason why cable companies and telephone companies cannot provide both telephone and television service, but the structure of the current marketplace is a product of the fact that these two have been kept apart. As the regulatory barriers between sectors of the communication industry fall, the impacts upon all communication technologies will be profound.

The impact of organizational structure and the political system (through regulation) illustrates the importance of using the umbrella perspective discussed in Chapter 1 to analyze technologies. In almost every technology discussed, at least one of the other four elements of the umbrella (software, organizational infrastructure, social systems, and individual user) was more important than the hardware.

The authors of each chapter have worked to ensure that all information is current as of the time the manuscripts were submitted (late spring 1998). It is certain that a few items will be out of date by the time this book is printed, and more items will become dated every week. The reader is therefore encouraged to use the online updates provided on the *Communication Technology Update* home page (http://www.tfi.com/ctu) as well as study the bibliographies of each chapter to determine the best sources for up-to-date information.

The authors of these chapters have one thing in common: They all enjoy the process of observing changes in the spectrum of communication technologies. This author likens the "playing field" of communication technologies to a spectator sport, with the "scores" reported on a daily basis in the business section of the newspaper. It is my hope that this book will provide you with the introduction you need to follow this "sport," and give you the most important tool you will need in the next millennium—the ability to understand and predict the future of communication technologies so that you can take advantage of the wealth of opportunities they will provide.

Glossary

A

Adaptive Transform Acoustic Coding (ATRAC). A method of digital compression of audio signals used in the MD (minidisc) format. ATRAC ignores sounds out of the range of human hearing to eliminate about 80% of the data in a digital audio signal.

Addressability. The ability of a cable system to individually control its converter boxes, allowing the cable operator to enable or disable reception of channels for individual customers instantaneously from the home office.

ADSL (asymmetrical digital subscriber line). A system of compression and transmission that allows broadband signals up to 6 Mb/s to be carried over twisted pair copper wire for relatively short distances.

Advanced television (ATV). Television technologies that offer improvement in existing television systems.

Agent. Any being in a virtual environment.

Algorithm. A specific formula used to modify a signal. For example, the key to a digital compression system is the algorithm that eliminates redundancy.

AM (amplitude modulation). A method of superimposing a signal on a carrier wave in which the strength (amplitude) of the carrier wave is continuously varied. AM radio and the video portion of NTSC TV signals use amplitude modulation.

Analog. A continuously varying signal or wave. As with all waves, analog waves are susceptible to interference, which can change the character of the wave.

ANSI (American National Standards Institute). An official body within the United States delegated with the responsibility of defining standards.

ASCII (American Standard Code for Information Interchange). Assigns specific letters, numbers, and control codes to the 256 different combinations of 0s and 1s in a byte.

Aspect ratio. In visual media, the ratio of the screen width to height. Ordinary television has an aspect ratio of 4:3, while high-definition television is "wider" with an aspect ratio of 16:9 (or 5.33:3).

Asynchronous. Occurring at different times. For example, electronic mail is asynchronous communication because it does not require the sender and receiver to be connected at the same time.

ATM (Asynchronous Transfer Mode). A method of data transportation whereby fixed length packets are sent over a switched network. Speeds of up to over 2 Gb/s can be achieved, making it suitable for carrying voice, video, and data.

Augmented reality. The superimposition of virtual objects on physical reality. For example, a technician could use augmented reality to display a three-dimensional image of a schematic diagram on a piece of equipment to facilitate repairs.

Available bit rate (ABR). An ATM service type in which the ATM network makes a "best effort" to meet the transmitter's bandwidth requirements. ABR differs from other best effort service types by employing a congestion feedback mechanism that allows the ATM network to notify the transmitters that they should reduce their rate of data transmission until the congestion decreases. Thus, ABR offers a qualitative guarantee that the transmitter's data can get to the intended receiver without experiencing unwanted cell loss.

Avatar. An animated character representing a person in a virtual environment.

AVI (audio-video interleaved). Microsoft's video driver for Windows.

B

Bandwidth. A measure of capacity of communications media. Greater bandwidth allows communication of more information in a given period of time.

Basic rate interface ISDN (BRI-ISDN). The basic rate ISDN interface provides two 64-Kb/s channels (called B channels) to carry voice or data and one 16-Kb/s signaling channel (the D channel) for call information.

Bit. A single unit of data, either a one or a zero, used in digital data communications. When discussing digital data, a small "b" refers to bits, and a capital "B" refers to bytes.

Bridge. A type of switch used in telephone and other networks that connects three or more users simultaneously.

Broadband. An adjective used to describe large-capacity networks that are able to carry several services at the same time, such as data, voice, and video.

Buy rate. The percentage of subscribers purchasing a pay-per-view program divided by the total number of subscribers who can receive the program.

Byte. A compilation of bits, seven bits in accordance with ASCII standards and eight bits in accordance with EBCDIC standards.

C

CATV (community antenna television). One of the first names for local cable television service, derived from the common antenna used to serve all subscribers.

C-band. Low-frequency (1 GHz to 10 GHz) microwave communication. Used for both terrestrial and satellite communication. C-band satellites use relatively low power and require relatively large receiving dishes.

CCD (charge coupled device). A solid-state camera pickup device that converts an optical image into an electrical signal.

CCITT (International Telegraph and Telephone Consultative Committee). CCITT is the former name of the international regulatory body that defines international telecommunications and data communications standards. It has been renamed the Telecommunications Standards Sector of the International Telecommunications Union.

CDDI (copper data distributed interface). A subset of the FDDI standard targeted toward copper wiring.

CD-I (compact disc-interactive). A proprietary standard created by Philips for interactive video presentations including games and education.

CDMA (code division multiple access). A spread spectrum cellular telephone technology, which digitally modulates signals from all channels in a broad spectrum.

CD-ROM (compact disc-read only memory). The use of compact discs to store text, data, and other digitized information instead of (or in addition to) audio. One CD-ROM can store up to 700 megabytes of data.

CD-RW (compact disc-rewritable). A special type of compact disc that allows a user to record and erase data, allowing the disc to be used repeatedly.

Cell. The area served by a single cellular telephone antenna. An area is typically divided into numerous cells so that the same frequencies can be used for many simultaneous calls without interference.

Central processing unit (CPU). The "brains" of a computer, which uses a stored program to manipulate information.

Circuit-switched network. A type of network whereby a continuous link is established between a source and a receiver. Circuit switching is used for voice and video to ensure that individual parts of a signal are received in the correct order by the destination site.

Coaxial cable. A type of "pipe" for electronic signals. An inner conductor is surrounded by a neutral material, which is then covered by a metal "shield" that prevents the signal from escaping the cable.

Codec (COmpression/DECompression). A device used to compress and decompress digital video signals.

COFDM (coded orthogonal frequency division multiplexing). A flexible protocol for advanced television signals that allows simultaneous transmission of multiple signals at the same time.

Common carrier. A business, including telephone and railroads, which is required to provide service to any paying customer on a first-come, first-served basis.

Compression. The process of reducing the amount of information necessary to transmit a specific audio, video, or data signal.

Constant bit rate (CBR). A data transmission that can be represented by a non-varying, or continuous, stream of bits or cell payloads. Applications such as voice circuits generate CBR traffic patterns. CBR is an ATM service type in which the ATM network guarantees to meet the transmitter's bandwidth and quality-of-service requirements.

Cookies. A file used by a Web browser to record information about a user's computer, including Web sites visited, which is stored on the computer's hard drive.

Cyberspace. The artificial worlds created within computer programs.

D

DCC (digital compact cassette). A digital audio format resembling an audiocassette tape. DCC was introduced by Philips (the creator of the common audio "compact cassette" format) in 1992, but never established a foothold in the market.

Desktop publishing (DTP). The process of producing printed materials using a desktop computer and laser printer. Commonly used peripherals in DTP include scanners, modems, and color printers.

Digital audio broadcasting (DAB). Radio broadcasting that uses digital signals instead of analog to provide improved sound quality.

Digital audiotape (DAT). An audio recording format that stores digital information on 4mm tape.

Digital signal. A signal that takes on only two values, off or on, typically represented by "0" or "1." Digital signals require less power but (typically) more bandwidth than analog, and copies of digital signals are *exactly* like the original.

Digital video compression. The process of eliminating redundancy or reducing the level of detail in a video signal in order to reduce the amount of information that must be transmitted or stored.

Direct broadcast satellites (DBS). High-powered satellites designed to beam television signals directly to viewers with special receiving equipment.

Domain name system (DNS). The protocol used for assigning addresses for specific computers and computer accounts on the Internet.

Downlink. Any antenna designed to receive a signal from a communication satellite.

DSL (digital subscriber line). A method of transmitting digital signals over ordinary copper telephone lines. Common varieties of DSL include ADSL, HDSL, IDSL, SDSL, and RADSL. The generic term xDSL is used to represent all forms of DSL.

DVD (digital video or versatile disc). A plastic disc similar in size to a compact disc that has the capacity to store up to 20 times as much information as a CD, using both sides of the disc.

DVD-RAM. A type of DVD that can be used to record and play back digital data (including audio and video signals).

DVD-ROM. A type of DVD that is designed to store computer programs and data for playback only.

DVI (digital video interactive). An interactive video system that uses a computer to record and/or play back compressed video, text, and audio.

E

E-1. The European analogue to T-1, a standard, high-capacity telephone circuit capable of transmitting approximately 2 Mb/s, the equivalent of 30 voice channels.

EDL (edit decision list). In video editing, a sequential list of all of the video and audio pieces which must be assembled to create a program. Also refers to a computer-generated library of numerical information that instructs an edit controller to operate videotape recorders to assemble a program.

Electromagnetic spectrum. The set of electromagnetic frequencies that includes radio waves, microwave, infrared, visible light, ultraviolet rays, and gamma rays. Communication is possible through the electromagnetic spectrum by radiation and reception of radio waves at a specific frequency.

E-mail. Electronic mail or textual messages sent and received through electronic means.

Extranet. A special type of Intranet that allows selected users outside an organization to access information on a company's Intranet.

F

FDDI-I (fiber data distributed interface I). A standard for 100 Mb/s LANs using fiber optics as the network medium linking devices.

FDDI-II (fiber data distributed interface II). Designed to accommodate the same speeds as FDDI-I, but over a twisted-pair copper cable.

Federal Communications Commission (FCC). The U.S. federal government organization responsible for the regulation of broadcasting, cable, telephony, satellites, and other communication media.

Fiber-in-the-loop (FITL). The deployment of fiber optic cable in the local loop, the area between the telephone company's central office and the subscriber.

Fiber optics. Thin strands of ultrapure glass or plastic which can be used to carry light waves from one location to another.

Fiber-to-the-curb (FTTC). The deployment of fiber optic cable from a central office to a platform serving numerous homes. The home is linked to this platform with coaxial cable or twisted-pair (copper wire). Each fiber carries signals for more than one residence, lowering the cost of installing the network versus fiber-to-the-home.

Fiber-to-the-home (FTTH). The deployment of fiber optic cable from a central office to an individual home. This is the most expensive broadband network design, with every home needing a separate fiber optic cable to link it with the central office.

Fixed satellite services (FSS). The use of geosynchronous satellites to relay information to and from two or more fixed points on the earth's surface.

FM (frequency modulation). A method of superimposing a signal on a carrier wave in which the frequency on the carrier wave is continuously varied. FM radio and the audio portion of an NTSC television signal use frequency modulation.

Footprint. The coverage area of a satellite signal, which can be focused to cover specific geographical areas.

Frame. One complete still image that makes up a part of a video signal.

FTP (file transfer protocol). An Internet application that allows a user to download and upload programs, documents, and pictures to and from databases virtually anywhere on the Internet.

Full-motion video. The projection of 20 or more frames (or still images) per second to give the eye the perception of movement. Broadcast video in the United States uses 30 frames per second, and most film technologies use 24 frames per second.

Fuzzy logic. A method of design that allows a device to undergo a gradual transition from on to off, instead of the traditional protocol of all-or-nothing.

G

Geosynchronous orbit (GEO, also known as geostationary orbit). A satellite orbit directly above the equator at 22,300 miles. At that distance, a satellite orbits at a speed that matches the revolution of the earth so that, from the earth, the satellite appears to remain in a fixed position.

Gigabyte. 1,000,000,000 bytes or 1,000 megabytes (see *Byte*).

Global positioning system (GPS). Satellite-based services that allow a receiver to determine its location within a few meters anywhere on earth.

Gopher. An early Internet application that assisted users in finding information on and accessing the resources of remote computers.

Graphical user interface (GUI). A computer operating system that is based upon icons and visual relationships rather than text. Windows and the Macintosh computer use GUIs because they are more user friendly.

Groupware. A set of computer software applications that facilitates intraorganizational communication, allowing multiple users to access and change files, send and receive e-mail, and keep track of progress on group projects.

GSM (Group standard mobile). A type of digital cellular telephony used in Europe that uses time-division multiplexing to carry multiple signals in a single frequency.

H

Hardware. The physical equipment related to a technology.

HFC (hybrid fiber/coax). A type of network that includes a fiber optic "backbone" to connect individual nodes and coaxial cable to distribute signals from an optical network interface to the individual users (up to 500 or more) within each node.

High-definition television (HDTV). Any television system that provides a significant improvement in existing television systems. Most HDTV systems offer more than 1,000 scan lines, in a wider aspect ratio, with superior color and sound fidelity.

Hologram. A three-dimensional photographic image made by a reflected laser beam of light on a photographic film.

http (hypertext transfer protocol). The first part of an address (URL) of a site on the Internet, signifying a document written in hypertext markup language (HTML).

Hypertext. Documents or other information with embedded links that enable a reader to access tangential information at programmed points in the text.

Hypertext markup language (HTML). The computer language used to create hypertext documents, allowing connections from one document or Internet page to numerous others.

I

Image stabilizer. A feature in camcorders that lessens the shakiness of the picture either optically or digitally.

Instructional Television Fixed Service (ITFS). A microwave television service designed to provide closed-circuit educational programming. Underutilization of these frequencies led wireless cable (MMDS) operators to obtain FCC approval to lease these channels to deliver television programming to subscribers.

Interactive TV. A television system in which the user interacts with the program in such a manner that the program sequence will change for each user.

Interexchange carrier. Any company that provides interLATA (long distance) telephone service.

Interlaced scanning. The process of displaying an image using two scans of a screen, with the first providing all the even-numbered lines and the second providing the odd-numbered lines.

Intranet. A network serving a single organization or site that is modeled after the Internet, allowing users access to almost any information available on the network. Unlike the Internet, Intranets are typically limited to one organization or one site, with little or no access to outside users.

IP (Internet protocol). The standard for adding "address" information to data packets to facilitate the transmission of these packets over the Internet.

ISDN (Integrated Services Digital Network). A planned hierarchy of digital switching and transmission systems synchronized to transmit all signals in digital form, offering greatly increased capacity over analog networks.

ISO (International Organization of Standardization). Develops, coordinates, and promulgates international standards that facilitate world trade.

ITU (International Telecommunications Union). A U.N. organization that coordinates use of the spectrum and creation of technical standards for communication equipment.

J

JPEG (Joint Photographic Experts Group). A committee formed by the ISO to create a digital compression standard for still images. Also refers to the digital compression standard for still images created by this group.

K

Killer app. Short for "killer application," this is a function of a new technology that is so strongly desired by users that it results in adoption of the technology.

Kilobyte. 1,000 bytes (see *Byte*).

Ku-band. A set of microwave frequencies (12 GHz to 14 GHz) used exclusively for satellite communication. Compared to C-band, the higher frequencies produce shorter waves, requiring smaller receiving dishes.

L

Laser. From the acronym for "light amplification by stimulated emission of radiation." A laser usually consists of a light-amplifying medium placed between two mirrors. Light not perfectly aligned with the mirrors escapes out the sides, but light perfectly aligned will be amplified. One mirror is made partially transparent. The result is an amplified beam of light that emerges through the partially transparent mirror.

LATA (local access transport area). The geographical areas defining local telephone service. Any call within a LATA is handled by the local telephone company, but calls between LATAs must be handled by long distance companies, even if the same local telephone company provides service in both LATAs.

LEO (low earth orbit). A satellite orbit between 400 and 800 miles above the earth's surface. The close proximity of the satellite reduces the power needed to reach the satellite, but the fact that these satellites complete an entire orbit in a few hours means that a large number of satellites must be used in a LEO satellite system in order to have one overhead at all times.

Liner notes. The printed material that accompanies a CD or record album, including authors, identification of musicians, lyrics, pictures, and commentary.

LMDS (local multipoint distribution service). A new form of wireless technology similar to MMDS that uses frequencies above 28 GHz.

Local area network (LAN). A network connecting a number of computers to each other or to a central server so that the computers can share programs and files.

Local exchange carrier (LEC). A local telephone company.

M

Mb/s. Megabits per second.

Megabyte. 1,000,000 bytes or 1,000 kilobytes (see *Byte*).

Microcell. The area, typically a few hundred yards across, served by a single transmitter in a PCS network. The use of microcells allows the reuse of the same frequencies many times in an area, allowing more simultaneous users.

MIDI (musical instrument digital interface). An international standard for representing music in digital form. Music can be directly input from a computer keypad and stored to disc or RAM, then played back through a connected instrument. Conversely, a song can be played by the performer on an instrument interfaced with the computer.

MIPS (millions of instructions per second). This is a common measure of the speed of a computer processor.

MMDS (multichannel, multipoint distribution systems). A service similar to cable television that uses microwaves to distribute the signals instead of coaxial cable. MMDS is therefore better suited to sparsely-populated areas than cable.

Mobile satellite services (MSS). The use of satellites to provide navigation services and to connect vehicles and remote regions with other mobile or stationary units.

Modem (MOdulator-DEModulator). Enables transmission of a digital signal, such as that generated by a computer, over an analog network, such as the telephone network.

Monochromatic. Light or other radiation with one single frequency or wavelength. Since no light is perfectly monochromatic, the term is used loosely to describe any light of a single color over a very narrow band of wavelengths.

MPEG (Moving Pictures Experts Group). A committee formed by the ISO to set standards for digital compression of full-motion video. Also stands for the digital compression standard created by the committee that produces VHS-quality video.

MPEG-1. An international standard for the digital compression of VHS-quality, full-motion video.

MPEG-2. An international standard for the digital compression of broadcast-quality, full-motion video.

MTSO (mobile telephone switching office). The "heart" of a cellular telephone network, containing switching equipment and computers to manage the use of cellular frequencies and connect cellular telephone users to the landline network.

Multicast. The transmission of information over the Internet to two or more users at the same time.

Multimedia. The combination of video, audio, and text in a single platform or presentation.

Multiple system operator (MSO). A cable company that owns and operates many local cable systems.

Multiplexing. Transmitting several messages or signals simultaneously over the same circuit or frequency.

Must-carry. A set of rules requiring cable operators to carry all local broadcast television stations.

N

Nanometer. One billionth of a meter. Did you know that "nano" comes from the Greek word "dwarf?"

National information infrastructure (NII). A Clinton administration initiative to support the private sector construction and maintenance of a "seamless web" of communication networks, computers, databases, and consumer electronics that will put vast amounts of information at users' fingertips.

Near video on demand (NVOD). A pay-per-view service offering movies on up to eight channels each with staggered start times so that a viewer can watch one at any time. NVOD is much more flexible than PPV, and costs far less than VOD to implement.

Newbies. New users of an interactive technology, usually identified because they have not yet learned the etiquette of communication in a system.

Nodes. Routers or switches on a broadband network that provide a possible link from point A to point B across a network.

NTSC (National Television Standards Committee). The group responsible for setting the U.S. standards for color television in the 1950s.

O

OCR (optical character recognition). Refers to computer programs that can convert images of text to text.

Octet. A byte, more specifically, an eight-bit byte. The origins of the octet trace back to when different networks had different byte sizes. Octet was coined to identify the eight-bit byte size.

Operating system. The program embedded in most computers that controls the manner in which data are read, processed, and stored.

Overlay. The process of combining a graphic with an existing video image.

P

Packet switched networks. A network that allows a message to be broken into small "packets" of data that are sent separately by a source to the destination. The packets may travel different paths and arrive at different times, with the destination site reassembling them into the original message. Packet switching is used in most computer networks because it allows a very large amount of information to be transmitted through a limited bandwidth.

PCN (personal communications network). Similar to PCS, but incorporating a wider variety of applications including voice, data, and facsimile.

PCS (personal communications services). A new category of digital cellular telephone service which uses much smaller service areas (microcells) than ordinary cellular telephony.

Peripheral. An external device that increases the capabilities of a communication system.

Personal digital assistants (PDAs). Extremely small computers (usually about half the size of a notebook) designed to facilitate communication and organization. A typical PDA accepts input from a special pen instead of a keyboard, and includes appointment and memo applications. Some PDAs also include fax software and a cellular telephone modem to allow faxing of messages almost anywhere.

Photovoltaic cells. A device that converts light energy to electricity.

Pixel. The smallest element of a computer display. The more pixels in a display, the greater the resolution.

Point-to-multipoint service. A communication technology designed for broadcast communication, where one sender simultaneously sends a message to an unlimited number of receivers.

Point-to-point service. A communication technology designed for closed-circuit communication between two points such as in a telephone circuit.

POTS (plain old telephone service). An acronym identifying the traditional function of a telephone network to allow voice communication between two people across a distance.

PPV (pay-per-view). A television service in which the subscriber is billed for individual programs or events.

Primary-rate interface ISDN (PRI-ISDN). The primary rate ISDN interface provides twenty-three 64-Kb/s channels (called B channels) to carry voice or data and one 16-Kb/s signaling channel (the D channel) for call information.

Progressive scanning. A video display system that sequentially scans all the lines in a video display.

PTT. A government organization that offers telecommunications services within a country. (The initials refer to the antecedents of modern communication: the post [mail], telephone, and telegraph.)

Q

QuickTime. A computer video playback system that enables a computer to automatically adjust video frame rates and image resolution so that sound and motion are synchronized during playback.

R

RBOC (regional Bell operating company). One of the seven local telephone companies formed upon the divestiture of AT&T in 1984. The seven were Bell Atlantic, NYNEX, Pacific Telesis, BellSouth, SBC Corporation, U S WEST, and Ameritech.

Retransmission consent. The right of a television station to prohibit retransmission of its signal by a cable company. Under the 1992 Cable Act, U.S. television stations may choose between must-carry and retransmission consent.

RF (radio frequency). Electromagnetic carrier waves upon which audio, video, or data signals can be superimposed for transmission.

Router. The central switching device in a computer network that directs and controls the flow of data through the network.

S

SCMS (Serial Copy Management System). A method of protecting media content from piracy that allows copies to be made of a specific piece of content, but will not allow copies of copies.

SCSI (small computer system interface) [pronounced "scuzzy"]. A type of interface between computers and peripherals that allows faster communication than most other interface standards.

SDTV (standard-definition television). Digital television transmissions that deliver approximately the same resolution and aspect ratio of traditional television broadcasts, but do so in a fraction of the bandwidth through the use of digital video compression.

SMPTE (Society of Motion Picture and Television Engineers). The industry group responsible for setting technical standards in most areas of film and television production.

Software. The messages transmitted or processed through a communications medium. This term also refers to the instructions (programs) written for programmable computers.

SONET (Synchronous Optical Network). A standard for data transfer over fiber optic networks used in the United States that can be used with a wide range of packet- and circuit-switched technologies.

Spot beam. A satellite signal targeted at a small area, or footprint. By concentrating the signal in a smaller area, the signal strength increases in the reception area.

Switched virtual connection (SVC). A logical connection between endpoints established by the ATM network on demand after receiving a connection request from the source. It is defined in the ATM Forum UNI specification and transmitted using the Q.2931 signaling protocols.

Synchronous transmission. The transmission of data at a fixed rate, based on a master clock, between the transmitter and receiver.

T

T-1. A standard for physical wire cabling used in networks. A T-1 line has the bandwidth of 1.54 Mb/s.

T-3. A standard for physical wire cabling that has the bandwidth of 44.75 Mb/s.

TCP/IP (transmission control protocol/Internet protocol). A method of packet-switched data transmission used on the Internet. The protocol specifies the manner in which a signal is divided into parts, as well as the manner in which "address" information is added to each packet to ensure that it reaches its destination and can be reassembled into the original message.

TDMA (time division multiple access). A cellular telephone technology that sends several digital signals over a single channel by assigning each signal a periodic slice of time on the channel. Different TDMA technologies include North America's Interim Standard (IS) 54, Europe's global system for mobile communications (GSM), and a version developed by InterDigital Corporation. These systems differ in circuits per channel, timing, and channel width.

Telecommuting. The practice of using telecommunications technologies to facilitate work at a site away from the traditional office location and environment.

Teleconference. Interactive, electronic communication among three or more people at two or more sites. Includes audio-only, audio and graphics, and videoconferencing.

Teleport. A site containing multiple satellite uplinks and downlinks, along with microwave, fiber optic, and other technologies to facilitate the distribution of satellite signals.

Terabyte. 1,000,000,000,000 bytes or 1,000 gigabytes (see *Byte*).

Time division multiplexing (TDM). The method of multiplexing where each device on the network is provided with a set amount of link time.

Transponder. The part of a satellite that receives an incoming signal from an uplink and retransmits it on a different frequency to a downlink.

TVRO (television receive only). A satellite dish used to receive television signals from a satellite.

Twisted pair. The set of two copper wires used to connect a telephone customer with a switching office. The bandwidth of twisted pair is extremely small compared with coaxial cable or fiber optics.

U

UHF (ultra high frequency). Television channels numbered from 14 through 83.

Universal service. In telecommunications policy, the principle that an interactive telecommunications service must be available to everyone within a community in order to increase the utility and value of the network for all users.

Uplink. An antenna that transmits a signal to a satellite for relay back to earth.

URL (uniform resource locator). An "address" for a specific page on the Internet. Every page has a URL that specifies its server and file name.

V

Variable bit rate (VBR). A data transmission that can be represented by an irregular grouping of bits or cell payloads followed by unused bits or cell payloads. Most applications other than voice circuits generate VBR traffic patterns.

Vertical blanking interval (VBI). In an NTSC television signal, the portion of the signal that is not displayed on a television receiver. Some of the lines in the VBI contain "sync" information that is used to identify the beginning of a new picture. Some of the blank lines in the VBI can be used to carry data such as closed captions.

Vertical integration. The ownership of more than one function of production or distribution by a single company, so that the company, in effect, becomes its own customer.

VFW (video for Windows). Software that lets users create, edit, and play digital video on Windows-based computers.

VHF (very high frequency). Television channels numbered from two through 13.

Videoconference. Interactive, audio/visual communication among three or more people at two or more sites.

Video dialtone. A set of rules enabling telephone companies to transmit video programming through the telephone network.

Video on demand (VOD). A pay-per-view television service in which a viewer can order a program from a menu and have it delivered instantly to the television set, typically with the ability to pause, rewind, etc.

Videophone. A telephone that provides both sound (audio) and picture (video).

Videotext (also known as *videotex*). An interactive computer system using text and/or graphics that allows access to a central computer using a terminal or personal computer to engage in data retrieval, communication, transactions, and/or games.

Virtual reality (VR). A cluster of interactive technologies that gives users a compelling sensation of being inside a circumambient environment created by a computer.

VRML (virtual reality markup language). A computer language that provides a three-dimensional environment for traditional Internet browsers, resulting in a simple form of virtual reality available over the Internet.

VSAT (very small aperture terminal). A satellite system that uses relatively small satellite dishes to send and receive one- or two-way data, voice, or even video signals.

W

Wide area network (WAN). A network that interconnects geographically distributed computers or LANs.

Wireless cable. See *MMDS*.

WORM (Write once, read many). A technique that allows recording of information on a medium only once, with unlimited playback.

X

X.25 data protocol. A packet switching standard developed in the mid-1970s for transmission of data over twisted-pair copper wire.

xDSL. See DSL.